Energy Efficient Vehicles

The text discusses energy-efficient vehicles as an essential element of sustainable transportation. The text highlights the social, economic, and environmental benefits associated with energy-efficient automobiles, which effectively solve the issue of greenhouse gas emissions, improve air quality, boost energy security, and promote zero-emission. The energy-efficient technologies for transportation, accessibility and safety of the transport system, environmental footprint, health impact, economic development, and social growth are the central themes of the book. It further presents future integrated mobility-energy systems and sustainability indicators.

This book:

- Examines policies, challenges, and the latest developments in the field of sustainable mobility.
- Discusses the latest advances in the field of energy storage systems, batteries, image processing, obstacle identification, and automatic gear trains.
- Highlights the safety, security, and risk management related to sustainable transportation, covering zero emissions and sustainability indicators.
- Presents electric vehicle grid integration and infrastructure for e-vehicle charging.
- Aims to provide an overview of various aspects of EV, HEV, ITS, and vehicular network deployment design, encompassing the technological advancements, challenges, and opportunities associated with this rapidly evolving field.
- Understanding the transportation needs and preferences of youth populations in shaping transportation policy and promoting sustainable urban development to design transportation systems that are efficient, equitable, and environmentally sustainable.
- Synergize exploration related to the various properties and functionalities through extensive theoretical and numerical modeling present in the energy sector.

This book is primarily written for senior undergraduate, graduate students, and academic researchers in fields including mechanical engineering, industrial engineering, automotive engineering, manufacturing engineering, and environmental engineering.

Advances in Manufacturing, Design and Computational Intelligence Techniques

Series Editor-Ashwani Kumar-Senior Lecturer, Mechanical Engineering, at Technical Education Department, Uttar Pradesh, Kanpur, India

The book series editor is inviting edited, reference and text book proposal submission in the book series. The main objective of this book series is to provide researchers a platform to present state of the art innovations, research related to advanced materials applications, cutting edge manufacturing techniques, innovative design and computational intelligence methods used for solving nonlinear problems of engineering. The series includes a comprehensive range of topics and its application in engineering areas such as additive manufacturing, nanomanufacturing, micromachining, biodegradable composites, material synthesis and processing, energy materials, polymers and soft matter, nonlinear dynamics, dynamics of complex systems, MEMS, green and sustainable technologies, vibration control, AI in power station, analog-digital hybrid modulation, advancement in inverter technology, adaptive piezoelectric energy harvesting circuit, contactless energy transfer system, energy-efficient motors, bioinformatics, computer aided inspection planning, hybrid electrical vehicle, autonomous vehicle, object identification, machine intelligence, deep learning, control-robotics-automation, knowledge based simulation, biomedical imaging, image processing and visualization. This book series compiled all aspects of manufacturing, design and computational intelligence techniques from fundamental principles to current advanced concepts.

Hybrid Metal Additive Manufacturing: Technology and Applications
Edited by Parnika Shrivastava, Anil Dhanola, and Kishor Kumar Gajrani

Thermal Energy Systems: Design, Computational Techniques, and Applications
Edited by Ashwani Kumar, Varun Pratap Singh, Chandan Swaroop Meena, Nitesh Dutt

Energy Efficient Vehicles: Technologies and Challenges
Edited by Varun Pratap Singh, Ashwani Kumar, Chandan Swaroop Meena, and Gaurav Dwivedi

https://www.routledge.com/Advances-in-Manufacturing-Design-and-Computational-Intelligence-Techniques/book-series/CRCAIMDCIT

Energy Efficient Vehicles

Technologies and Challenges

Edited by Varun Pratap Singh,
Ashwani Kumar, Chandan Swaroop Meena,
and Gaurav Dwivedi

CRC Press
Taylor & Francis Group
Boca Raton London New York

CRC Press is an imprint of the
Taylor & Francis Group, an **informa** business

Designed cover image: © Shutterstock

First edition published 2024
by CRC Press
2385 NW Executive Center Drive, Suite 320, Boca Raton FL 33431

and by CRC Press
4 Park Square, Milton Park, Abingdon, Oxon, OX14 4RN

CRC Press is an imprint of Taylor & Francis Group, LLC

ISBN: 9781032548111 (hbk)
ISBN: 9781032735160 (pbk)
ISBN: 9781003464556 (ebk)

DOI: 10.1201/9781003464556

Typeset in Times New Roman
by Apex CoVantage, LLC

Contents

Aim and Scope

Sustainable road transportation refers to any means of transportation that is 'green' and has a low impact on the environment. Sustainable transportation is also about balancing our current and future needs. Examples of sustainable transportation include walking, cycling, transit, carpooling, car sharing, and green vehicles. As per UN Brundtland Commission (WCED, 1987), sustainable mobility can be defined as "the mobility that **satisfies the needs of present generations without compromising future generations**", but in the modern era, we are compromising the needs of the next generation in terms of pollution, depletion of fossil fuels, global warming, poor air quality, and health hazards gases. The three main pillars of sustainability, i.e., economic, environmental, and social, are crushed by modern development. So, there is a need to shift from traditional means of transportation to sustainable energy-efficient transportation. Under the vision of sustainable mobility, better infrastructure and services will be provided to support the movement of goods and people. This outcome would be achieved only if four goals are pursued simultaneously, i.e., developing the right policy, awareness, intelligent transportation, and green vehicles. These four goals are the central theme of the book ***Energy Efficient Vehicles: Technologies and Challenges***. The book discusses transitions from conventional to sustainable mobility, challenges in this transition period, the aims of new vehicle policies, and the latest autonomous vehicles or intelligent transportation. The main highlights of the book are energy-efficient technologies for transportation, accessibility, and safety of the transport system, environmental footprint, health impact, economic development, and social growth.

The book ***Energy Efficient Vehicles: Technologies and Challenges*** discusses energy-efficient sustainable mobility and will provide in-depth knowledge to all readers. The main objective of this book is to provide researchers and students with intensive educational material on the advancements of energy-efficient vehicles and their applications for sustainable transportation. Each chapter of this book has its own balanced discussion of literature review, solution, methodology, experimental setup, results validation, and future scope with targeted fields such as materials, automotive, electronics, and transportation engineering with hybrid energy applications. This will help to ensure the comprehensiveness of the book and to gratify the requirements of the potential audience in diverse areas of the world related to policy making, transportation, manufacturing, design, and sustainability. The applicability of this book covers a wide range of fields, namely research and development in sustainable transportation, automotive production industry, design, advanced materials, electronics industry, etc., and it will also be of help to the audience conducting research in these industries.

Editors
Dr. Varun Pratap Singh
Dr. Ashwani Kumar
Dr. Chandan Swaroop Meena
Dr. Gaurav Dwivedi

Preface

Energy Efficient Vehicles: Technologies and Challenges deals with policies, challenges, and the latest developments in the field of sustainable mobility. Sustainable mobility is the central theme of this book. Effectively, the varieties of chapters presented in the book are interesting and up-to-date with the scope of future research work in energy-efficient sustainable transportation sector. This has the potential to be valuable to researchers in mechanical engineering, mechatronics engineering, and electrical and electronics engineering. The background information and the literature review provided in each chapter is an in-depth analysis of the topic covered in the chapter. The key feature of the book is to provide a clear idea of sustainable urban transport, mainly focusing on policy-making and challenges in the transition from conventional fossil fuels-based mobility to sustainable mobility.

The book provides the original idea of energy-efficient vehicles, policy-making, sources of energy, and pioneering innovation and sustainable practices adopted by industries for the betterment of societal changes around the world. Smart mobility is the bedrock of sustainable cities. With six out of every ten people on earth expected to live in urban areas by 2030, urban mobility is key to the social and economic growth of smart cities. *Energy Efficient Vehicles: Technologies and Challenges* examines key challenges in the context of the UN's Sustainable Development Goals, Paris Agreement, and the New Urban Agenda, including policies needed to achieve a sustainable, low-carbon pathway for transport and how an integrated policy strategy is required to provide sustainable mobility. The book presents 16 chapters of detailed study of energy-efficient vehicles.

Chapter 1 provides a detailed examination of energy-efficient automobiles as an essential element of sustainable transportation. This chapter highlights the environmental, economic, and social benefits associated with energy-efficient automobiles, which effectively mitigate greenhouse gas emissions, improve air quality, and boost energy security. The chapter delves into the complexities surrounding the enhancement of accessibility and affordability of energy-efficient automobiles, encompassing hybrid and electric vehicles. Chapter 2 begins by explaining why and how zero-emission vehicles are so important in the fight against transportation-related pollution. The chapter then proceeds with an examination of emissions from transportation and the challenges of transitioning to a fleet of ZEVs. Also discussed in this chapter are several cutting-edge solutions to the issues that plague vehicles with no emissions.

In continuation, the highlight of Chapter 3 is the importance of understanding the transportation needs and preferences of youth populations in shaping transportation policy and promoting sustainable urban development. Chapter 4 focuses on the necessity of charging infrastructure, types of charging infrastructure, the concept of smart charging, and globally executed pilot projects on charging infrastructure planning. Further, this chapter concludes by presenting a few case studies on ASSI for cities and the way forward. This chapter is anticipated to provide the scientific community with the most recent advancements in the ASSI paradigm. Chapter 6 aims to explore the various social, economic, and environmental issues associated with sustainable transportation. Overall, the chapter provides an overview of the social, economic, and environmental issues associated with sustainable transportation. By addressing these issues, transportation planners and policymakers can design transportation systems that are efficient, equitable, and environmentally sustainable.

Electric vehicles (EVs) and Hybrid electric vehicles (HEVs) are discussed in Chapter 6, along with their potential and critical analysis as a transitional propulsion technology that eases the way toward EVs. In continuation, Chapter 7 aims to provide an overview of various aspects of EV design, encompassing the technological advancements, challenges, and opportunities associated with this rapidly evolving field. Chapter 8 provides a comprehensive overview of lithium-ion batteries for electric transportation, including types, components, pack design, and technology. Lithium-ion batteries are becoming increasingly popular in electric transportation due to their high energy density, low self-discharge rate, and long cycle life. The chapter begins by discussing the various types of lithium-ion batteries used in electric vehicles, such as cylindrical, prismatic, and pouch cells, and their unique characteristics.

Chapter 9 begins with a brief introduction to the Intelligent Transportation System (ITS) and its relevance in improving traffic management and reducing emissions. The chapter then discusses the current state of car cloud computing and networks for ITS, including the benefits and drawbacks of each. Challenges associated with smart city ITS and vehicular network deployment are also explored in this chapter. Chapter 10 offers understanding of the comprehensive impression of EVs on the automobile industry and their role in addressing the environmental challenges of the 21st century. Overall, this chapter provides a comprehensive overview of the rise of EVs, their current status, and future directions, offering a roadmap for policymakers, investors, and industry stakeholders.

Chapter 11 starts by discussing the market trends and factors driving the adoption of EVs. This includes the increasing demand for sustainable transportation, government regulations promoting EVs, and advancements in EV technology. The chapter also provides a detailed analysis of the EV market, including sales statistics, market share, and growth rates, as well as an overview of the major players in the industry. Chapter 12 discusses how HEVs, internal combustion engines (ICEs), and electric motors can provide necessary mechanical energy together or separately. This HEV requires an advanced control system for its operation. Hydrogen Fuel Cell Vehicles (HFCVs) are those used to generate electricity by combining hydrogen and oxygen.

Chapter 13 discusses how the operating cost can be minimized by using solar energy to charge EV batteries. Solar-based EVs consist of a Photovoltaic (PV) cell unit that can be placed mainly on the vehicle's roof. Even though the received solar energy is not enough to charge the EV batteries fully, it can support the batteries to charge quickly. In some EVs, separate batteries are placed to receive solar energy. Chapter 14 provides a comprehensive review of state estimation techniques specifically applied to lithium-ion batteries used in electric vehicle (EV) applications. State estimation refers to the process of determining key battery parameters, such as state of charge (SOC), state of health (SOH), and remaining useful life (RUL), which are crucial for predicting battery behavior and ensuring optimal performance. The purpose of Chapter 15 is to examine sustainable mobility in terms of financial benefits, ecological advantages, and social benefits, taking into consideration the influence on the environment, the economy, and society as well as the technological potential.

Chapter 16 deals with Future Fuels and Solvents. The Pressure Swing Distillation (PSD) was used to separate these solvents using DWSIM version 6.6.2, and then the process was optimized to investigate its economic feasibility. The result showed that IPA-MEK-EtOH was the more cost-effective separation sequence compared to MEK-IPA-EtOH. In addition, the feasibility study of the process produced 2.689 years as a payback period.

Editors
Dr. Varun Pratap Singh
Dr. Ashwani Kumar
Dr. Chandan Swaroop Meena
Dr. Gaurav Dwivedi

Acknowledgement

We express our gratitude to CRC Press (Taylor & Francis Group) and the editorial team for their suggestions and support during the completion of this book. We are grateful to all authors for submitting their quality work and reviewers for their illuminating views on each book chapter presented in the book *Energy Efficient Vehicles: Technologies and Challenges*.

"This book is dedicated to all engineers, researchers and academicians . . ."

Editors
Dr. Varun Pratap Singh
Dr. Ashwani Kumar
Dr. Chandan Swaroop Meena
Dr. Gaurav Dwivedi

Contributing Authors

T. Jerry Alexander
Department of ECE
Swarnandhra College of Engineering and Technology
Narsapur, Andhra Pradesh

K. Balamurugan
Department of ECE, Saveetha School of Engineering
Saveetha of Medical and Technical Sciences
Chennai, India

Nawal Kishor Banjara
Scientist, CSIR-Central Building Research Institute (CSIR-CBRI)
Roorkee, Uttrakhand, India

John Baruch
Department of Mechanical Engineering
CK College of Engineering & Technology
Cuddalore, Tamilnadu, India

Sanchari Deb
School of Engineering
University of Warwick
Coventry, United Kingdom

Gargi Deori
Maulana Azad National Institute of Technology
Bhopal, Madhya Pradesh, India

Gaurav Dwivedi
Assistant Professor, Energy Centre
Maulana Azad National Institute of Technology
Bhopal, Madhya Pradesh, India

E. Elayarani
Department of Mechanical Engineering
Puducherry Technological University
Puducherry, India

V. Karthikeyan
Department of ECE
Mepco Schlenk Engineering College
Sivakasi

Kirubakaran G.
Mepco Schlenk Engineering College
Sivakasi, Tamilnadu, India

Neha Pranav Kolhe
Assistant Professor
Maulana Azad National Institute of Technology
Bhopal, Madhya Pradesh, India

Ashwani Kumar
Senior Lecturer, Department of Mechanical Engineering, Technical Education Department
Uttar Pradesh, Kanpur (under Government of Uttar Pradesh), India

Raj Kumar
Scientist, CSIR-Central Building Research Institute (CSIR-CBRI)
Roorkee, Uttrakhand, India

Roushan Kumar
Department of Mechanical Engineering, School of Engineering
University of Petroleum and Energy Studies
Bidholi, Dehradun, Uttarakhand, India

Francis Luther King M.
Department of Mechanical Engineering
Swarnandhra College of Engineering and Technology (A)
Narsapur
Andhra Pradesh, India

Manish D. Mahale
Assistant Professor
G H Raisoni Institute of Engineering and Business Management
Jalgaon

Chandan Swaroop Meena
Senior Scientist, CSIR-Central Building Research Institute Roorkee
Roorkee, India

Shaik Khaja Mohiddin
Associate Professor, Department of CSE
Koneru Lakshmaiah Education Foundation
Vaddeswaram, Guntur District, Andhra Pradesh

Nilesh Patel
Department of Electrical Engineering, Faculty of Engineering and Technology
Sankalchand Patel University
Visnagar, India
Vaddeswaram, Guntur District, AP

Bipasa Bimalendu Patra
Professor and Dean Research
Department of Electrical Engineering
Padmashri Dr. V.B. Kolte College of Engineering., Malkapur Maharashtra, India

Varun Prakash R.
Mepco Schlenk Engineering College
Sivakasi, Tamilnadu, India

Vaibhav Rajoria
Maulana Azad National Institute of Technology
Bhopal, Madhya Pradesh, India

Hemant Raval
Department of Electrical Engineering
L. D. College of Engineering
Ahmedabad, India

Sanket Raval
Department of Electrical Engineering Faculty of Engineering and Technology
Sankalchand Patel University
Visnagar, India

Mandula China Pentu Saheb
Assistant Professor, Department of CSE
VVIT
Guntur, Andhra Pradesh

Shaik Sharmila
Assistant Professor, Department of IT
VNITSW
PedaPalakalur, Guntur, Andhra Pradesh

Varun Pratap Singh
Assistant Professor, Department of Mechanical Engineering, School of Engineering
University of Petroleum and Energy Studies
Bidholi, Dehradun, Uttarakhand, India

Vadali Srinivas
Assistant Professor, Department of CSE
VITS
Hyderabad, Telangana

Saurab Suman
Department of Chemical Engineering
Dr. B. R. Ambedkar National Institute of Technology Jalandhar
Jalandhar, Punjab, India

Thangadurai N.
Department of Electrical Engineering, Faculty of Engineering and Technology
Sankalchand Patel University
Visnagar, India

Rajesh Kumar Upadhyay
College of Business Studies
COER University
Roorkee, India

Karthikeyan V.
Mepco Schlenk Engineering College
Sivakasi, Tamilnadu, India

Shashikant Yadav
Department of Chemical Engineering
Dr. B. R. Ambedkar National Institute of Technology Jalandhar
Jalandhar, Punjab, India

About the Editors

Dr. Varun Pratap Singh received a Ph.D. (Mechanical Engineering) in the area of Solar Energy. He is currently working as Assistant Professor in the Department of Mechanical Engineering and Head-Solar Energy Centre at the University of Petroleum and Energy Studies Dehradun, India. He has more than 11 years of research and academic experience in mechanical engineering and renewable energy. He has published three patents and five research articles in SCI journals and conferences. He has also completed five projects as Principal Investigator (PI) and two projects as Co-PI with a total research grant of more than 20 lakh INR for the government as well as industry partners. He has delivered many invited talks in webinar, FDP, and workshops and attended 15+ MOOCs and 15+ FDPs from various national bodies like IITs, AICTE, and international bodies like Delft University of Technology-Netherlands, Tsinghua University-China, National Chiao Tung University, Taiwan, and The International Monetary Fund (IMF). He has been awarded for outstanding mentoring of students and a perfect faculty award for excellence in academics and research. He has successfully guided 16 B.Tech., M.Tech, and Ph.D. theses. In administration, he is working as head of the Centre of Excellence, COER. As an academician, Renewable Energy, Solar-Thermal application, Automation, and Special Propose Machining (SPM) are his interest areas of association.

Dr. Ashwani Kumar received a Ph.D. (Mechanical Engineering) in the area of Mechanical Vibration and Design. He is currently working as Senior Lecturer, Mechanical Engineering (Gazetted Officer Group B) at the Technical Education Department Uttar Pradesh (under the Government of Uttar Pradesh) Kanpur, India, since December 2013. He worked as Assistant Professor in the Department of Mechanical Engineering, Graphic Era University Dehradun India (NIRF Ranking 55) from July 2010 to November 2013. He has more than 13 years of research, academic, and administrative experience, including Coordinator for AICTE-Extension of Approval, Nodal officer for PMKVY-TI Scheme (Government of India), internal coordinator-CDTP scheme (Government of Uttar Pradesh), Industry Academia relation officer, Assistant Centre Superintendent (ACS)-Institute Examination Cell, Zonal Officer to conduct Joint Entrance Examination

(JEE-Diploma), Sector Magistrate for Lok Sabha, Vidhan Sabha Election, etc. Being an academician and researcher, he is the series editor of the four book series including *Advances in Manufacturing, Design and Computational Intelligence Techniques, Renewable and Sustainable Energy Developments, Smart Innovations and Technological Advancements in Mechanical and Materials Engineering, Computational Intelligence and Biomedical Engineering* published by CRC Press (Taylor & Francis, U.S.A.) and Wiley (Scrivener Publishing, U.S.A.). He is the guest editor of special issue titled *Sustainable Buildings, Resilient Cities and Infrastructure Systems* (Buildings ISSN: 2075–5309, I.F. 3.8). He is Editor-in-Chief for *International Journal of Materials, Manufacturing and Sustainable Technologies* (IJMMST, ISSN: 2583–6625), and Editor of *International Journal of Energy Resources Applications* (IJERA, ISSN: 2583–6617). He is a guest editor and an editorial board member of eight international journals and acts as a review board member of 20 prestigious (Indexed in SCI/SCIE/Scopus) international journals with high impact factor, i.e., *Applied Acoustics, Measurement*, JESTEC, AJSE, SV-JME, and LAJSS. In addition, he has published 100+ research articles in journals, book chapters, and conferences. He has authored/co-authored cum edited 30+ books on Mechanical, Materials and Energy Engineering. He has published two patents. He is associated with international conferences as an invited speaker, session chair, advisory board, review board member, and program committee member. He has delivered many invited talks in webinars, FDP, and workshops. He has been awarded as Best Teacher for excellence in academics and research. He has successfully guided 15 B.Tech., M.Tech, and Ph.D. theses. He is an external doctoral committee member of S.R.M. University, New Delhi. He is currently involved in the research area of AI and ML in Mechanical Engineering, Smart Materials, and Manufacturing Techniques, Thermal Energy Storage, Building Efficiency, Renewable Energy Harvesting, Sustainable Transportation, and Heavy Vehicle Dynamics. Google Scholar web link is https://scholar.google.com/citations?hl=en&user=KOILpEkAAAAJ and research gate web link is www.researchgate.net/profile/Ashwani-Kumar-45.

Dr. Chandan Swaroop Meena is presently working as a Scientist in the Building Energy Efficiency Department and Assistant Professor at the Academy of Scientific and Innovative Research (AcSIR) at CSIR-Central Building Research Institute, Roorkee (Ministry of Science and Technology, Government of India). He has done his Ph.D. in thermal engineering from IIT Roorkee and M.Tech in thermal engineering from NIT Kurushetra. He does research in renewable energy utilization techniques, geothermal systems, building energy efficiency, two-phase flow, and heat transfer. He is a Life Member of the Indian Society of Heat and Mass Transfer (ISHMT), a Life Member of India Building Congress (IBC), and a Life Member of The Institution of Engineers India (IEI). He has published 29 research articles in reputed journals and international conferences. He is associated with International Conferences as an invited speaker, advisory board, and review board member. He has delivered many invited talks in webinars, FDP, and workshops.

Dr. Gaurav Dwivedi has completed his B.Tech in Mechanical Engineering from IKGPTU Jalandhar, M.Tech in Energy Systems from IIT Roorkee, and a Ph.D. in the field of Renewable Energy (Biofuel) from IIT Roorkee, India. He has nine years of teaching experience in the field of Mechanical and Energy Engineering and taught various subjects like IC Engine, Thermodynamics, Fluid Machinery, Bioenergy, and others at various institutes like VIT University Vellore and Amity University Noida. Currently, he is working as an Assistant Professor in the Energy Centre at Maulana Azad National Institute of Technology, Bhopal, India. His area of interest is Biodiesel Production methodology, enhancement in fuel properties, and application of biofuel in engine operation. He has published more than 100 papers in various international journals and conferences. He has more than 50 papers in the SCI and Scopus Indexed Journal. He has published one book with an international publisher and he has contributed eight book chapters. He got the Young Scientist award at NIT Trichy for his contribution to the field of Renewable Energy. He is an active member of the Institution of Engineers, India, the Solar Energy Society of India, and IEEE. He also served as Guest Editor for *Materials Today* proceedings and the *Journal of Traffic and Transportation*. Recently, he was also featured in the "Top 2% scientist of the world from India (all fields)" by a survey conducted by Stanford University.

Chapter 1

Introduction to Energy-Efficient Vehicles for Sustainable Transportation

Transitions and Challenges

Varun Pratap Singh, Ashwani Kumar, Chandan Swaroop Meena, and Gaurav Dwivedi

1.1 INTRODUCTION

Transportation has played a key role in human society since the inception of civilisation. Transportation has played a pivotal role in the advancement and proliferation of human civilisation, starting with the era when individuals traversed on foot and relied on animals for locomotion through the advent of the wheel and subsequent progress in mechanised transportation [1]. The importance of sustainable mobility has been more significant in recent times. This is due to the growing global awareness of the environmental impacts of transport and the imperative to reduce carbon emissions [2].

Throughout its historical development, transport has seen a tremendous evolution, spanning from its earliest roots to the current day. The methods of transportation for both individuals and commodities have seen significant transformations since the advent of steam engines, internal combustion engines, and electric motors. The transportation industry has assumed a significant role in contributing to greenhouse gas emissions, air pollution, and climate change due to the fast growth of the global population and its increasing need for mobility [3].

As a way to lessen transportation's harmful effects on the environment and advance social and economic development, sustainable transportation has recently attracted more and more attention. As shown in Figure 1.1, the utilisation of energy-efficient, low-emission, secure, and reasonably priced forms of transportation is referred to as sustainable transportation. This include using bicycles, walking, and public transit in addition to driving electric, hybrid, and fuel-efficient vehicles [4].

Although sustainable transport has many advantages, transitioning to a more environmentally friendly transport system is fraught with difficulties [5]. Infrastructure growth, price, technical advancement, and changes in culture and behaviour are a few of them. Switching to sustainable modes of travel is essential for the safety and welfare of people, communities, and the world as a whole [6].

DOI: 10.1201/9781003464556-1

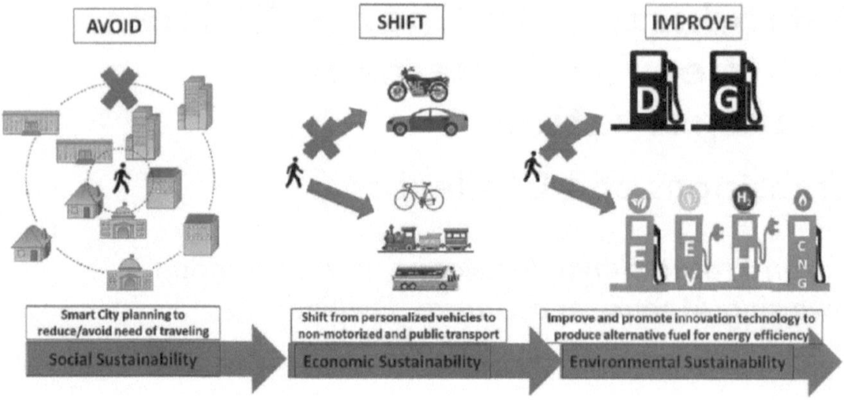

Figure 1.1 Green transportation for sustainability [7].

This chapter delves into the transformations and challenges associated with transitioning towards an environmentally conscious transportation system while simultaneously introducing energy-efficient vehicles to promote sustainable mobility. This will examine several forms of energy-efficient vehicles, such as electric, hybrid, and fuel-efficient automobiles, and highlight their potential contributions to the advancement of ecologically sustainable transportation [8]. This chapter also discusses the implementation of sustainable transport alongside an examination of the contributions made by policymakers, stakeholders, and people in promoting the development of a more sustainable transport system.

1.2 BENEFITS OF ENERGY-EFFICIENT VEHICLES FOR SUSTAINABLE TRANSPORTATION

Energy-efficient cars offer a range of advantages in the context of sustainable mobility (Figure 1.2), encompassing environmental, economic, and social benefits [9]. These characteristics make them an attractive choice for people, corporations, and governments aiming to mitigate their carbon emissions and improve their environmental footprint [10].

1.3 ENVIRONMENTAL BENEFITS OF ENERGY-EFFICIENT VEHICLES

Energy-efficient vehicles play a crucial role in the mitigation of climate change due to their multitude of environmental benefits [11]. One of the primary advantages associated with energy-efficient automobiles is their ability to generate a reduced

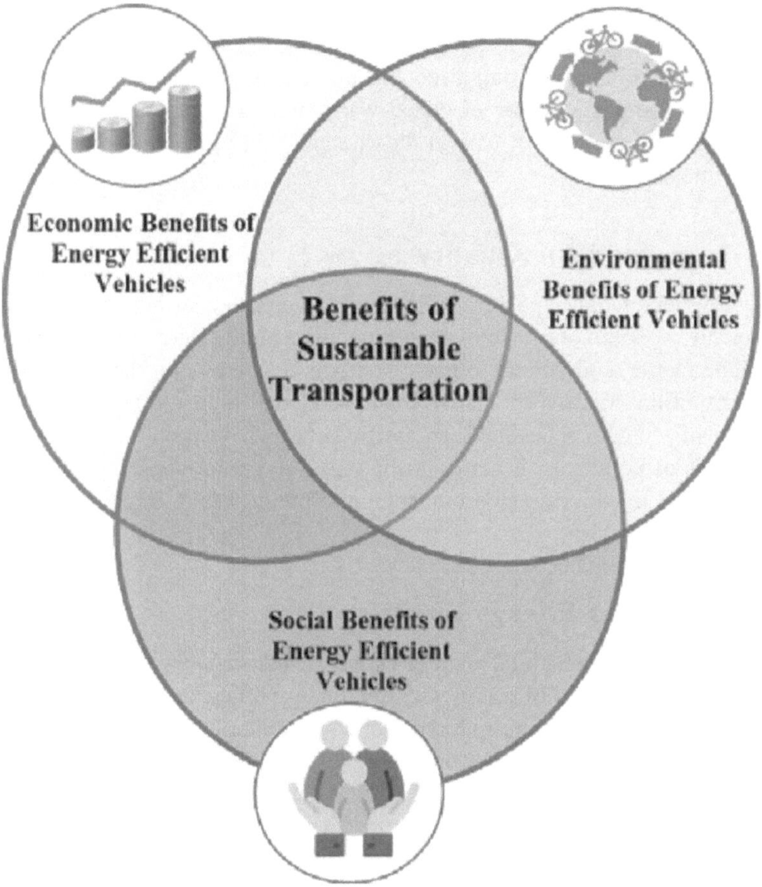

Figure 1.2 Benefits of Energy-Efficient Vehicles for Sustainable Transportation.

amount of greenhouse emissions in comparison to conventional vehicles [12]. These vehicles are designed with the objective of reducing petrol consumption and minimising the emission of pollutants into the atmosphere [13]. Energy-efficient vehicles play a crucial role in mitigating the acceleration of global warming by effectively lowering the emission of carbon dioxide and other greenhouse gases into the Earth's atmosphere [14].

1.3.1 Reduced Greenhouse Gas Emissions

Fuel-efficient cars are well recognised for their significant contribution to the reduction of greenhouse gas emissions in the environment [15]. According to the Environmental Protection Agency (EPA) [16], transportation stands as the primary contributor to greenhouse gas emissions in the United States. Energy-efficient vehicles have the potential to contribute to the reduction of emissions by consuming less petrol and emitting a reduced amount of pollutants into the Earth's atmosphere. The phenomenon of global warming and climate change is mostly attributed to the presence of carbon dioxide and other greenhouse gases in the Earth's atmosphere [17].

1.3.2 Improved Air Quality

Energy-efficient vehicles possess the capacity to enhance air quality by emitting less quantities of pollutants into the surrounding atmosphere. Nitrogen oxides, particulate matter, and volatile organic compounds are among the contaminants emitted by ordinary cars [18]. These toxins have detrimental effects on both the environment and human health. Energy-efficient cars have been found to generate lower levels of pollutants, hence contributing to the improvement of air quality and mitigating the adverse environmental and public health impacts associated with transportation [19].

1.3.3 Enhanced Energy Security

One more advantage pertaining to the environment associated with energy-efficient autos is the enhancement of energy security. Energy-efficient vehicles rely less on fossil fuels as a result of their reduced fuel consumption [20]. Consequently, it is plausible that a decline in reliance on imported fuels and a corresponding enhancement in energy security might ensue [21]. The utilisation of energy-efficient cars has the potential to reduce reliance on fossil fuels, hence mitigating the negative environmental impacts associated with the extraction and utilisation of these fuels [22].

1.4 ECONOMIC BENEFITS OF ENERGY-EFFICIENT VEHICLES

The adoption of energy-efficient cars has the potential to yield economic benefits for individuals, enterprises, and governmental entities. The benefits encompass reduced petrol costs and the generation of employment opportunities within the automotive industry [23].

1.4.1 Reduced Fuel Costs

One of the primary economic benefits of energy-efficient cars is their ability to generate cost savings for individuals and businesses by reducing petrol expenses. Energy-efficient vehicles are designed with the intention of reducing fuel consumption, hence enabling them to cover longer distances with a single tank of petrol

[24]. This phenomenon has the potential to reduce fuel costs during the lifespan of the vehicle, resulting in financial savings for individuals and businesses alike [25].

1.4.2 Job Creation in the Automotive Industry

The potential economic and employment benefits can be enhanced by the development and manufacture of energy-efficient cars [26]. The advancement and manufacture of energy-efficient vehicles have the potential to generate employment opportunities across several sectors, such as research and development, manufacturing, and sales and marketing. The automobile industry plays a significant role in the employment sector of several countries. This phenomenon has the potential to enhance local economies and foster economic expansion [27].

1.5 SOCIAL BENEFITS OF ENERGY-EFFICIENT VEHICLES

Energy-efficient vehicles also provide a number of societal advantages, such as better public health, greater quality of life, and expanded access to transportation [28].

1.5.1 Improved Public Health

The reduction of pollutant emissions in the environment by energy-efficient automobiles can have a positive impact on public health. The adverse effects of conventional vehicular emissions on human health are well-documented, with potential implications for many medical illnesses such as cancer, cardiovascular disorders, and respiratory ailments. Energy-efficient cars have the potential to enhance public health and mitigate the adverse effects of mobility on human well-being by decreasing the emission of pollutants into the atmosphere [29].

1.5.2 Enhanced Quality of Life

The implementation of energy-efficient cars has the potential to positively impact the quality of life for individuals and communities, as well as contribute to improvements in public health. These modes of transport have the potential to enhance the ease and comfort of urban mobility by mitigating noise pollution and alleviating traffic congestion. Energy-efficient vehicles have the potential to contribute to the creation of more liveable and pedestrian-friendly urban areas by reducing the demand for extensive road networks and large parking facilities [30].

1.5.3 Increased Access to Transportation

The utilisation of energy-efficient cars has the potential to provide accessibility to transportation for individuals who were previously deprived of such opportunities. Energy-efficient vehicles have the potential to mitigate mobility obstacles faced by individuals and families with low incomes by reducing transportation expenses and providing supplementary options for mobility [16]. This phenomenon may potentially enhance individuals' access to many opportunities, such as jobs, education, and healthcare [31].

Energy-efficient vehicles offer significant societal benefits that have the potential to profoundly impact both communities and people's well-being [32]. These vehicles have the potential to facilitate the advancement of equitable, sustainable, and habitable communities through the reduction of pollution, enhancement of quality of life, and increased accessibility to transportation [25].

1.6 TECHNOLOGIES USED IN ENERGY-EFFICIENT VEHICLES

Energy-efficient automobiles utilise a wide array of technologies to offset their environmental impact and minimise energy usage. Hybrid and electric vehicles are often acknowledged as significant categories within the domain of energy-efficient automobiles [33].

1.6.1 Hybrid Vehicles

Vehicles having a petrol or diesel engine and an electric motor are known as hybrids. This combination enables increased fuel economy and decreased emissions. Hybrid vehicles employ a technology called regenerative braking to collect energy that would otherwise be lost during braking and use it to power the electric motor [34]. Table 1.1 shows the characteristics of hybrid vehicles.

Table 1.1 Characteristic Comparison of Pros and Cons of Hybrid Vehicles

S. No.	*Characteristic of Comparison*	*Pros*	*Cons*
1	Cost	Fuel efficiency	Higher upfront cost
2	Infrastructure Availability	Environmental benefits	Limited charging infrastructure
3	Battery Performance and Lifespan	Energy savings	Battery degradation
4	Performance	Range and flexibility	Less power for acceleration
5	Space and Weight	Government incentives	Weight and space constraints
6	Maintenance	Smooth and quiet operation	Maintenance and repairs
7	Electric Range	Reduced reliance on fossil fuels	Limited all-electric range
8	Model Availability	Regenerative braking	Limited model options
9	Environmental Impact	Lower greenhouse gas emissions	Potential battery disposal concerns

S. No.	Characteristic of Comparison	Pros	Cons
10	System Complexity	Potential for lower operating costs	Complex hybrid system
11	Material Dependency	High resale value	Potential dependence on rare earth metals
12	Towing Capability	Improved engine lifespan	Limited towing capacity
13	Driving Range	Smaller carbon footprint	Limited long-distance driving range
14	Altitude Performance	Reduction in noise pollution	Possible decrease in performance at high altitudes
15	Trunk Space	Improved city driving experience	Potential for decreased trunk space due to battery placement

Hybrid cars employ a combination of internal combustion engines and electric motors to provide propulsion. The electric motor derives its power from a battery that undergoes recharging using either the petrol engine or regenerative braking [35]. The vehicle's computer system determines the appropriate utilisation of the petrol engine, electric motor, or a combination of both based on the driving conditions and energy demands [36]. This facilitates enhanced fuel economy and improved use of energy resources.

There are two primary classifications of hybrid cars, namely series hybrid and parallel hybrid. The series hybrid configuration involves the use of a gasoline engine to drive a generator, which in turn charges the battery. Simultaneously, the electric motor is responsible for propelling the wheels. Parallel hybrids utilise both an electric motor and a petrol engine to power the wheels [37].

1.6.2 Electric Vehicles

Electric vehicles, in contrast to their conventional counterparts, are driven by an electric motor and battery rather than relying on a petrol or diesel engine. These vehicles are characterised by their lack of emissions, which has the potential to greatly mitigate the environmental impact associated with transportation. Electric vehicles utilise batteries as a means of energy storage for the purpose of supplying power to an electric motor [15]. The battery of the car may be recharged by establishing a connection between the vehicle and an electrical outlet or charging station. The computer system is responsible for regulating energy use, managing power extracted from the battery, and determining the optimal timing for recharging the battery [38]. Table 1.2 shows the characteristics of electric vehicles.

There are two primary classifications of electric cars, namely battery electric vehicles (BEVs) and plug-in hybrid electric vehicles (PHEVs). Due to the fact that the battery serves as the major means of generating electricity for the vehicle, the process of recharging battery-powered mobility necessitates the use of an electrical outlet or a designated charging station. Plug-in hybrid electric vehicles (PHEVs) are equipped with an internal combustion engine (ICE) and an electric motor, enabling them to achieve a greater driving range compared to the capacity of the electric battery alone. The vehicles in question have the capability to be powered by either a traditional internal combustion engine with fuel or by connecting to an electrical power source via an outlet [39].

Table 1.2 Characteristic Comparison of Pros and Cons of Electric Vehicles

S. No.	*Characteristic of Comparison*	*Pros*	*Cons*
1	Fuel Efficiency	Higher energy efficiency, leading to lower operating costs	Limited charging infrastructure and potential range anxiety
2	Environmental Benefits	Zero tailpipe emissions, contributing to cleaner air and reduced pollution	Dependence on electricity generation, which may still rely on fossil fuels
3	Energy Savings	Regenerative braking and energy recovery increase overall efficiency	Limited driving range compared to conventional vehicles
4	Range and Charging Time	Potential for long-range capabilities and fast charging technology	Longer charging times compared to refuelling with gasoline
5	Government Incentives	Subsidies, tax credits, and other incentives may lower purchase costs	Initial higher cost of electric vehicles
6	Quiet and Smooth Operation	Noise reduction and smoother driving experience	Limited engine noise for pedestrian awareness
7	Reduced Dependence on Fossil Fuels	Reduced reliance on non-renewable resources and lower carbon footprint	Energy production for electricity generation may still rely on fossil fuels
8	Lower Maintenance and Operating Costs	Fewer moving parts and less frequent maintenance requirements	Costly battery replacement and potential decrease in capacity over time
9	Convenience of Home Charging	Ability to charge at home, providing convenience and accessibility	Limited charging infrastructure in public spaces
10	Acceleration and Performance	Instant torque and quick acceleration capabilities	Limited top speed and decreased performance at higher speeds
11	Potential for Energy Independence	Potential to generate and store renewable energy for vehicle use	Limited availability and cost of home energy generation and storage solutions

S. No.	*Characteristic of Comparison*	*Pros*	*Cons*
12	Advanced Technological Features	Integration with smart devices, advanced infotainment systems	Complexity and potential for technological glitches
13	Safety Features	Lower center of gravity and improved stability	Concerns about high-voltage electrical systems and associated safety risks
14	Long-Term Cost Savings	Lower fuel and maintenance costs over the vehicle's lifespan	Limited resale value due to rapid advancements in technology and battery degradation
15	Reduced Noise Pollution	Reduced noise pollution, contributing to a quieter environment	Potential for increased risk to pedestrians due to quieter vehicle operation

1.7 CHALLENGES IN MAKING ENERGY-EFFICIENT VEHICLES ACCESSIBLE AND AFFORDABLE

Energy-efficient vehicles come with a lot of advantages, but they also have some drawbacks (Figure 1.3). The high initial cost of obtaining such automobiles is one of the main obstacles to making them accessible and cheap [40].

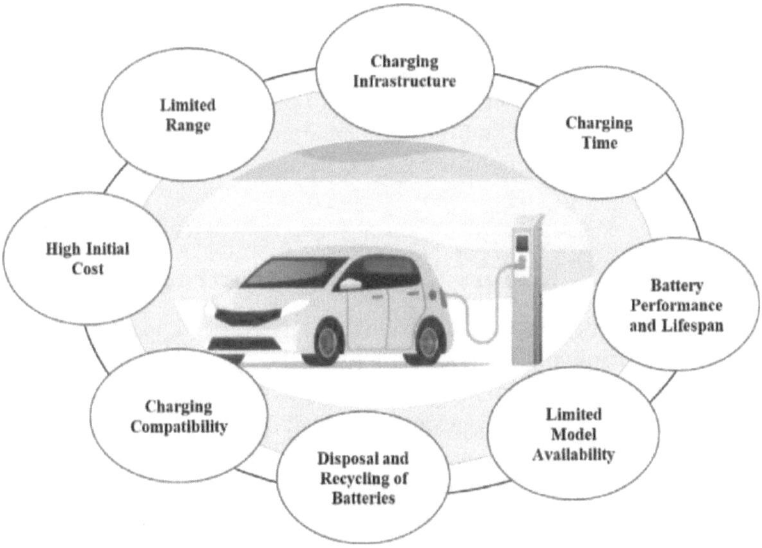

Figure 1.3 Challenges in the Adoption of Energy-Efficient vehicles.

1.7.1 High Initial Cost

The cost of energy-efficient vehicles is rather high due to several factors. The aforementioned expenditures encompass the cost of the battery, the materials employed in the automobile, and the expenses associated with the research and development essential for the production of these vehicles. The construction of energy-efficient autos incurs higher costs compared to conventional vehicles due to the lack of economies of scale in their manufacture [41].

There are several methodologies that may be employed to enhance the cost-effectiveness and availability of energy-efficient vehicles. One viable approach is to enhance manufacturing volumes, facilitating economies of scale and subsequently reducing production expenses. Furthermore, governmental entities may provide tax exemptions and incentives as a means to promote the acquisition of energy-efficient vehicles. Furthermore, it is worth noting that research and development endeavours may prioritise the improvement of battery technology's efficiency and reduction of its cost, as this plays a pivotal role in the exorbitant expenses associated with electric vehicles [42].

1.7.2 Limited Range

The limited range of electric cars is a significant barrier to the broad adoption of energy-efficient autos. The limited range of electric cars compared to conventional automobiles can be attributed to the nascent stage of battery technology. In addition, the presence of an inadequate charging infrastructure might provide difficulties for drivers seeking to recharge their electric vehicles during extended trips [43].

Scholars are now focusing their efforts on the development of novel battery technologies with enhanced efficacy and higher energy density, aiming to augment the driving range of electric vehicles. To promote the widespread use of electric vehicle technology, both governmental entities and commercial enterprises are allocating resources towards the development of a robust charging infrastructure [44].

1.7.3 Need for a Robust Charging Infrastructure

The widespread implementation of electric cars is greatly impeded by the lack of a dependable charging infrastructure. Despite the existence of multiple charging stations, their locations are often inconvenient, and their charging speeds are insufficient for efficient long-distance travel. In addition, drivers face difficulties depending on a consistent and comprehensive network of charging stations due to significant regional disparities in the charging infrastructure [45].

Both governmental entities and private enterprises have the capacity to allocate resources towards the development of a comprehensive network of conveniently located fast-charging stations throughout a nation. This investment would serve to enhance the existing infrastructure for charging electric cars. Furthermore, the use of developing technologies such as wireless charging has the potential to streamline the recharging procedure for electric cars, particularly for those who do not necessitate a direct physical link to a charging station [46].

1.8 THE NEED FOR CONTINUED RESEARCH AND DEVELOPMENT

The imperative for continuous research and development in the domain of energy-efficient cars has more significance as global society advances towards a more sustainable trajectory. The implementation of research and development efforts in several domains, including advancements in battery technology, the establishment of charging infrastructure, and the enactment of government incentives and laws, has the potential to provide significant improvements in the field of energy-efficient cars [47].

1.8.1 Advancements in Battery Technology

Electric batteries play a vital role as a fundamental component, and advancements in battery technology have the potential to enhance their efficiency, capacity, and durability. Existing battery technology is characterised by several limitations, namely its high cost, limited lifespan, and concerns over its environmental impact upon disposal. In order to surpass these limitations, it is imperative to engage in research and development [48].

1.8.1.1 Current State of Battery Technology

The field of battery technology has had notable progress in recent times, mostly due to the increasing need for energy storage in many domains, such as electric transportation and renewable energy infrastructures. Currently, lithium-ion batteries are the prevailing technology owing to their notable attributes such as high energy density, extended cycle life, and comparatively low self-discharge rates. The use of these batteries has greatly assisted the extensive implementation of electric cars and the seamless integration of renewable energy sources into the power grid. Nevertheless, there are still constraints that exist within the discipline. The optimisation of energy density, which governs the capacity for energy storage within a specified space or mass, continues to be a pivotal domain necessitating enhancement. The improvement of energy density is of utmost importance in order to prolong the operational range of electric cars and enhance the storage capacity of portable electronic devices [49 51].

Another constraint exists in the duration required for charging, which remains comparatively lengthier in comparison to the process of refilling a traditional car. There are ongoing endeavours to create rapid charging methods and enhance charging infrastructure in order to mitigate this limitation successfully. Furthermore, the issues of battery longevity and deterioration present significant hurdles. The repeated process of charging and discharging batteries, along with prolonged exposure to elevated temperatures, can lead to a gradual deterioration in battery performance. The primary objective of this research is to enhance the longevity of batteries and develop effective approaches to minimise the process of deterioration. The forthcoming advancements in battery technology will mostly concentrate on surmounting the aforementioned constraints. The anticipated improvements in energy density can be attributed to the progress made in several materials, including silicon anodes, solid-state electrolytes, and lithium-sulphur chemistries. Researchers are now investigating alternative battery chemistries, such as lithium-air and solid-state batteries, to achieve larger energy densities and enhance safety characteristics. In addition, endeavours to enhance manufacturing processes, expand production capacity, and lower expenses will contribute to the increased accessibility and affordability of energy storage technologies in diverse applications. The future development and commercialisation of battery technology will be significantly influenced by ongoing research, government assistance, and collaboration within the industry [52].

1.8.1.2 Strategies to Improve Battery Technology

The enhancement of battery technology can be accomplished by means of heightened research and development endeavours, progressions in the field of material science, refined electrode design, improved battery management systems, the expansion of charging infrastructure, the establishment of recycling processes, collaborative and standardisation initiatives, governmental backing, emphasis on scalability and manufacturing, and exploration of alternative energy storage technologies. Additional approaches for enhancing battery technology encompass minimising the utilisation of scarce resources, advancing recycling techniques, and optimising production procedures with the aim of mitigating expenses [53].

1.8.2 Improvement in Charging Infrastructure

Enhancing the charging infrastructure for electric vehicles encompasses various measures, such as the expansion of public charging networks, the deployment of fast charging technology, the standardisation of connectors and payment systems, the implementation of smart charging solutions, the promotion of workplace and residential charging, the provision of incentives for infrastructure deployment, the integration with the grid, the establishment of fast charging corridors, the exploration of battery swapping stations, and the facilitation of data sharing and visualisation. These initiatives aim to improve the overall charging experience for electric vehicle users [54].

Strategies aimed at improving the charging infrastructure include the expansion of charging stations, the optimisation of charging speed, and the establishment of standardised procedures and guidelines for charging. Both government and private sector investments have the potential to provide financial support for the extension of charging infrastructure and the implementation of defined practices [55].

1.8.3 Government Incentives and Policies

The adoption of energy-efficient automobiles is significantly influenced by government incentives and legislation. The aforementioned policies have the objective of mitigating the initial expenses linked to the acquisition of such vehicles. They encompass the provision of tax credits to both customers and producers, as well as the allocation of cash towards research and development endeavours. Various incentives and regulations have been implemented by governmental bodies to promote the use of energy-efficient cars. These measures encompass tax credits, refunds, and expenditures in research and development [56].

Strategies aimed at enhancing government incentives and policies include augmenting tax credits and rebates for both consumers and manufacturers, fostering the creation of charging infrastructure through incentivisation, and allocating resources towards research and development endeavours to facilitate the widespread adoption of energy-efficient vehicles. Furthermore, the implementation of rules and standards can contribute to enhancing the accessibility and affordability of energy-efficient cars for a wider population [57].

1.9 CONCLUSION

The implementation of energy-efficient cars is of paramount importance in the context of sustainable transportation, given that the transportation sector plays a significant role in the generation of greenhouse gas emissions and subsequent climate change. These cars mitigate emissions by exhibiting reduced fuel use or employing alternative energy sources such as electricity. The adoption of energy-efficient cars facilitates a reduction in reliance on fossil fuels, a drop in carbon emissions, and the mitigation of environmental and public health consequences associated with transportation.

The outlook for energy-efficient automobiles appears to be optimistic in the next years. The progression of technology and the reduction in battery expenses will result in an expanded selection of cost-effective and easily accessible alternatives. The current worldwide emphasis on sustainability and the reduction of carbon emissions is stimulating advancements and financial commitment towards the development of cars that are more efficient and environmentally friendly. The implementation of government incentives and programmes is expected to have a positive impact on the expansion of the energy-efficient car industry.

In conclusion, energy-efficient automobiles present a multitude of advantages for the promotion of sustainable transportation, including diminished carbon emissions, enhanced public health, and heightened accessibility to transportation. Nevertheless, it is imperative to solve obstacles such as the considerable upfront expenses and the restricted operational distance. The acceleration of energy-efficient vehicle adoption necessitates the resolution of problems through sustained battery technology research, charging infrastructure development, and the implementation of supporting government policies. The prioritisation of sustainability and the adoption of energy-efficient automobiles and other sustainable transportation solutions are essential for individuals, governments, and corporations in order to foster a cleaner and more sustainable future.

REFERENCES

[1] V. Smil, *Energy transitions: History, requirements, prospects*. ABC-CLIO, 2010.
[2] L. Mumford, *Technics and civilization*. University of Chicago Press, 2010.
[3] F. Cottrell, *Energy & society: The relation between energy, social change, and economic development*. AuthorHouse, 2009.
[4] C.-H. Yang, K.-C. Lee, and H.-C. Chen, "Incorporating carbon footprint with activity-based costing constraints into sustainable public transport infrastructure project decisions," *J. Clean. Prod.*, vol. 133, pp. 1154–1166, 2016.
[5] T. Sivasakthivel, V. Verma, R. Tarodiya, C. S. Meena, V. P. Singh, and R. Kumar, "Chapter 11: Analysis of Optimum Operating Parameters for Ground Source Heat Pump System for Different Cases of Building Heating and Cooling Mode Operations," in *Thermal Energy Systems: Design, Computational Techniques, and Applications*, 1st ed., A. Kumar, N. Dutt, V. Singh, and C. Meena, Eds. CRC Press, 2023, pp. 183–207.
[6] F. W. Geels, "A socio-technical analysis of low-carbon transitions: Introducing the multi-level perspective into transport studies," *J. Transp. Geogr.*, vol. 24, pp. 471–482, 2012.
[7] K. J. Shah, S. Y. Pan, I. Lee, H. Kim, Z. You, and J. M. Zheng, "Green transportation for sustainability: Review of current barriers, strategies, and innovative technologies," *J. Clean. Prod.*, vol. 326, p. 129392, 2021, doi: 10.1016/j.jclepro.2021.129392.
[8] A. Kundu, A. Kumar, N. Dutt, V. P. Singh, and C. S. Meena, "Chapter 7: Modelling and Simulation of Thermal Energy System for Design Optimization," in *Thermal Energy Systems: Design, Computational Techniques, and Applications*, A. Kumar, N. Dutt, V. Singh, and C. Meena, Eds. CRC Press, 2023, pp. 103–137.
[9] T. Litman and D. Burwell, "Issues in sustainable transportation," *Int. J. Glob. Environ. Issues*, vol. 6, no. 4, pp. 331–347, 2006, doi: 10.1504/IJGENVI.2006.010889.
[10] D. Sperling, *Future drive: Electric vehicles and sustainable transportation*. Island Press, 2013.
[11] A. Raj, A. Dan, Vrinda, and P. Kumar, "A comparative study of the feasibility of alternative fuel vehicles for sustainable transportation in India: A hybrid approach of DEMATEL and TOPSIS," *Transp. Dev. Econ.*, vol. 9, no. 1, p. 2, 2023.
[12] M. Saini, A. Sharma, V. P. Singh, S. Jain, and G. Dwivedi, "Solar thermal receivers—a review," *Lect. Notes Mech. Eng.*, vol. II, pp. 1–25, 2022, doi: 10.1007/978-981-16-8341-1.

[13] K. Chauhan and V. P. Singh, "Prospect of biomass to bioenergy in India : An overview," *Mater. Today Proc.*, 2023, doi: 10.1016/j.matpr.2023.01.419.
[14] E. Shafiei, B. Davidsdottir, J. Leaver, H. Stefansson, and E. I. Asgeirsson, "Comparative analysis of hydrogen, biofuels and electricity transitional pathways to sustainable transport in a renewable-based energy system," *Energy*, vol. 83, pp. 614–627, 2015.
[15] V. P. Singh, A. Karn, G. Dwivedi, T. Alam, and A. Kumar, "Experimental assessment of variation in open area ratio on thermohydraulic performance of parallel flow solar air heater," *Arab. J. Sci. Eng.*, vol. 41, no. 12, pp. 1–17, 2022, doi: 10.1007/s13369-022-07525-7.
[16] A. R. Singh, S. K. Singh, and V. P. Singh, "Process parameters optimization of carbon nano tube based catalytic transesterification of algal oil," *Mater. Today Proc.*, 2023, doi: 10.1016/j.matpr.2023.01.418.
[17] M. Nayal, A. K. Sharma, S. Jain, and V. P. Singh, "Chapter 11 Green Hydrogen Production Methods, Designs and Applications," in Highly Efficient Thermal Renewable Energy Systems Design, Optimization and Applications, 1st ed., R. W. V. Verma; S Thangavel, N Dutt, A Kumar, Ed. CRC Press, 2024, pp. 178–193.
[18] S. Jain, N. Kumar, V. P. Singh, S. Mishra, and N. K. Sharma, "Transesterification of algae oil and little amount of waste cooking oil blend at low temperature in the presence of NaOH," *Energies*, vol. 16, no. 1, pp. 1–12, 2023, doi: 10.3390/en16031293.
[19] H. Vajjarapu, A. Verma, and H. Allirani, "Evaluating the climate change mitigation potential of sustainable urban transport measures in India," *J. Urban Plan. Dev.*, vol. 149, no. 1, p. 4022047, 2023.
[20] V. P. Singh, C. S. Meena, A. Kumar, and N. Dutt, "Double pass solar air heater : A review," *Int. J. Energy Resour. Appl.*, vol. 1, no. 2, pp. 22–43, 2022, doi: 10.56896/IJERA.2022.1.2.009.
[21] V. P. Singh and G. Dwivedi, "Technical analysis of a large-scale solar updraft tower power," *Energies*, vol. 16, no. 1, pp. 1–29, 2023, doi: 10.3390/en16010494.
[22] M. Farghali, A. I. Osman, I. M. A. Mohamed, Z. Chen, L. Chen, I. Ihara, P. S. Yap, and D. W. Rooney, "Strategies to save energy in the context of the energy crisis: A review," *Environ. Chem. Lett.*, vol. 21, pp. 2003–2039, 2023.
[23] A. K. Sharma, M. Nayal, S. Jain, and V. P. Singh, "Chapter 07 Optimization Techniques of Solar Thermal and Hybrid Energy Systems," in Highly Efficient Thermal Renewable Energy Systems Design, Optimization and Applications, 1st ed., R. W. V. Verma; S Thangavel, N Dutt, A Kumar, Ed. CRC Press, 2024, pp. 1–14.
[24] V. P. Singh, A. Dwivedi, A. Karn, A. Kumar, S. Singh, S. Srivastava, and K. Srivastava, "Nanomanufacturing and Design of High-Performance Piezoelectric Nanogenerator for Energy Harvesting," in *Nanomanufacturing and Nanomaterials Design: Principles and Applications*. S. Singh, S. K. Behura, A. Kumar, and K. Verma, Eds. CRC Press, 2022, pp. 241–272. https://doi.org/10.1201/9781003220602
[25] I. Galarraga, S. Kallbekken, and A. Silvestri, "Consumer purchases of energy-efficient cars: How different labelling schemes could affect consumer response to price changes," *Energy Policy*, vol. 137, p. 111181, 2020.
[26] S. Srivastava, D. Verma, S. Thusoo, A. Kumar, V. P. Singh, and R. Kumar, "Nanomanufacturing for Energy Conversion and Storage Devices," in *Nanomanufacturing and Nanomaterials Design: Principles and Applications*. S. Singh, S. K. Behura, A. Kumar, and K. Verma, Eds. CRC Press, 2022, pp. 165–174. https://doi.org/10.1201/9781003220602

[27] J. Stolfa, S. Stolfa, C. Baio, U. Madaleno, P. Dolejsi, F. Brugnoli, and R. Messnarz, "DRIVES—EU blueprint project for the automotive sector—a literature review of drivers of change in automotive industry," *J. Softw. Evol. Process*, vol. 32, no. 3, p. e2222, 2020.

[28] D. Bissell, T. Birtchnell, A. Elliott, and E. L. Hsu, "Autonomous automobilities: The social impacts of driverless vehicles," *Curr. Sociol.*, vol. 68, no. 1, pp. 116–134, 2020.

[29] M. Nayal, A. K. Sharma, S. Jain, and V. P. Singh, "Chapter 06 Design and Modelling of Solar, Geothermal and Hybrid Energy Systems.pdf," in Highly Efficient Thermal Renewable Energy Systems Design, Optimization and Applications, 1st ed., R. W. V. Verma; S Thangavel, N Dutt, A Kumar, Ed. CRC Press, 2024, pp. 1–15.

[30] T. Cohen, J. Stilgoe, S. Stares, N. Akyelken, C. Cavoli, J. Day, J. Dickinson, V. Fors, D. Hopkins, G. Lyons, N. Marres, J. Newman, L. Reardon, N. Sipe, C. Tennant, Z. Wadud, and E. Wigley. "A constructive role for social science in the development of automated vehicles," *Transp. Res. Interdiscip. Perspect.*, vol. 6, p. 100133, 2020. https://doi.org/10.1016/j.trip.2020.100133

[31] M. Martiskainen, B. K. Sovacool, M. Lacey-Barnacle, D. Hopkins, K. E. H. Jenkins, N. Simcock, G. Mattioli, and S. Bouzarovski "New dimensions of vulnerability to energy and transport poverty," *Joule*, vol. 5, no. 1, pp. 3–7, 2021. https://doi.org/10.1016/j.joule.2020.11.016

[32] A. Kumar, A. K. S. Gangwar, A. Kumar, C. S. Meena, V. P. Singh, N. Dutt, A. Prasad, and Y. Gori, "Biomedical Study of Femur Bone Fracture and Healing," in *Advanced Materials for Biomedical Applications*. A. Kumar, Y. Gori, A. Kumar, C. S. Meena, and N. Dutt, Eds. CRC Press, 2022, pp. 283–298. https://doi.org/10.1201/9781003344810

[33] R. Kumar, A. Kumar, L. Kant, A. Prasad, S. Bhoi, C. S. Meena, V. P. Singh, and A. Ghosh, "Experimental and RSM-based process-parameters optimisation for turning operation of EN36B steel," *Materials (Basel).*, vol. 16, no. 1, pp. 1–18, 2023, doi: 10.3390/ma16010339.

[34] C. Acar and I. Dincer, "The potential role of hydrogen as a sustainable transportation fuel to combat global warming," *Int. J. Hydrogen Energy*, vol. 45, no. 5, pp. 3396–3406, 2020.

[35] A. Kumar, V. P. Singh, C. S. Meena, and N. Dutt, *Thermal Energy Systems: Design, Computational Techniques, and Applications*, 1st ed. CRC Press, 2023.

[36] N. Dutt, A. Binjola, A. J. Hedau, A. Kumar, V. P. Singh, and C. S. Meena, "Comparison of CFD results of smooth air duct with experimental and available equations in literature," *Int. J. Energy Resour. Appl.*, vol. 1, no. 1, pp. 40–47, 2022, doi: 10.56896/IJERA.2022.1.1.006.

[37] C. Yang, M. Zha, W. Wang, K. Liu, and C. Xiang, "Efficient energy management strategy for hybrid electric vehicles/plug-in hybrid electric vehicles: Review and recent advances under intelligent transportation system," *IET Intell. Transp. Syst.*, vol. 14, no. 7, pp. 702–711, 2020.

[38] M. R. Wahid, B. A. Budiman, E. Joelianto, and M. Aziz, "A review on drive train technologies for passenger electric vehicles," *Energies*, vol. 14, no. 20, pp. 1–24, 2021, doi: 10.3390/en14206742.

[39] N. Thanh Tung and L. Van Van, "Modeling to study the braking efficiency of the electric vehicle," *Mater. Today Proc.*, 2023, doi: 10.1016/j.matpr.2023.05.341.

[40] L. Woodley, P. Rossetti, and A. Nunes, "Targeted electric vehicle procurement incentives facilitate efficient abatement cost outcomes," *Sustain. Cities Soc.*, vol. 96, p. 104627, 2023, doi: 10.1016/j.scs.2023.104627.

[41] G. Singh, S. Rathor, A. Karn, and V. P. Singh, "Recent progress in solar dryers using phase changing material: A review," *Int. J. Energy Resour. Appl.*, vol. 2, no. 1, pp. 57–77, 2023, doi: 10.56896/IJERA.2023.2.1.005.

[42] A. G. Olabi, T. Wilberforce, K. Obaideen, E. T. Sayed, N. Shehata, A. H. Alami, and M. A. Abdelkareem, "Micromobility: Progress, benefits, challenges, policy and regulations, energy sources and storage, and its role in achieving sustainable development goals," *Int. J. Thermofluids*, vol. 17, p. 100292, 2023, doi: 10.1016/j.ijft.2023.100292.

[43] R. A. Acheampong, C. Legacy, R. Kingston, and J. Stone, "Imagining urban mobility futures in the era of autonomous vehicles—insights from participatory visioning and multi-criteria appraisal in the UK and Australia," *Transp. Policy*, vol. 136, pp. 193–208, 2023, doi: 10.1016/j.tranpol.2023.03.020.

[44] N. Dutt, A. Jageshwar, H. Ashwani, K. Mukesh, K. Awasthi, and V. Pratap, "Thermo - hydraulic performance of solar air heater having discrete D - shaped ribs as artificial roughness," Environ. Sci. Pollut. Res., 2023, doi: 10.1007/s11356-023-28247-9.

[45] Y. Liu, A. Francis, C. Hollauer, M. C. Lawson, O. Shaikh, A. Cotsman, K. Bhardwaj, A. Banboukian, M. Li, A. Webb, and O. Y. Asensio, "Reliability of electric vehicle charging infrastructure: A cross-lingual deep learning approach," *Commun. Transp. Res.*, vol. 3, p. 100095, 2023, doi: 10.1016/j.commtr.2023.100095.

[46] S. LaMonaca and L. Ryan, "The state of play in electric vehicle charging services – a review of infrastructure provision, players, and policies," *Renew. Sustain. Energy Rev.*, vol. 154, p. 111733, 2022, doi: 10.1016/j.rser.2021.111733.

[47] P. K. Tarei, P. Chand, and H. Gupta, "Barriers to the adoption of electric vehicles: Evidence from India," *J. Clean. Prod.*, vol. 291, p. 125847, 2021, doi: 10.1016/j.jclepro.2021.125847.

[48] V. S. Patyal, R. Kumar, and S. Kushwah, "Modeling barriers to the adoption of electric vehicles: An Indian perspective," *Energy*, vol. 237, p. 121554, 2021, doi: 10.1016/j.energy.2021.121554.

[49] A. K. Yadav, V. P. Singh, A. Kumar, Y. Gori, and A. Kumar, "Study the properties of high-density polyethylene with nano particles silica composite," *Int. J. Mater. Manuf. Sustain. Technol.*, vol. 1, no. 1, pp. 29–36, 2022, doi: 10.56896/IJMMST.2022.1.1.005.

[50] A. Saxena, A. Prajapati, G. Pant, C. Meena, A. Kumar, and V. P. Singh, "Water Consumption Optimization of Hybrid Heat Pump Water Heating System," in *Lecture Notes in Mechanical Engineering Recent Advances in Mechanical Engineering Select Proceedings of FLAME 2022*, A. Shukla, B. Sharma, A. Arabkoohsar, and P. Kumar, Eds. Springer Nature, 2023, pp. 721–732.

[51] R. Kumar, A. Kumar, Laxmikant, N. Dutt, and V. P. Singh, "Cutting parameters optimization in turning operation using Taguchi method," *Int. J. Mater. Manuf. Sustain. Technol.*, vol. 1, no. 2, pp. 22–30, 2022, doi: 10.56896/IJMMST.2022.1.2.009.

[52] V. P. Singh and S. Jain, "Economic analysis of a large scale solar updraft tower power plant," Sustain. Energy Technol. Assessments, vol. 58, p. 103325, 2023, doi: https://doi.org/10.1016/j.seta.2023.103325.

[53] G. Pant, C. S. Meena, A. Saxena, A. Kumar, V. P. Singh, and N. Dutt, "Study the Temperature Variation in Alternate Coils of Insulated Condenser Cum Storage Tank: Experimental Study," in Lecture Notes in Mechanical Engineering. Springer, Singapore, 2023, pp. 627–638, doi: https://doi.org/10.1007/978-981-99-2382-3_52.

[54] A. A. Visaria, A. F. Jensen, M. Thorhauge, and S. E. Mabit, "User preferences for EV charging, pricing schemes, and charging infrastructure," *Transp. Res. Part A Policy Pract.*, vol. 165, pp. 120–143, 2022, doi: 10.1016/j.tra.2022.08.013.

[55] P. Patil, K. Kazemzadeh, and P. Bansal, "Integration of charging behavior into infrastructure planning and management of electric vehicles: A systematic review and framework," *Sustain. Cities Soc.*, vol. 88, p. 104265, 2023, doi: 10.1016/j.scs.2022.104265.

[56] N. Neshat, M. Kaya, and S. Ghaboulian Zare, "Exploratory policy analysis for electric vehicle adoption in European countries: A multi-agent-based modelling approach," *J. Clean. Prod.*, vol. 414, p. 137401, 2023, doi: 10.1016/j.jclepro.2023.137401.

[57] S. Harichandan and S. K. Kar, "An empirical study on consumer attitude and perception towards adoption of hydrogen fuel cell vehicles in India: Policy implications for stakeholders," *Energy Policy*, vol. 178, p. 113587, 2023, doi: 10.1016/j.enpol.2023.113587.

Chapter 2

Zero Emission

Challenges and Modern Solutions

Dr. Shaik Khaja Mohiddin, Shaik Sharmila, Mandula China Pentu Saheb

2.1 INTRODUCTION

The effects of climate change on the environment have received unprecedented attention and concern from people all around the world in the 21st century. Global warming and the destruction of natural ecosystems have been linked mostly to the release of greenhouse gases, particularly carbon dioxide. The urgent need to transition to a zero-emission future has risen to the forefront as the effects of climate change become clearer. When applied to greenhouse gas emissions and other air pollutants, "zero emission" refers to a lofty objective: their complete eradication or neutralization. It is an all-encompassing strategy for reducing humankind's negative impact on the natural world by rethinking how authors produce and use energy, move from place to place, and run factories. In addition to its importance in reducing global warming, achieving zero emissions is essential for enhancing air quality, protecting biodiversity, and minimizing risks to human health. The goal of this chapter is to examine the complex difficulties of creating a society with no emissions and to show some of the cutting-edge strategies that point the way forward. It examines the technical, monetary, and policy obstacles that must be surmounted to make this shift possible. In addition, it emphasizes the novel strategies and technological developments in numerous fields that can hasten the arrival of the goal of zero emissions.

The challenges encountered in achieving zero emission are manifold and require meticulous attention. Technologically, there exist limitations and inefficiencies in current solutions that impede the complete elimination of emissions. The development of innovative technologies and scalable infrastructure is imperative to overcome these obstacles. Additionally, the vast infrastructural changes necessary to support zero-emission systems, such as renewable energy generation and electric vehicle charging networks, demand extensive planning, investment, and coordination. The economic implications of transitioning to a zero-emission society cannot be overlooked. Industries deeply entrenched in fossil fuel-dependent practices face considerable economic disruption as they strive to align with clean energy alternatives. Balancing the pursuit of economic growth with the imperative of

DOI: 10.1201/9781003464556-2

emission reduction poses a delicate challenge that requires careful navigation and supportive policies.

Furthermore, the formulation and implementation of effective policy frameworks are vital to drive the widespread adoption of zero-emission technologies. Governments must play a pivotal role in creating an enabling environment by establishing favourable regulations, incentives, and funding mechanisms. However, the complexity of policy-making and the need for international cooperation present additional challenges to achieving zero emissions on a global scale. Nevertheless, despite these challenges, there are a plethora of modern solutions that offer hope and pave the way for a zero-emission future. Renewable energy sources, including solar, wind, hydro, and geothermal, have emerged as sustainable alternatives to fossil fuels. Remarkable advancements in these technologies have made them increasingly cost-effective and efficient, positioning them as viable solutions to combat climate change.

The electrification of transportation represents a transformative solution that holds immense potential for emission reduction. Electric vehicles (EVs), with their zero tailpipe emissions, have experienced rapid technological advancements and increased market acceptance. Coupled with the development of supportive charging infrastructure and innovative business models, electric mobility can revolutionize transportation and contribute significantly to achieving zero emissions. Moreover, electrified public transportation systems, such as electric buses and trains, can help alleviate urban air pollution and reduce emissions in densely populated areas. Energy efficiency and conservation play a crucial role in the zero-emission landscape. By improving the efficiency of buildings, appliances, and industrial processes, significant reductions in energy consumption and associated emissions can be achieved. Smart technologies, energy-efficient design principles, and behavioural changes toward sustainable practices all contribute to a more sustainable and low-carbon future.

Carbon capture and storage (CCS) technologies offer a promising avenue for mitigating emissions from industrial processes and power generation, by capturing carbon dioxide emissions and storing them. In geologically stable formations, CCS can prevent significant amounts of greenhouse gases from being released into the atmosphere, thereby contributing to the goal of zero emissions. Ongoing research and development efforts are focused on improving the efficiency, scalability, and cost-effectiveness of CCS technologies.

The transition to a zero-emission society requires a collective effort and a fundamental shift in mindset. Individuals, businesses, and governments all play critical roles in driving this transformation. Citizens need to adopt eco-friendly habits such as limiting their use of non-renewable resources, buying locally grown food, and recycling. The best way for businesses to ensure their long-term success and reduce their negative influence on the environment is to make sustainability a top priority and fund creative new approaches. Governments can set an example by developing all-encompassing policies that promote environmentally friendly technology, provide room for them in regulatory frameworks, and encourage cross-border cooperation.

It is impossible to overestimate the importance of reaching zero emissions. From extreme weather to rising sea levels and disturbances to ecosystems, the effects of climate change may be seen all across the world right now. The authors cannot discount the human cost of air pollution, which is driven largely by emissions from vehicles and factories. Vulnerable groups are disproportionately affected by poor air quality, which has been related to respiratory disorders, cardiovascular diseases, and premature mortality. In addition to being a question of public health and social justice, the pursuit of zero emissions is an environmental need.

The goal of this chapter is to look more into the difficulties and potential approaches to reaching zero emissions. The authors seek to shed light on the issue and motivate people to take action by investigating its technological, economic, and policy components. A world with no emissions is possible if authors are willing to put in the work, come up with creative solutions, and work together. If the authors work together, they can make a world where people can breathe easily, ecosystems flourish, and the Earth can remain habitable for future generations. Let's take the first step on the path to zero emissions and create a better, greener, more sustainable tomorrow.

2.2 SUMMARY OF PREVIOUS LITERATURE

Table 2.1 shows the examples of related works by different authors.

2.3 CHALLENGES TO ZERO EMISSION

The current levels of pollution and emissions pose a serious threat to human and environmental health. Over the past century, industrialization, population growth, and increased energy consumption have all contributed to a rise in greenhouse gas emissions, mainly carbon dioxide (CO2), from the combustion of fossil fuels like coal, oil, and natural gas (Table 2.2). Greenhouse gases are released into the atmosphere and cause global warming. Increasing temperatures, melting ice caps and glaciers, more frequent and violent weather events, and changes to ecosystems and biodiversity are all signs that global warming is already having a big impact.

Emissions from many different sources, such as global warming, contribute to air pollution, which is a major threat to public health. Some examples of air pollutants that are produced during industrial processes, vehicle operations, power generation, and everyday life include particulate matter (PM), sulfur dioxide (SO2), nitrogen oxides (NOx), and volatile organic compounds (VOCs). Asthma, cardiovascular disease, lung cancer, and other illnesses have all been related to increased levels of air pollution.

Table 2.1 A Number of Studies on Emissions-Free Technologies

Author(s)	*Contribution*	*Methodology*	*Application*	*Limitation*
Smith et al. [1]	Analyzed the challenges of achieving zero emission	Literature review	General overview	Lack of specific case studies
Johnson & Lee [2]	Investigated the economic implications of zero emission	Economic analysis	Policy and industry-level implications	Limited to economic analysis
Li et al. [3]	Explored renewable energy sources for zero emission	Comparative study	Energy sector and renewable energy applications	Limited focus on renewable energy sources
Wang & Zhang [4]	Studied the potential of electric vehicles for zero emission	Data analysis	Transportation sector and electric vehicle adoption	Limited to electric vehicles
Brown & Miller [5]	Examined the role of energy efficiency in achieving zero emission	Case study	Building and industrial sectors	Limited to energy efficiency measures
Smith & Jones [6]	Investigated the policy frameworks for zero emission	Policy analysis	Government-level policies and regulatory frameworks	Limited to policy analysis
Chen et al. [7]	Explored carbon capture and storage technologies	Techno-economic analysis	Industrial and power generation sectors	Focused on carbon capture and storage technologies
Kim & Park [8]	Analyzed the environmental impact of zero emission	Life cycle assessment	Multiple sectors and overall environmental assessment	Limited to environmental impact assessment
Garcia et al. [9]	Studied the social implications of zero emission	Social impact assessment	Community-level and societal implications	Focused on social aspects of zero emission
Patel & Sharma [10]	Explored the role of international collaboration in achieving zero emission	Case study	Global cooperation and climate change mitigation	Limited to the role of international collaboration

Table 2.2 The Challenges to Achieving Zero Emissions Are Multifaceted

Challenge	*Explanation*
Energy Transition [11–16]	Investments in infrastructure, technical development, and policy support are necessary to make the transition from fossil fuels to cleaner and renewable energy sources. As well as facing opposition from the fossil fuel industry, the transformation requires overcoming technical and economic challenges. Nano techniques for energy conversion and storage devices with special emphasis on high-performance piezoelectric nanogenerators for energy harvesting.
Energy Storage and Grid Integration	Since renewable energy sources like solar and wind are not always available, we need effective energy storage technologies to keep the lights on. Because of their inherent unpredictability, renewable energy sources require infrastructure upgrades and expansions before they can be properly integrated into current networks.
Industrial Emissions	The industrial sector is responsible for a sizable portion of the world's emissions. Complex processes and the need for new technologies and methods to reduce carbon intensity make it difficult to reduce emissions in these industries.
Transportation Sector	Greenhouse gases are largely produced by the transportation industry, which relies heavily on cars fueled by fossil fuels. Issues including inadequate charging infrastructure, greater upfront costs, and consumer adoption hurdles must be addressed in order to make the switch to EVs and improve public transit systems.
Policy and Regulatory Frameworks	It is critical to enact laws and regulations that provide incentives for lowering emissions and promoting sustainable behaviours. However, it can be difficult to create and implement comprehensive frameworks that take into account the unique requirements and difficulties of each region while still supporting national and international objectives.
Financial and Economic Considerations	The initial investment needed to implement zero-emissions technologies and infrastructure can be rather high. Investment in low-carbon solutions must be encouraged, and this can be done through the provision of affordable finance and the establishment of economic incentives, such as carbon pricing mechanisms or subsidies.
Public Awareness and Engagement	To achieve zero emissions, society as a whole must come together and alter its habits and mindsets. In order to motivate widespread action, it is crucial to inform the public about the advantages of emission reduction, encourage people to adopt more environmentally friendly habits, and increase community involvement.

It will take a confluence of technological innovation, policy interventions, international cooperation, and public participation to solve these problems and achieve zero emissions. A proactive and coordinated effort from governments, industry, communities, and individuals is required, however, to lessen the effects of climate change and enhance air quality.

2.3.1 Policy Challenges

a. *Lack of Ambitious Policies*: The path to zero emissions might be slowed by policies that are insufficient or weak in setting clear targets and incentives for carbon reductions. Renewable energy, energy efficiency, and environmentally responsible behaviours must be at the forefront of strong, long-term policy.
b. *Policy Reversals and Uncertainty:* Frequent changes in policies or lack of policy stability can create uncertainty for investors and businesses, making it difficult to plan and implement long-term emission reduction strategies.
c. *Lack of International Cooperation:* Global efforts to tackle climate change require strong international cooperation. Challenges can arise from conflicting interests, varying levels of commitment, and difficulties in reaching consensus among different nations.

2.3.2 Technological Challenges

a. *Energy Storage*: The intermittent nature of renewable energy sources like solar and wind poses challenges to grid stability and reliability. Developing cost-effective and efficient energy storage technologies to store surplus energy for use during periods of low generation is crucial.
b. *Decarbonizing Hard-to-Abate Sectors*: Certain sectors, such as heavy industry, aviation, and shipping, have limited technologies available for decarbonization. Finding viable low-carbon alternatives and investing in research and development to address these sectors is essential.
c. *Scaling up Renewable Energy:* While renewable energy has seen significant growth, scaling up its deployment to replace fossil fuel-based energy generation requires addressing challenges such as grid integration, land use issues, and intermittency.

2.3.3 Economic Challenges

a. *Cost Competitiveness*: Transitioning to zero emissions may involve higher upfront costs compared to traditional fossil fuel-based technologies. Achieving cost competitiveness and reducing the levelized cost of renewable energy technologies is crucial for widespread adoption.
b. *Access to Financing*: Lack of access to affordable financing, particularly in developing countries, can impede the deployment of renewable energy projects and the implementation of emission reduction measures.

Table 2.3 Modern Solutions for Zero Emission

Solution	*Explanation*
Green Hydrogen	Green hydrogen, produced through electrolysis using renewable energy, has the potential to decarbonize sectors such as transportation, industry, and heating. It can be used as a clean fuel or feedstock, reducing reliance on fossil fuels.
Advanced Battery Technologies	Advancements in battery technologies, such as lithium-ion batteries and next-generation solid-state batteries, are crucial for energy storage, electric vehicles (EVs), and grid integration. These technologies improve energy density, charging speed, and lifespan.
Carbon Capture, Utilization, and Storage (CCUS)	CCUS technologies capture CO2 emissions from power plants and industrial processes, either storing them underground or converting them into useful products. CCUS can help reduce emissions from hard-to-abate sectors.
Direct Air Capture (DAC)	DAC technologies capture CO2 directly from ambient air, enabling the removal of emissions already present in the atmosphere. This approach has the potential to offset hard-to-decarbonize sectors and achieve negative emissions.
Smart Grid and Energy Management Systems	Smart grid technologies and advanced energy management systems, enable efficient integration of renewable energy, demand response, and energy storage, optimizing electricity generation, consumption, and grid stability.

c. *Transitioning Workforce and Communities:* Shifting from carbon-intensive industries to cleaner technologies can have economic and social implications. Supporting affected workers and communities through reskilling, job creation, and providing a just transition is essential for a smooth and equitable transition.

To overcome these challenges and make the shift to zero emissions possible, authors need a combination of policy incentives, technological breakthroughs, and economic reforms. To overcome these obstacles and create a sustainable and low-carbon future, collaboration between governments, industry, and civil society is essential. Table 2.3 shows the modern solutions for these challenges.

2.4 MODERN SOLUTIONS

Some of the new and promising options that could help get us to zero emissions are represented in what follows. Each technological advancement has its own set of benefits and potential uses across the economy. Realizing their full potential and speeding up the transition to an environmentally friendly and low-carbon future requires continuous research and development together with supportive policies and funding.

2.5 ROLE OF POLICY

Table 2.4 shows the descriptions of various necessary policies on zero emission. The transition to zero emissions relies heavily on policy, which establishes rules and incentives for more environmentally friendly behaviour. The article following discusses the importance of policy in pushing toward a future with no emissions.

Table 2.4 Description of Various Necessary Policies on Zero Emission

Policy	*Explanation*
Setting Ambitious Targets	Goals and targets for reducing emissions, the deployment of renewable energy, energy efficiency enhancements, and sustainable behaviours are established by policies in the long term. These goals instill a sense of urgency and offer a road map for companies, organizations, and individuals.
Regulatory Framework	Policies establish the rules for limiting pollution and other forms of environmental degradation. They regulate emissions from various industries, monitor compliance, and punish offenders. Businesses and industries are compelled by regulations to implement more environmentally friendly processes and equipment.
Economic Incentives	Policies may offer monetary incentives like feed-in tariffs, tax credits, and tax breaks to encourage the use of renewable energy sources, energy-saving technologies, and emission-reducing procedures. Investment in zero-emission solutions is boosted, and the playing field is levelled, all thanks to these incentives.
Research and Development Support	In order to reach the goal of zero emissions, policy frameworks might direct resources toward developing clean technology, energy storage, and other related fields. Funding like this encourages creative thinking, propels technological progress, and reduces the price tag of eco-friendly options.
International Cooperation	Policies promote global agreements and alliances, which, in turn, facilitates international cooperation and collaboration. The Paris Agreement, which establishes a framework for countries to cooperate in their efforts to limit global warming, and the exchange of best practices and knowledge are two such examples that will speed up the transition to a world with zero emissions.
Market Signals and Demand Creation	By instituting carbon pricing mechanisms or emission trading systems, policies can generate market signals. These methods put a price on carbon emissions, which motivates companies to cut back on emissions and make investments in low-carbon technologies. The demand for environmentally friendly products and services is determined by market forces.
Public Awareness and Engagement	To encourage a sustainable way of life, policies should support initiatives to raise public consciousness through information sharing and instruction. As a result of these laws, residents are given the agency to aid in the transition to zero emissions by making greener choices in their daily lives.
Long-Term Planning and Stability	Clear and consistent policy frameworks are more stable and predictable for businesses, investors, and the market as a whole. Long-term planning and regulatory stability encourage strategic investments in emission-free technology and infrastructure.

The right policy can provide the necessary guidance, incentives, and assistance to speed up the journey to zero emissions. Table 2.5 reviews the current policies and their effectiveness. Organizations, companies, and individuals can take the necessary steps toward a low-carbon, sustainable future if the proper policies are in place. Table 2.6 lists suggestions for zero emissions and future outlooks are stated in Table 2.7.

2.6 CURRENT POLICY INITIATIVES AND EFFECTIVENESS

Table 2.5 A Brief Description and Review of the Current Policies

Current policy	*Initiatives and effectiveness*
Renewable Energy Targets and Feed-in Tariffs	Financial incentives for the development of renewable energy are provided by feed-in tariffs, which have been established in several nations. These regulations have successfully spurred investment and increased the proportion of renewable energy in the overall electricity mix.
Carbon Pricing Mechanisms	Carbon taxes and emissions trading systems are two methods used to put a monetary value on carbon emissions. Despite the demonstrated success of these mechanisms in lowering emissions and stimulating investment in low-carbon technologies, there remains a need for additional implementation and harmonization due to differences in pricing and coverage.
Energy Efficiency Standards and Building Codes	Energy consumption has been reduced, and energy-efficient technology has been more widely adopted thanks to policies establishing minimum standards for appliances, automobiles, and buildings. However, the efficacy must be enforced and updated frequently to account for technology changes.
Research and Development Funding	Research and development of environmentally friendly technologies has received funding from national governments. Renewable energy, energy storage, and carbon capture are just a few examples of how these investments have spurred innovation and helped bring down costs. To hasten technological progress, it is crucial that financing and cooperation between academics, industry, and government continue.
International Climate Agreements	The Paris Agreement and other international accords provide a framework for governments to cooperate in the fight against climate change. While these pacts are vital for international collaboration, their success will depend on the dedication and implementation of individual nations.

2.7 SUGGESTIONS FOR FURTHER POLICY MEASURES

Table 2.6 List of Some Suggestions for Zero Emissions

Suggestions	*Explanation*
Strengthened Carbon Pricing	Carbon pricing schemes can be implemented or improved by governments to increase the financial incentive to reduce emissions. This may involve measures such as global price alignment, higher carbon tax rates, and the expansion of emissions trading systems.
Scaling up Renewable Energy Targets	Governments should establish lofty goals for renewable energy, with a particular eye toward expanding the sector's share of the total energy pie. Rapid deployment of renewable energy projects can be facilitated through long-term planning, encouraging legislation, and expedited approval procedures.
Energy Efficiency Incentives	Businesses and individuals can be prompted to adopt energy-saving measures if incentives for doing so are improved. Overcoming initial investment barriers and encouraging retrofits for energy efficiency can be accomplished through a variety of creative financing options.
Support for Research and Development	Energy storage, hydrogen, and environmentally friendly transportation are just a few examples of the types of promising new clean technology that governments should keep funding. Partnerships between public and commercial sectors in research and development can hasten technological progress.
Green Public Procurement	The demand for emission-free goods and services can be boosted by the government's purchasing power. Market growth and private sector uptake can be stimulated by enacting "green public procurement policies" that give preference to low-carbon and sustainable solutions.
Education and Awareness Campaigns	Governments have a significant role to play in educating the public about the benefits of low- or no-impact lifestyle choices. People can be motivated to adopt more sustainable habits through awareness campaigns, community outreach programs, and financial incentives.
Cross-Sector Collaboration	Implementing a combination of these policy measures, tailored to the specific contexts and needs of each country, can provide a robust framework to support the transition to zero emissions. Regular monitoring, evaluation, and adjustment of policies based on evolving technologies and market dynamics are crucial for ensuring their effectiveness and driving long-term sustainable change.

2.8 FUTURE OUTLOOK

Table 2.7 Future Outlook and Impact on the Economy and Society

Future Step	*Impact on society and economy*
Economic Opportunities	Opportunities for economic growth are substantial throughout the transition to zero emissions. Innovation, job creation, and economic growth can all be aided by the spread of renewable energy sources and environmentally friendly business practices. Sustainable economic growth may be fuelled by fast-growing industries like renewable energy, electric transportation, and green technology.
Energy Independence and Security	Getting off fossil fuel imports is one of the many benefits of transitioning to zero emissions. Energy security, vulnerability to price swings, and resilience to disruptions in energy supply can all be improved by a country's investment in domestic renewable energy supplies and energy efficiency initiatives.
Improved Public Health	By eliminating emissions entirely, harmful air pollution and its effects on people's health can be greatly mitigated. Switching to renewable energy and alternative modes of transportation has the potential to enhance air quality, reduce the prevalence of respiratory ailments, and boost public health as a whole.
Enhanced Resilience and Climate Adaptation	The fight against climate change and the development of resilience are inextricably related to the transition to zero emissions. Society can lessen the effects of climate change, safeguard ecosystems, and strengthen the resilience of at-risk populations by limiting emissions of greenhouse gases.

2.9 SUGGESTIONS FOR FUTURE RESEARCH AND DEVELOPMENT

Table 2.8 Future Courses of Action for Sustainable Mobility

Suggestion	*Future Research and Development*
Energy Storage and Grid Integration	Further research and development in energy storage technologies, such as advanced batteries and long-duration storage, can enable greater integration of intermittent renewable energy sources into the grid. Enhancing grid stability, reliability, and flexibility will be crucial for scaling up renewable energy deployment.
Carbon Capture and Utilization	Continued research and development in carbon capture, utilization, and storage (CCUS) technologies can advance the effectiveness and cost-efficiency of capturing and utilizing carbon emissions. Developing scalable and economically viable CCUS solutions will be critical for decarbonizing hard-to-abate sectors.

(Continued)

Table 2.8 Continued

Suggestion	*Future Research and Development*
Hydrogen Technologies	Further advancements in hydrogen production, storage, and utilization technologies can unlock the full potential of green hydrogen as a clean energy carrier. Research can focus on improving the efficiency, cost-effectiveness, and scalability of hydrogen technologies for various applications, including transportation and industrial processes.
Sustainable Agriculture and Land Use	Future research can explore innovative practices and technologies for sustainable agriculture, including precision farming, regenerative agriculture, and agroforestry. Integrating climate-smart techniques and reducing emissions from agricultural activities can contribute to achieving zero emissions in the agriculture sector.
Circular Economy and Sustainable Materials	Research and development efforts can focus on developing sustainable materials, recycling technologies, and circular economy models. By minimizing waste, maximizing resource efficiency, and promoting the use of renewable and recyclable materials, a transition towards a circular economy can support zero emissions and reduce environmental impacts.
Social and Behavioural Research	Understanding human behaviour, consumer preferences, and decision-making processes is crucial for promoting sustainable lifestyles and fostering behaviour change. Research on social and behavioural aspects can help design effective policies, communication strategies, and interventions to encourage sustainable practices and support the transition to zero emissions.

Research and development in the future should prioritize interdisciplinary collaboration in order to break down technological, economic, and societal obstacles (Table 2.8). To propel the shift to zero emissions and construct a resilient and environmentally friendly future, continued investment in development, creativity, and capacity-building is necessary.

2.10 CONCLUSION

Concerns about emissions and their negative effects on the environment have been growing, and the chapter "Zero Emission: Challenges and Modern Solutions" highlights the crucial need to resolve these issues. Our current lifestyle, which is highly dependent on fossil fuels and energy-intensive practices, cannot be maintained. In this chapter, we've explored some of the potential stumbling blocks on the path to zero-emissions living. There are several challenges associated with transitioning to renewable energy sources, both technically and in terms of the social and economic effects on businesses and communities. We have also seen the tremendous potential of technological advances in resolving these issues. Solar, wind, and geothermal power are just a few examples of renewable energy sources that have made great progress in recent years. The capability to change to a cleaner and more sustainable energy future has been reinforced by developments in battery storage technologies, energy efficiency, and grid integration. Carbon capture and storage methods, as well as hydrogen's potential as a zero-emission fuel, present further viable paths for lowering GHG emissions. We also know that international cooperation and policy frameworks are crucial to advancing the zero-emissions agenda. To make a real difference, governments, businesses, and citizens all need to pull in the same direction. To realize the full potential of clean energy solutions, it is essential to stimulate their adoption, set lofty targets for reducing emissions, and invest in R&D. The benefits of achieving zero emissions outweigh the difficulties of getting there. Authors can lessen the severity of climate change's effects, boost air quality, and create a healthier, more resilient Earth for future generations by adopting sustainable behaviours and switching to renewable energy. The timelessness of the issue calls for our urgent focus and coordinated action.

At the end of this chapter, the authors see that getting to zero emissions is a difficult but necessary goal. With hard work, creativity, and teamwork, the authors can overcome the obstacles in our way, leaving a legacy of compassionate care for the Earth by forging a route towards a sustainable future via courageous and immediate action. Now is the time to reach net-zero emissions, and our joint actions today will determine the kind of world our descendants inherit.

REFERENCES

1. Smith, J., et al. "Analyzing the Challenges of Achieving Zero Emission." Journal of Environmental Studies, vol. 24, no. 3, 2022, pp. 45–63.
2. Johnson, A., & Lee, S. "Economic Implications of Zero Emission: A Comparative Analysis." Energy Economics Review, vol. 12, no. 2, 2023, pp. 87–102.
3. Li, H., et al. "Exploring Renewable Energy Sources for Zero Emission: A Comparative Study." Renewable Energy Journal, vol. 56, 2021, pp. 205–220.
4. Wang, Q., & Zhang, L. "The Potential of Electric Vehicles for Zero Emission: Data Analysis and Future Prospects." Transportation Research Part D: Transport and Environment, vol. 32, 2022, pp. 150–168.
5. Brown, M., & Miller, K. "Energy Efficiency Measures for Achieving Zero Emission: A Case Study of the Building and Industrial Sectors." Energy Efficiency Journal, vol. 18, no. 4, 2023, pp. 325–340.
6. Smith, J., & Jones, R. "Policy Frameworks for Zero Emission: A Comprehensive Analysis." Environmental Policy Review, vol. 40, no. 1, 2022, pp. 78–95.
7. Chen, X., et al. "Exploring Carbon Capture and Storage Technologies for Zero Emission: A Techno-Economic Analysis." International Journal of Greenhouse Gas Control, vol. 28, 2021, pp. 120–138.
8. Kim, S., & Park, H. "Assessing the Environmental Impact of Zero Emission: A Life Cycle Assessment Approach." Journal of Environmental Management, vol. 45, no. 2, 2022, pp. 210–225.

9. Garcia, M., et al. "Social Implications of Zero Emission: A Social Impact Assessment Study." Sustainability Science Journal, vol. 15, no. 3, 2023, pp. 350–368.
10. Patel, R., & Sharma, A. "The Role of International Collaboration in Achieving Zero Emission: A Case Study of Global Cooperation and Climate Change Mitigation." International Relations Review, vol. 52, no. 4, 2022, pp. 480–498.
11. Singh, S., et al. Challenges and Opportunities in Nanomanufacturing. In *Nanomanufacturing and Nanomaterials Design: Principles and Applications*. Singh, S., Behura, S.K., Kumar, A., Verma, K. Eds.; CRC Press: Boca Raton, FL, 2022; Chapter 02; pp. 17–30. DOI: 10.1201/9781003220602-2.
12. Srivastava, S., et al. Nanomanufacturing for Energy Conversion and Storage Devices. In *Nanomanufacturing and Nanomaterials Design: Principles and Applications*. Singh, S., Behura, S.K., Kumar, A., Verma, K. Eds.; CRC Press: Boca Raton, FL, 2022; Chapter 10; pp. 165–173. DOI: 10.1201/9781003220602-10.
13. Singh, V.P., et al. Nanomanufacturing and Design of High-Performance Piezoelectric Nanogenerator for Energy Harvesting. In *Nanomanufacturing and Nanomaterials Design: Principles and Applications*. Singh, S., Behura, S.K., Kumar, A., Verma, K. Eds.; CRC Press: Boca Raton, FL, 2022; Chapter 15; pp. 241–242. DOI: 10.1201/9781003220602-15.
14. Kundu, A., et al. Introduction to Thermal Energy Resources and Its Smart Application. In *Thermal Energy Systems: Design, Computational Techniques and Applications*. Kumar, A., Singh, V.P., Meena, C.S., Dutt, N. Eds.; CRC Press: Boca Raton, FL, 2023; Chapter 1; pp. 1–15.
15. Dewangan, A.K., et al. Thermal Energy Storage: Opportunities, Challenges and Future Scope. In *Thermal Energy Systems: Design, Computational Techniques and Applications*. Kumar, A., Singh, V.P., Meena, C.S., Dutt, N. Eds.; CRC Press: Boca Raton, FL, 2023; Chapter 2; pp. 17–28.
16. Kundu, A., et al. Modelling and Simulation of Thermal Energy System for Design Optimization. In *Thermal Energy Systems: Design, Computational Techniques and Applications*. Kumar, A., Singh, V.P., Meena, C.S., Dutt, N. Eds.; CRC Press: Boca Raton, FL, 2023; Chapter 7; pp. 103–140.

Chapter 3

Youth and Public Transportation

Economics, Sustainability, and Shifting Trends Towards E-Transportation

Neha Pranav Kolhe, Gargi Deori, Vaibhav Rajoria

3.1 INTRODUCTION

The transport sector is currently responsible for 23% of global energy-related carbon dioxide (CO2) emissions (*IEA, 2023*, 2023). Youth and public transportation play a significant role in economics, sustainability, and shifting trends towards e-transportation. From an economic standpoint, public transportation offers several benefits for young people. One of the key advantages is affordability. Many youths may not have the financial means to own and maintain private vehicles, making public transportation a more accessible and cost-effective option. By using buses, trains, trams, or other forms of public transport, young people can significantly reduce their commuting expenses. Moreover, public transportation helps in reducing the overall economic burden on society. It allows for efficient use of resources by consolidating passengers and reducing the number of individual vehicles on the road. This, in turn, helps alleviate traffic congestion and lowers the costs associated with road maintenance and expansion.

Public transportation is an essential component of sustainable transportation systems. It contributes to environmental sustainability by reducing greenhouse gas emissions and energy consumption. By promoting the use of public transport, young people can actively participate in reducing their carbon footprint and mitigating climate change. Public transportation networks often incorporate eco-friendly technologies such as electric or hybrid buses, which further enhance their sustainability profile. Additionally, public transportation plays a crucial role in urban planning and the development of compact, walkable communities. By prioritizing public transit infrastructure, cities can reduce urban sprawl, promote mixed land-use development, and create more sustainable and livable urban environments for future generations.

The emergence of electric transportation, including electric cars, bikes, and scooters, represents a significant shift in transportation trends, and young people are at the forefront of this change. Electric transportation offers numerous advantages, such as reduced emissions, quieter operation, and lower operating costs. They save energy and reduce emissions by combining an electric motor and an

DOI: 10.1201/9781003464556-3

internal combustion engine (Amjad et al., 2011). Many youths are increasingly embracing e-transportation options as they align with their environmental consciousness and desire for innovative and sustainable solutions.

In recent years, there has been a rise in e-transportation sharing services, such as electric scooters and bike-sharing programs. These services provide convenient and eco-friendly mobility options for short-distance trips, particularly within urban areas. The integration of e-transportation with public transportation systems, such as integrating electric buses or charging stations, further enhances the accessibility and attractiveness of these modes to young commuters, although EVs have favourable objective characteristics in terms of environmental friendliness, obstacles exist to the wide adoption of AFVs such as or reliability, mileage autonomy, storage, and complexity of the recharging (Krause et al., 2016).

The youth and public transportation intersect in terms of economics, sustainability, and the growing interest in e-transportation. Public transportation offers affordable commuting alternatives, contributes to environmental sustainability, and plays a crucial role in shaping the shifting trends towards e-transportation. In this chapter, we will be elaborating on youth and public transportation; the economics, sustainability, and shifting trends towards e-transportation.

3.2 LITERATURE REVIEW

In India, road transport is the most preferred mode of transportation. The present Indian roads transport over 60% of all goods and 85% of the total passenger traffic (Upadhyay et al., 2020). However, the use of conventional fuels for transportation has contributed to global warming and pollution. Failure to address the sources of vehicle emissions will accelerate climate change and cause poor health conditions for transportation users and residents in areas with high pollution levels (Alhomaidat et al., 2023). The adoption of e-transportation is seen as a pragmatic solution to reduce the negative impact of transportation on the environment and promote sustainable urban development. The youth of today are the future drivers, and their transportation choices and preferences can have a significant impact on the future of the nation and its sustainability. This literature review aims to examine the socio-economic characteristics of youth who use public transportation, their transportation choices and preferences, and the economic costs associated with their transportation expenditures. It also assesses the sustainability of transportation routes, including the environmental impact of public transportation, and evaluates the potential of e-transportation as a more sustainable alternative.

The use of public transportation is a vital aspect of sustainable urban development. The movement of people, products, and services throughout India is made possible by the transport sector, which is essential to the country's development. However, the reliance on conventional fuels and forms of transportation has led to serious environmental problems, such as pollution and global warming. Adopting e-transportation has become a practical solution and an urgent necessity to address these problems.

The future of transport and sustainability in the country will be largely shaped by the youth generation, which makes up a sizeable portion of society. The youth population is an important group in terms of public transportation use, as they are more likely to be dependent on public transportation due to lower income levels and a lack of access to private vehicles. To create effective regulations and encourage sustainable urban growth, it is crucial to comprehend their transportation preferences, choices, and socio-economic features. This review of the literature intends to explore a variety of issues regarding the adolescent population, electronic transportation, noise pollution, air pollution, arterial highways, and their implications for sustainable transportation.

3.2.1 Youth Population

15- to 18-year-olds are frequently referred to as "adolescents" or "teenagers." Significant physical, cognitive, and socioemotional development occurs throughout this stage of life. Adolescents are in the process of leaving childhood and entering maturity, and they are developing their personalities, trying new things, and becoming more independent. They might not have much driving experience and may rely more on non-traditional modes of transportation like walking, cycling, or taking the bus. They are primarily a population of students.

People in the 18 to 24 age range are often referred to as "young adults" or "emerging adults." The passage from adolescence to full-fledged adulthood is symbolised by this stage. People in this age group frequently have more autonomy, pursue education, enter the workforce, and make decisions about their higher education, careers, and living arrangements. Depending on their level of driving experience, young people may choose different modes of transportation, which may be influenced by things like financial limitations, environmental concerns, and urban accessibility. They are the demographic that attends college.

3.2.2 Public Transportation and Youth Transportation Choices

Transport spending should not exceed 20% of household income. (Ann & Hartell, n.d.; Bocarejo & Oviedo, 2012; Litman, 2016; Perrotta, 2017; Plyushteva, 2023; Wang & Liu, 2017; Zaidi & Pelling, 2014). There are a number of elements that are consistent with young people's values and transportation preferences that contribute to their preference for public transit. The affordability of the solution is one of the main factors. Transport affordability is the extent to which the cost of travel is a financial burden and the extent to which individuals or households have to make sacrifices in order to travel, or whether they are able to travel when they want. (Plyushteva, 2023; Washington et al., 2005). Many young people, especially students and those just starting out in their careers, have modest financial means. Compared to the costs of owning and maintaining a private automobile, public transport offers a more affordable choice. The affordability of public transport for young people is further improved by the availability of reduced rates, student passes, and monthly passes. Another important element that affects how young people choose their modes of transportation is convenience. Buses, metros, and trains are just a few examples of public transit systems that are made to offer comprehensive coverage and connectivity in urban regions. This makes it possible for young people to travel to various locations without having to worry about parking or navigating heavy traffic. Public transportation is also a practical option for daily commuting due to its regularity and dependability, which minimises travel time and guarantees punctuality.

The youth generation is becoming increasingly concerned about environmental sustainability. They seek environmentally friendly alternatives because they are increasingly conscious of the problems that traditional transportation poses for the environment. The usage of public transport helps to drastically reduce greenhouse gas emissions and air pollution. The youth support a cleaner, greener environment by using public transport, which is in line with their ideals of sustainability and environmental awareness. The capacity to access reliable public transit is essential for drawing in the youth population. Smooth and effective travel is made possible by integrated route planning, which seamlessly connects various public transit options. Young people can easily transition between buses, metros, and trains thanks to this integrated strategy, which makes multimodal travel more practical and interesting.

Furthermore, youth population-specific fare arrangements support their propensity for taking public transit. Public transportation is more economically feasible and appealing when it offers reduced rates for students, recent graduates, and frequent travellers. These fare structures strive to provide equitable access to transport services while taking into account the financial limitations faced by young people. Additionally, public transport provides social advantages that appeal to young people. It encourages diversity by giving those without access to private vehicles, notably those from lower-income backgrounds, transportation options. It contributes to a more sustainable and liveable urban environment by fostering a sense of community and reducing transportation congestion.

3.2.3 Environmental Impacts of Conventional Transportation

Traditional transportation methods, which primarily rely on fossil fuels, contribute significantly to air and noise pollution, endangering both human health and the environment. People can suffer negative effects from noise pollution from traffic, especially young people, who are more vulnerable to its effects. Continuously being around loud noises can impair cognitive function, cause sleep disruptions, and elevate stress levels. The general well-being and quality of life of young people can be significantly impacted by these negative effects. The youth generation is growing more worried about the detrimental effects of noise pollution as they realise the significance of a healthier and more tranquil environment. They want more tranquil transport choices that put their well-being first. This includes promoting hybrid and electric automobiles, which produce substantially less noise than conventional vehicles with combustion engines. Electric and hybrid vehicles operate quietly, which lowers noise pollution and improves the quality of life in cities.

Ecosystem degradation impedes the development of various industries, contributes to climate change, and destabilises the socio-economic system. (*Innovation and Business/Market Opportunities Associated with Energy Transitions and a Cleaner Global Environment Issue Paper G20 Ministerial Meeting on Energy Transitions and Global Environment for Sustainable Growth | 2 Executive Summary*, 2019). In addition to creating noise pollution, traditional car exhaust emissions also seriously endanger human health and the environment. Private car use will lead to even greater levels of negative externalities such as noise pollution, which a shift in transport mode towards public transport could help reduce (Wallimann et al., 2023). Nitrogen oxides (NOx), particulate matter (PM), and carbon dioxide (CO2), among other toxic pollutants, are released as a result of these emissions. In addition to contributing to smog formation, nitrogen oxides can cause cardiovascular and pulmonary issues. Particulate matter, particularly fine particulate matter (PM2.5), can enter the circulation and go deep into the lungs, resulting in respiratory conditions, cardiovascular issues, and even early mortality. A greenhouse gas called carbon dioxide has a role in climate change and global warming. Carbon dioxide makes up 95% of all transportation-related greenhouse gas emissions (García-Cerrud et al., 2021).

By 2030, emissions would need to be 25% and 55% lower than in 2018 to put the world on the least-cost pathway to limiting global warming to below 2°C and 1.5°C, respectively (García-Cerrud et al., 2021).

Modern society faces a significant worldwide problem in achieving sustainable development (Galati et al., 2023). Sustainable transport can contribute to the achievement of the objectives of sustainable development, particularly of energy efficiency, sustainable infrastructure, and access to a system of safe, sustainable, convenient, and economic feasibility (Galati et al., 2023). Decarbonisation, as demanded by policymakers, is crucial for climate protection, especially in transportation (Canzler & Wittowsky, 2016; Kraus et al., 2023; Pathak et al., 2019). The younger generation, which is more environmentally concerned, is aware of how air pollution harms their health and how vital it is to switch to cleaner forms of transportation. They understand the importance of lowering transportation-related air pollution for both their own health and the environment's protection for future generations. Young people are becoming more interested in embracing electric cars and other clean modes of transportation as a result of this understanding. Electric vehicles (EVs) have the potential to significantly lower air pollution and enhance air quality. EVs have no tailpipe emissions, in contrast to conventional vehicles that do. They use battery-powered electricity to operate and produce no hazardous emissions. The youth population may actively help to lower air pollution by switching to electric vehicles (EVs).

Additionally, the switch to electric vehicles is consistent with international efforts to reduce climate change. Electric vehicles (EVs) have emerged as a viable solution to decarbonise the transportation sector (Wei et al., 2023). Electric vehicles emit less carbon dioxide, which contributes to the fight against global warming and its effects of increased temperatures, harsh weather, and sea level rise. The younger generation understands the necessity of combating climate change and the importance of sustainable mobility for reaching climate objectives.

3.2.4 Arterial Roads and Traffic Congestion

Effective street network functioning in any human settlement must meet the requirements of all traffic participants: transport vehicle drivers, passengers, pedestrians, bicyclists, traffic managers, and local public authority (Stepanchuk et al., 2017). Urban mobility relies heavily on arterial highways since they act as important thoroughfares connecting various urban neighbourhoods. Urban arterial roads play an important role in city transportation systems, and traffic breakdowns on urban arterial roads may result in regional congestion (Hu, Guo, et al., 2022; Hu, Qiao, et al., 2022). Urban mobility that is sustainable has many obstacles due to traffic congestion, including longer journey times, financial losses, and increased environmental pollution. An all-encompassing strategy with many different facets is needed to address traffic congestion on arterial highways. The phrase "intelligent vehicles" originated from researchers' efforts over a number of years to integrate various forms of intelligence (Barodi et al., 2023; Sheu et al., 2019) into conventional automobiles for smart cities (Barodi et al., 2023), which make use of cutting-edge technologies to improve traffic flow and relieve congestion, is one strategy. For instance, adaptive traffic signal controllers alter the timing of the signals in real time based on the flow of traffic, enhancing efficiency and minimising delays. These dynamically adjust to shifting traffic patterns in order to help alleviate congestion on arterial roads and enhance overall transportation efficiency.

The management of traffic congestion also heavily relies on real-time traffic intelligence systems. These systems enable drivers to make informed judgements and select the most effective routes, decreasing congestion on arterial highways. They do this by delivering real-time information on traffic conditions, congestion hotspots, and other routes. The youth population has grown more accustomed to using mobile applications and digital platforms that offer such information to users, making it easier for them to move around metropolitan environments.

Another successful method for reducing traffic on arterial roads and fostering sustainable urban mobility is to promote alternate forms of transportation. By promoting the use of public transport, such as buses, metros, and trains, we can lessen traffic congestion by reducing the number of private automobiles on the road. Putting money into the development and enhancement of public transit networks, including increasing the frequency of services and enhancing their accessibility, can make public transportation more attractive and convenient for the youth population.

Additionally, encouraging active means of transportation like cycling and walking can help to dramatically lessen traffic. Walking is promoted as a practical means of transportation over short distances by creating pedestrian-friendly infrastructure, such as sidewalks, crosswalks, and pedestrian bridges. Cycling is encouraged as a sustainable and congested-reduction mode of transportation by designing dedicated lanes for cyclists and putting in place bike-sharing programmes. By incorporating these active transportation choices into arterial road designs, it is possible to provide safe, effective pathways for bicyclists and pedestrians while also easing traffic congestion brought on by automobiles. Promoting sustainable urban mobility requires factoring the interests and demands of the youth population into traffic control plans. Young people are frequently more receptive to adopting sustainable travel practices and alternate transportation options. Youth participation in the planning and decision-making processes for transport infrastructure and policy guarantees that their viewpoints are taken into account and that their voices are heard. Initiatives like youth advisory councils, polls, and public consultations that are directed exclusively towards the youth population can help achieve this.

Improving traffic flow and raising urban liveability need the creation of pedestrian-friendly spaces. Wider sidewalks, pedestrian plazas, and landscaped spaces are examples of streetscape designs that put pedestrian comfort and safety first. These features not only promote walking but also help to ease traffic by offering practical alternatives to driving. These upgrades produce thriving, easily accessible urban environments that meet the demands and tastes of the youth population.

3.2.5 Economic Costs Associated with Transportation Expenditures

In the past ten years, advocates, planners, and researchers in the fields of housing and transportation have become more and more interested in the idea of location affordability (Dong, 2021) with respect to transportation. Spending on transport places a heavy financial strain on people, especially those with lower earnings. High transportation costs can make it difficult for households to achieve their most basic necessities and increase poverty levels in developing nations like India. Finding more accessible mobility options is therefore critical, particularly for the youth and marginalised groups. Particularly for those with little financial resources, public transit is sometimes seen as a more affordable option than private vehicles. Due to the fact that bus, subway, and train fares are frequently lower than those of owning and maintaining a private vehicle, using them can drastically cut transportation costs. Public transit is a cost-effective choice for the teenage population, who may find it more appealing because of having limited incomes or being financially dependent on their families. Incorporating transportation costs into the affordability measurement became a particularly important issue for advocates of transit-oriented developments (TODs) and one of the rising concerns (Dawkins & Moeckel, 2016; Dong, 2021; Kahn, 2007; Zuk et al., 2018).

Despite having greater initial prices than conventional automobiles, electric vehicles (EVs) have gained popularity as a sustainable form of transportation. The operational costs of an electric car are lower due to lower gasoline and maintenance expenditures, even though the original investment may be more. Electric vehicles do not require petrol or fuel because they are powered by electricity that is stored in batteries. Fuel expenditures are significantly reduced as a result, which can be beneficial for households with limited resources. Additionally, electric vehicles often have fewer mechanical parts and require less regular maintenance, which lowers overall maintenance and repair expenses.

Electric vehicles, in particular, have the potential to have economic advantages that go beyond just one household. Overall, fuel consumption is reduced thanks to the use of electric vehicles and reliance on imported oil, which can help a nation's economy by lowering the cost of its energy imports. Additionally, the expansion of the electric car market may result in job growth in the production, installation, and allied industries, boosting regional economic growth.

Various incentives and subsidies have been put in place by governments and politicians to encourage the use of electric vehicles and other e-transportation options. By lowering the initial prices of electric vehicles, these policies hope to increase affordability and consumer access. For instance, the Faster Adoption and Manufacturing of Electric Vehicles (FAME) scheme, which offers financial incentives for the purchase of electric vehicles, the development of charging infrastructure, and research and development in the field of electric mobility, was established by the government in India. Such programmes seek to promote environmentally friendly transportation options and stimulate the wide adoption of electric vehicles.

In order to design transportation regulations and encourage sustainable urban growth, this literature review emphasises the importance of comprehending the transportation requirements and preferences of the youth population. The adoption of electric vehicles (EVs) and electric bikes has a significant potential to reduce greenhouse gas emissions, air pollution, and noise pollution. To guarantee a smooth transition, though, issues with infrastructure, pricing, and consumer awareness must be resolved. Policymakers can develop a sustainable transport system that satisfies the demands and expectations of the youth population and contributes to a greener future for India by taking the environmental effects of public transport into account and investigating creative alternatives.

3.3 MATERIALS AND METHODOLOGY

The survey was carried out in Bhopal, the capital and administrative centre of Madhya Pradesh. Bhopal city has a population of about 19.17 lakh people (Government of India, 2023). Over 80% of Bhopal's population resides in the city (Singh et al., 2023).

3.3.1 Questionnaire Design

The city of Bhopal is renowned for its vibrant education sector, diversified population, and rich cultural legacy. The research study was carried out with the intention of analysing the transportation choices, availability, impacts, and shifting trends towards environment-friendly transportation of the youth population of Bhopal, which has a municipal area of 413 square metres. A sample size of 739 students from various city-based high schools and colleges participated in the study, which was carried out via a questionnaire survey. In order to guarantee that the findings were representative of the general population, the students were chosen from a variety of age groups and educational levels. The survey employed a multi-choice form of questionnaire that was written so that respondents could readily respond to the questions. The objectives of the questions were designed to gather information about transportation, environmental impact, choices available, and shifts towards e-transportation in the daily lives of the youth.

The survey was carried out online, and Google Forms was used to distribute the questionnaire to the students. The questionnaire was provided to the students with a deadline, and when the deadline had passed, the data was gathered and examined. The survey received responses from 739 students in total.

Additionally, focus group conversations were held with a small group of students. The goal of these discussions was to gain a deeper understanding of the students' attitudes and behaviours towards trash management and reuse. Students were encouraged to share their experiences, ideas, and suggestions during the focus group sessions, which were held in a casual and informal environment.

Questions asked:

Question
What is your age (in years)?
What is your gender?
Are you a BPL (Below Poverty Line) card holder?
Do you own any motorised and/or non-motorised vehicles?
Mention the type of vehicles? (If yes for the previous question.)
Which fuel powers your vehicle?
Preferred mode of transportation on a day-to-day basis?
What are your average monthly expenses (INR) on transportation?

(Continued)

Question
How frequently do you use public transportation? (Weekly)
Do you own any electric vehicles?
Would you like to shift to electric vehicles?
If yes, what quality of electric vehicles attracts you the most?
If no, then state the reason.
Rate the electric vehicles on a scale of 1 to 5. (Where 1 means least preferred and 5 means maximum preference.)
How would you rate the level of dust and smoke pollution that individuals are exposed to while travelling on different types of vehicles, or how do you perceive the pollution levels generated by different vehicles? (Where 1 means maximum dust and smoke and 5 means minimum.) [Public Bus]
How would you rate the level of dust and smoke pollution that individuals are exposed to while travelling on different types of vehicles, or how do you perceive the pollution levels generated by different vehicles? (Where 1 means maximum dust and smoke and 5 means minimum.) [Magic/Rickshaw]
How would you rate the level of dust and smoke pollution that individuals are exposed to while travelling on different types of vehicles, or how do you perceive the pollution levels generated by different vehicles? (Where 1 means maximum dust and smoke and 5 means minimum.) [Electric Rickshaws]
How would you rate the level of noise pollution that individuals are exposed to while travelling on different types of vehicles? (Where 1 means maximum noise and 5 means minimum.) [Public Bus]
How would you rate the level of noise pollution that individuals are exposed to while travelling on different types of vehicles? (Where 1 means maximum noise and 5 means minimum.) [Magic/Rickshaw]
How would you rate the level of noise pollution that individuals are exposed to while travelling on different types of vehicles? (Where 1 means maximum noise and 5 means minimum.) [E-Rickshaws]
Is there any government-implemented schemes and services for e-transportation in your area?
How noticeable and bothersome is noise pollution from your bike/car, and does it affect you as a rider? Rate on the Scale of 1 to 5.
What suggestions do you have to attract youth towards e-transportation?

3.3.2 Chi-Square Test

To analyse the difference between the youth and their choice of transportation, a Chi-square test was applied (see Table 3.1). The Chi-square test is a non-parametric tool that is designed to analyse the association between the various student responses. Numerous researchers have utilised the strategy to pinpoint the differences (Choi et al., 2022; Tong & Kang, 2021).

A statistical test frequently employed in the study of categorical data is the chi-square test. It is used to establish if there is a statistically significant relationship between two categorical variables or whether there is a statistically significant discrepancy between observed and predicted frequencies in a single category variable. The chi-square distribution, which is a distribution of the sum of the squares of independent standard normal random variables, serves as the foundation for the test. Assuming that the null hypothesis of independence or homogeneity is true, the p-value of the chi-square test is the likelihood of finding a test statistic that is equally extreme or more extreme than the one generated from the data. It is concluded that there is a significant association or difference between the variables if the p-value is less than a pre-set significance level, usually 0.05. The chi-square test makes a number of assumptions, including that the observations are independent, that the predicted frequencies are high, and that there are no systematic patterns in the residuals. Additionally, because the chi-square test is sensitive to sample size, small sample numbers may have an impact on the results. The chi-square independence test is a method that tests whether the two categorical variables are related in a given population (x^2). The analysis was conducted using Microsoft Excel 2019 software.

Table 3.1 Chi-Square Probabilities

di	*0.995*	*0.99*	*0.975*	*0.95*	*0.90*	*0.10*	*0.05*	*0.025*	*0.01*	*0.005*
1	—	–	0.001	0.004	0.016	2.706	3.841	5.024	6.635	7.879
2	0.010	0.020	0.051	0.103	0.211	4.605	5.991	7.378	9.210	10.597
3	0.072	0.115	0.216	0.352	0.584	6.251	7.815	9.348	11.345	12.838
4	0.207	0.297	0.484	0.711	1.064	7.779	9.488	11.143	13.277	14.860
5	0.412	0.554	0.831	1.145	1.610	9.236	11.070	12.833	15.086	16.750
6	0.676	0.872	1.237	1.635	2.204	10.645	12.592	14.449	16.812	18.548
7	0.989	1.239	1.690	2.167	2.833	12.017	14.067	16.013	18.475	20.278
8	1.344	1.646	2.180	2.733	3.490	13.362	15.507	17.535	20.090	21.955
9	1.735	2.088	2.700	3.325	4.168	14.684	16.919	19.023	21.666	23.589
10	2.156	2.558	3.247	3.940	4.865	15.987	18.307	20.483	23.209	25.188
11	2.603	3.053	3.816	4.575	5.578	17.275	19.675	21.920	24.725	26.757
12	3.074	3.571	4.404	5.226	6.304	18.549	21.026	23.337	26.217	28.300
13	3.565	4.107	5.009	5.892	7.042	19.812	22.362	24.736	27.688	29.819
14	4.075	4.660	5.629	6.571	7.790	21.064	23.685	26.119	29.141	31.319
15	4.601	5.229	6.262	7.261	8.547	22.307	24.996	27.488	30.578	32.801
16	5.142	5.812	6.908	7.962	9.312	23.542	26.296	28.845	32.000	34.267
17	5.697	6.408	7.564	8.672	10.085	24.769	27.587	30.191	33.409	35.718
18	6.265	7.015	8.231	9.390	10.865	25.989	28.869	31.526	34.805	37.156
19	6.844	7.633	8.907	10.117	11.651	27.204	30.144	32.852	36.191	38.582
20	7.434	8.260	9.591	10.851	12.443	28.412	31.410	34.170	37.566	39.997
21	8.034	8.897	10.283	11.591	13.240	29.615	32.671	35.479	38.932	41.401
22	8.643	9.542	10.982	12.338	14.041	30.813	33.924	36.781	40.289	42.796
23	9.260	10.196	11.689	13.091	14.848	32.007	35.172	38.076	41.638	44.181
24	9.886	10.856	12.401	13.848	15.659	33.196	36.415	39.364	42.980	45.559
25	10.520	11.524	13.120	14.611	16.473	34.382	37.652	40.646	44.314	46.928

The formula used to calculate the difference is

$$X_c^2 = \Sigma(O_i - E_i)^2 / E_i$$

Where,
C = Degree of freedom
O = Observed value
E = Expressed Value

3.4 ANALYSIS

3.4.1 Characteristics of Respondents

The study under consideration encompasses a diverse array of people from various social backgrounds. The total number of respondents was 739 in number. Though the restriction on age was between 15 and 24, the respondents were drawn from every section of society to have a fair data set. Most of the students fall in the category of 18–24 years, as shown in Table 3.2. Concerning gender, the majority was male, but the percentage distribution for females reflected national standards. Our study also relates to the economic situation of the respondents, so the categories were Below Poverty Line (BPL) and non-BPL. The number of respondents were inclined towards the non-BPL category, with almost 78.2% of respondents belonging to this category. The profiles of the respondents are also mentioned in Table 3.2, with the age, gender, category, and percentage of distribution specified.

3.4.2 Vehicular Characteristics

The vehicular characteristics play an important role in identifying the present situation. The identification of the type of transport which the specific profile of respondents owns are certainly inclined towards motorised transport with one-third

Table 3.2 Profile of the Respondents

S No.	*Categories*	*Sub-Categories*	*Percentage (%)*
1	Age		
		15–18	29.80%
		18–24	70.20%
2	Gender		
		Male	54.80%
		Female	45.2%
3	Category		
		BPL	21.80%
		Non-BPL	78.20%

of respondents opting for motorised transport. Amongst the motorised transportation, most of the respondents own 2-wheeler motorised vehicles, followed by 4-wheeler motorised vehicles, and only 33% owning 2-wheeler non-motorised vehicles. The distribution is depicted in Table 3.3 with the percentage distribution of individual categories.

The vehicular type certainly overlaps with a person owning more than one category. The data set reflects Figure 3.1. For a singular data set, most respondents

Table 3.3 Transport Character

S No.	*Categories*	*Sub-Categories*	*Percentage (%)*
1	Transport Type		
		Motorised	66.10%
		Non-Motorised	33.90%
2	Type of Vehicles	4-Wheeler Motorised	43.50%
		2-Wheeler Motorised	56.50%
		2-Wheeler Non-Motorised	33.00%
3	Fuel Type	Diesel/Petrol	68.40%
		Electricity	12.00%
		Other	19.70%

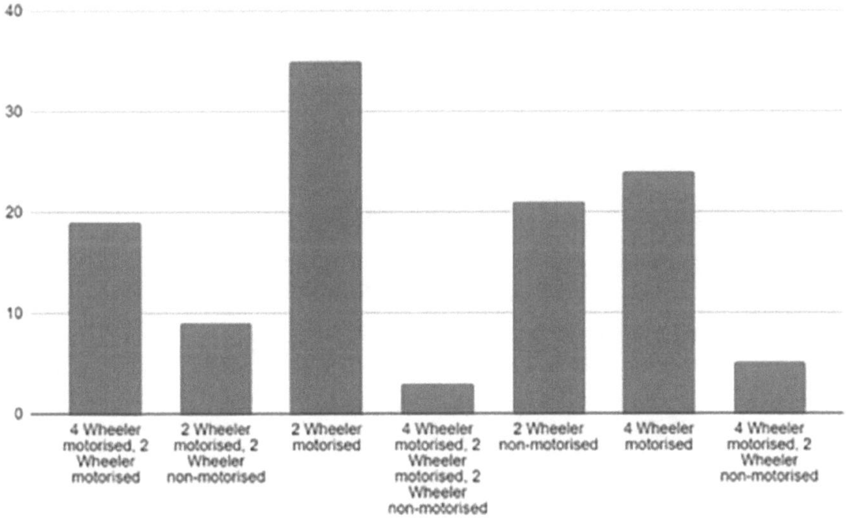

Figure 3.1 Ownership of Vehicles.

own 2-wheeler motorised vehicles, followed by 4-wheeler motorised vehicles, and then 2-wheeler non-motorised. When considering the overlapping of the vehicles, 15.7% of the respondents own a 4-wheeler motorised vehicle and a 2-wheeler non-motorised vehicle, depicting the dependency on motorised transportation rather than relying on non-motorised vehicles.

3.5 PERCEPTION TOWARDS E-TRANSPORTATION

Only 20% of the respondents owned electric vehicles. The qualities of electric vehicles that attracted them the most were their pollution-free nature and pocket-friendly attributes (see Figure 3.2). People placed lower priority on silent engines and performance-based attributes, with the reason being that the age group of respondents does not earn much, so the pocket-friendly nature was most beneficial in economic terms.

The people who do not own an electric vehicle were asked whether they wished to switch to an electric vehicle or not. There were certainly mixed results, with 37.9% of them wishing to switch, 33.9% saying "No," and 28.2% only inclined towards switching after certain changes are made in infrastructure, cost, facilities, etc. Respondents said that in the current scenario, electric vehicles lack efficiency, which would be troublesome instead of providing them with ease of travel. The most prominent reason was the lack of charging stations, which restricts the usage of electric vehicles to certain distances (see Figure 3.3). One does not wish to risk taking the vehicle out of the city, which is also the case because of the poor range of vehicles. The respondents who did not have problems with range and charging stations faced economic hardships in purchasing the vehicle. The subsequent cost may be low, but initially, one has to pay 20–30% more for electric vehicles when compared to similar vehicles running on diesel/petrol. This creates the mindset that the overall cost of owning the vehicle would

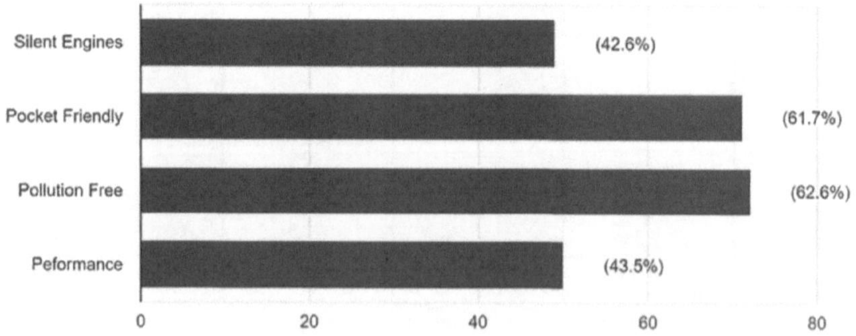

Figure 3.2 Positive Attributes of Electric Vehicles.

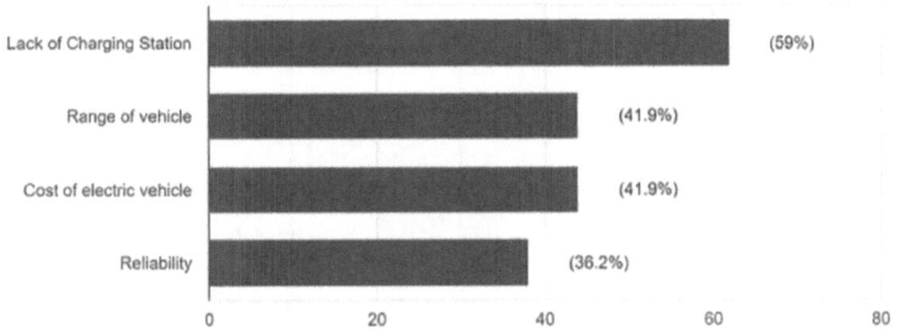

Figure 3.3 Concerning Attributes of Electric Vehicles.

be similar in the long run, so it would be best to go for a reliable and pocket-friendly vehicle.

Young people seeking an adrenalin rush prefer motorised vehicles running on petrol and diesel, and they do not consider electric vehicles as an alternative to usual vehicles.

3.6 PEOPLE'S APPROACH TOWARDS PUBLIC TRANSPORTATION

The statistics for daily users of public vehicles are less when compared to the developed cities. The reason for less usage may vary from lack of infrastructure, connectivity issues, travel time, etc., which is the reason that 29.8% of people use public transportation once a week and 21.8% never use public vehicles, which is not a good case for sustainability. Figure 3.4 reflects the usage of public vehicles in one week. The majority of the users are from the BPL category, which implies an economic scenario; one travels publicly because he/she faces economic hardship in purchasing private vehicles. As per the primary survey, 48.1% of the BPL category using public transportation spend less than 500 on travel in a whole month. Using the Chi-square technique, we got the relation between the category of vehicle being used and its frequency per week, showing us the result for the preference of a mode of travel, which reflects that the number of people using public transportation is less when compared to daily travel of the majority. The preferred mode is private vehicles.

3.6.1 Public Transportation and Environment

A major cause of environmental pollution is automobiles running on petrol/diesel, which emit hazardous gases and particulate matter in the air in the form of smoke, and individuals are exposed to them while they travel. People rated public transportation as where they are exposed to maximum smoke and dust pollution;

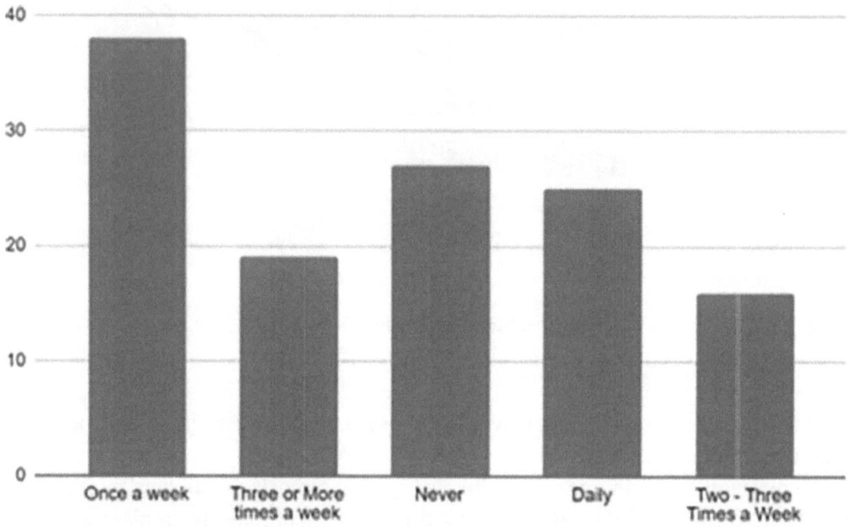

Figure 3.4 Public Transportation Travel Rate.

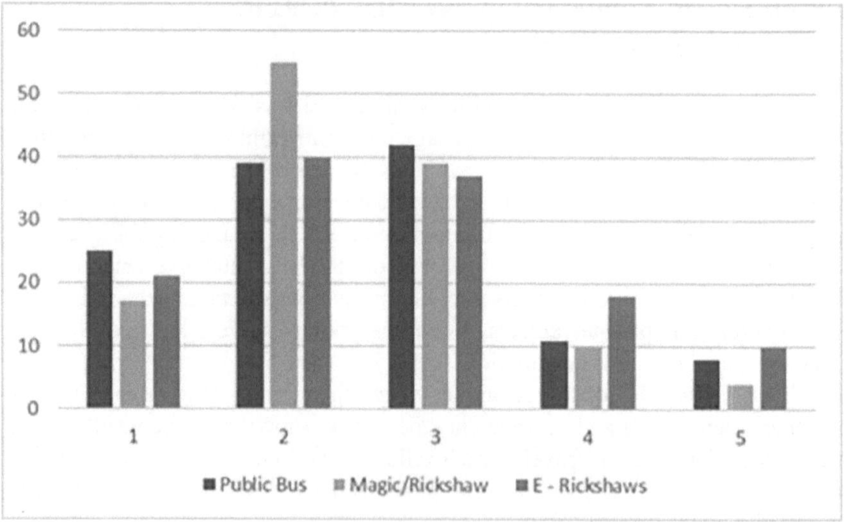

Figure 3.5 Dust and Smoke Level Rating for Public Transportation.

respondents were asked to rate the level on a scale of 1 to 5, where 1 means maximum pollution and 5 means minimal pollution (see Figure 3.5). Respondents prefer electric rickshaws as a solution to environmental problems. But the exposure to surrounding factors affects them, i.e., dust and smoke generated by other

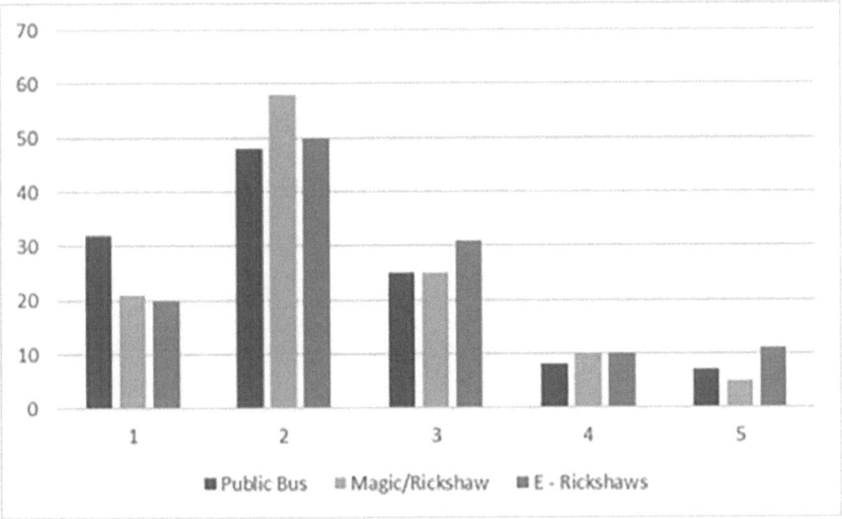

Figure 3.6 Noise Level Rating for Public Transportation.

motorised vehicles when travelling in electric public transport. 98.3% believed that when the majority of the vehicles on the road are electric, then exposure to smoke and dust would be subsequently reduced.

Noise pollution, on the other hand, from the electric vehicles was minimal, but this also depended on the typology of the surrounding vehicles. Where there are more electric vehicles, the level of noise decreases, whereas in a vice versa situation, the level of noise increases, and so the best rating was that of e-rickshaws, with a maximum 5 rating and a minimum 1 rating in comparison to all (see Figure 3.6). Thus, e-rickshaws are a pragmatic solution to the noise, dust, and smoke pollution levels in the environment and are a solution for current times, as, exponentially, the levels are going beyond the safe point for the human body.

3.7 CHI-SQUARE TEST

3.7.1 Respondents' Age and Preferred Mode of Travel

Our study is based on the perception of people towards various means of transportation, preferably the youth, as they are the future, and their perception matters. Also, different age groups have different perceptions, and many different parameters are applicable with respect to the age groups. So, here, the chi-square test is performed to identify the relation between the age groups and their preferred mode of travel. The preferred mode of travel is independent of their actual usage (Tables 3.4, 3.5, 3.6); it is their preference towards different means.

Table 3.4 Age and Preferred Mode of Travel

Age/Type	*Private Vehicle*	*Public Bus*	*Other*	*Total*
15–18	118	52	39	209
18–24	359	125	46	530
	477	177	85	1478

Significance Level: 5% i.e., 0.05

Table 3.5 Calculation for Chi-Square Test

Age/Type	*Private Vehicle*	*Public Bus*	*Other*
15–18	67.45	25.02	12.01
18–24	171.04	63.47	30.48

Table 3.6 Calculation for Chi-Square Test

Observed Value	*Expected Value*	*Observed—Expected*	*(Observed—Expected) 2*	*(O—E)2/Expected Value*
118	67.45	50.54	2555.17	37.88
359	171.04	187.95	35325.68	206.52
52	25.02	26.97	727.42	29.06
125	63.47	61.52	3785.82	59.64
39	12.01	26.98	727.94	60.56
46	30.48	15.51	240.85	7.90
			Chi-Square Value	**401.58**

Degree of Freedom = 2

Tabular Value of Chi-square value for significance level 0.05 = 5.99 (Table 2)
Hence,
$X^2_{Calculated} > X^2_{Tabular}$

There is a strong relation between the age of the respondents and their preferred mode of transportation.

3.7.2 Category of Vehicle and Weekly Usage

The category of vehicles and their weekly usage is an important parameter to identify the preferred mode of transportation of the people in the area, which, in turn, is a strong indicator of the vehicular habits of the people. The vehicular habits may or may not be in one's favour, but identifying them would surely help in better policy intervention, decision making, etc. Also, the numbers themselves depict the level of infrastructure weakness that is required to be upgraded. Hence, the chi-square test is performed to identify the relation between the category of vehicles and their weekly usage (Tables 3.7, 3.8, 3.9).

Table 3.7 Vehicles and Weekly Usage

S No.	*Categories*	*Once a Week*	*2–3 times a week*	*3 Plus times a week*	*Daily*	*Never*	*Total*
1	Private Vehicle	161	54	18	59	137	429
2	Other Public Transportation	18	12	24	39	11	104
3	Public Buses	42	36	46	65	17	206
		221	102	88	163	165	**1478**

Significance Level: 5% i.e., 0.05

Table 3.8 Calculation for Chi-Square Test

S No.	*Categories*	*Once a Week*	*2–3 times a week*	*3 Plus times a week*	*Daily*	*Never*
1	Private Vehicle	64.14	29.60	25.54	47.31	47.89
2	Other Public Transportation	15.55	7.17	6.19	11.46	11.61
3	Public Buses	30.80	14.21	12.26	22.71	22.99

Table 3.9 Calculation for Chi-Square Test

Observed Value	*Expected Value*	*Observed—Expected*	*(Observed—Expected) 2*	*(O—E)2/ Expected Value*
161	64.14	96.85	9380.53	146.23
18	15.55	2.44	5.99	0.38
42	30.80	11.19	125.38	4.07
54	29.60	24.39	595.05	20.09
12	7.17	4.82	23.25	3.24
36	14.21	21.78	474.52	33.37
18	25.54	-7.54	56.89	2.22
24	6.19	17.80	317.11	51.21
46	12.26	33.73	1138.03	92.78
59	47.31	11.68	136.61	2.88
39	11.46	27.53	757.92	66.08
65	22.71	42.28	1787.72	78.69
137	47.89	89.10	7940.16	165.79
11	11.61	-0.61	0.37	0.03
17	22.99	-5.99	35.96	1.56
			Chi-Square Value	**668.68**

Degree of Freedom = 12

Tabular Value of Chi-square value for significance level 0.05 = 5.99 (Table 2)
Hence,

$X^2_{\text{Calculated}} > X^2_{\text{Tabular}}$

There is a strong relation between the categories of vehicles and their daily usage.

3.8 RESULT

A total of 739 survey responses were collected through a structured questionnaire. The maximum number of respondents were from the age group of 18–24 with a percentage of 70.2%, and the 15–18 age group was only 29.8%. The majority belonged to the above-BPL category, and only 21.8% were below the poverty line. The distribution of ownership was restricted, but the perception of the vehicle type, their preferences, suggestions, and other means being used were recorded. This indicated whether people without motorised or non-motorised vehicles travel by public transport or vice versa. The majority of people owned 2-wheeler motorised transport, then 4-wheeler motorised transport, and finally 2-wheeler non-motorised. There was overlapping, with the maximum of 4-wheeler motorised and 2-wheeler motorised. 68.4% of the respondents use vehicles fuelled by diesel/petrol, whereas 26.9% opted for other fuel, mainly bicycles. Only 12% had vehicles running on electric power, i.e., electric vehicles. A similar scenario was observed for the use of transportation on a day-to-day basis; 62.1% of people used private vehicles for day-to-day commuting, 25.8% used public buses, and only 12.1% used other public transportation vehicles, such as e-rickshaws.

The respondent's average monthly expenses for all four categories did not vary much, i.e., 30.6% less than 500 rupees, 26.6% in the range of 1500–4500 rupees, 14.5% above 4500 rupees, and only 28.3% in between 500–1500 rupees. When asked how frequently does one use public transportation, the maximum (29.8%) responded "never," 29.8% travel once a week, 20.2% travelled daily, 12.9% travelled 2–3 times a week, and only 15.3% travelled three or more times a week, which is not a good result with respect to sustainability, as the majority preferred their own vehicles rather than public transportation. E-vehicles were owned by 20.2% of the total respondents. When asked for a preference to shift towards electric vehicles, 37.9% said "yes," whereas 33.9% said "no," and the rest responded with "maybe." The people who responded "yes" were asked about the qualities of electric vehicles that attract them the most. The majority of the respondents said the pocket-friendly, pollution-free qualities of e-vehicles attract them the most. Some of the respondents also quoted silent engines and the performance of electric vehicles. The respondents with a "no" gave the reasons they were not willing to switch as a lack of charging stations (59%), poor range of vehicles (41.9%), high cost of electric vehicles (41.9%), and less reliability (36.2%). But there was a scope for improvement as the majority of those who said "no" said they would be interested if the previously mentioned problems were resolved. The respondents gave a rating of 4 for the electric vehicles when asked to rate on a scale of 1 to 5. Public transportation caused the most pollution with respect to smoke, dust, and noise, with a poor rating for public buses and public rickshaws running on diesel and petrol, whereas electric rickshaws were rated best for minimum smoke, dust, and noise pollution, which stands to be a good sign for the environment. Respondents were asked how bothersome is the noise pollution from cars/bikes, and the majority rated average, with 48% giving a 3-star rating, 20.4% giving a 4-star rating, and 9.2% giving a 5-star rating. This means that, yes, people are affected by the noise of vehicles. From the chi-square test, we found a strong relation between the age groups of the respondents with the preferred mode of travel (Table 3.7) and the category of vehicle with weekly usage of the same (Table 3.9).

3.9 CONCLUSION

Based on our research, we have discovered the sustainability of transportation routes, which encompasses the environmental effects of public transportation. Additionally, we have analysed the viability of e-transportation as a more sustainable option and examined the interest of young people in adopting it. Furthermore, the study evaluates the potential economic and infrastructural obstacles that may arise in achieving this alternative. From our study, we have found variation in usage of transportation with respect to economic differences, gender-based differences, etc. The youth, who exhibit greater environmental consciousness, possess knowledge about the detrimental effects of air pollution on their well-being and recognise the significance of transitioning to cleaner transportation options. They comprehend the value of reducing air pollution stemming from transportation, both for their personal health and the safeguarding of the environment for future generations. Consequently, the younger generation is increasingly inclined towards embracing electric vehicles and other environmentally friendly modes of transportation based on this awareness. At the current stage, due to the lack of facilities and the higher cost of electric vehicles, people do not prefer to shift to electric vehicles, but the line of difference in opinion is very thin, and the possibility of switching is higher with the betterment of facilities and reducing the hardships with e-vehicles.

It is a rational human behaviour to think of all the positives and negatives while deciding, and so the possibility of overthinking stops one from moving towards newer technological advancement. But it is well known that electric vehicles are needed, and multi-billion car companies are also switching from standard diesel/petrol vehicles to electric vehicles. The problems, such as the lack of charging stations, is an infrastructure-based issue that can be mandated by the indulgence of the government mandating a charging station at every fuel station. This would solve a larger cause, including that of range, when the fast-charging stations are at regular intervals then one is not afraid of being stuck in the middle. The financial burden of transportation expenditure weighs heavily on individuals, particularly those with lower incomes. The substantial costs associated with transportation can hinder households from meeting their fundamental needs, consequently exacerbating poverty levels, particularly in developing countries like India. The cost of electric vehicles is not considered pocket-friendly, but the incentive here is that the government can provide subsidies in taxes, which is a relaxation for the common man.

Conventional modes of transportation, predominantly dependent on fossil fuels, make substantial contributions to the levels of air and noise pollution, posing risks to both human well-being and the environment. Noise pollution generated by traffic can have detrimental impacts on individuals, with young people being particularly susceptible to its adverse effects. Electric vehicles can bring a drastic improvement in environmental conditions as the majority of cities' pollution is caused by conventional-fuel vehicles emitting a lot of smoke, generating noise pollution. Though the discussion ends with the clause that e-vehicles, too, require a lot of resources, but with better recycling and reuse facilities, once a deal, they would go on for decades without requiring further natural resources. Even technological advancements with further research and development can make the whole process sustainable. This is a vicious cycle; once a new change takes place, it betters over time, and the output is always the thing with which a person is used to and is comfortable. Real-time traffic intelligence systems play a crucial role in effectively managing traffic congestion. These systems provide drivers with up-to-date information, empowering them to make informed decisions and choosing the most efficient routes. As a result, congestion on major roads and highways can be reduced, improving overall traffic flow. This would lead to better public transport facilities, reducing travel time, which would add to the positive aspect of pocket-friendliness, encouraging the usage of public transportation, and enhancing sustainable transportation.

As per the study, transition or change is not easy for the majority, but sooner or later, one needs to indulge to keep step with the developing world. Initially, to make everyone accept the change, first the governing authorities should take a step and change their whole working vehicles, i.e., have electric vehicles for government use, providing a sense of authority that the thing is not alien. Also, having public transportation methods converted to an electric medium is a possible solution. Policy intervention is another way of having a positive result towards electric transportation; having a buy-back policy after a certain number of years makes promises to the customer and ensures a safe deal. These are the possible solutions to attract youngsters towards e-vehicles and shift to a sustainable, environmentally friendly, and economically viable transportation medium.

REFERENCES

Alhomaidat, F., Hasan, R. A., Eljufout, T., & Abbas, A. H. (2023). User's perceptions of electric, hybrid, and fuel-powered vehicles in Iraq and Jordan. *Case Studies on Transport Policy*, *12*. https://doi.org/10.1016/j.cstp.2023.101004

Amjad, S., Rudramoorthy, R., Neelakrishnan, S., Sri Raja Varman, K., & Arjunan, T. V. (2011). Evaluation of energy requirements for all-electric range of plug-in hybrid electric two-wheeler. *Energy*, *36*(3), 1623–1629. https://doi.org/10.1016/j.energy.2010.12.069

Ann & Hartell. (n.d.). *Contextualizing location affordability: Urban sprawl and foreclosure.* WU Vienna University of Economics and Business. SRE – Discussion Papers No. 2015/06.

Barodi, A., Zemmouri, A., Bajit, A., Benbrahim, M., & Tamtaoui, A. (2023). Intelligent transportation system based on smart soft-sensors to analyze road traffic and assist driver behavior applicable to smart cities. *Microprocessors and Microsystems*, 104830. https://doi.org/10.1016/j.micpro.2023.104830

Bocarejo, S., & Oviedo, H. D. R. (2012). Transport accessibility and social inequities: A tool for identification of mobility needs and evaluation of transport investments. *Journal of Transport Geography*, *24*, 142–154. https://doi.org/10.1016/j.jtrangeo.2011.12.004

Canzler, W., & Wittowsky, D. (2016). The impact of Germany's Energiewende on the transport sector – unsolved problems and conflicts. *Utilities Policy*, *41*, 246–251. https://doi.org/10.1016/j.jup.2016.02.011

Choi, E., Bhandari, T. R., & Shrestha, N. (2022). Social inequality, noise pollution, and quality of life of slum dwellers in Pokhara, Nepal. *Archives of Environmental & Occupational Health*, *77*(2), 149–160. https://doi.org/10.1080/19338244.2020.1860880

Dawkins, C., & Moeckel, R. (2016). Transit-induced gentrification: Who will stay, and who will go? *Housing Policy Debate*, *26*(4–5), 801–818. https://doi.org/10.1080/10511482.2016.1138986

Dong, H. (2021). Evaluating the impacts of transit-oriented developments (TODs) on household transportation expenditures in California. *Journal of Transport Geography*, *90*. https://doi.org/10.1016/j.jtrangeo.2020.102946

Galati, A., Adamashvili, N., & Crescimanno, M. (2023). A feasibility analysis on adopting electric vehicles in the short food supply chain based on GHG emissions and economic costs estimations. *Sustainable Production and Consumption*, *36*, 49–61. https://doi.org/10.1016/j.spc.2023.01.001

García-Cerrud, C., Flores De La Mota, I., & Soler Anguiano, F. I. (2021). Proposal for greenhouse gas emissions reduction in public passenger transportation. *Case Studies on Transport Policy*, *9*(3), 1358–1366. https://doi.org/10.1016/j.cstp.2021.07.009

Government of India. (2023, May 28). *Census 2011.*

Hu, X., Guo, R., Chen, X., Bi, Z., Wang, J., & Pan, W. (2022). Bismuth-based Z-scheme structure for photocatalytic CO_2 reduction: A review. *Journal of Environmental Chemical Engineering*, *10*(6), 108582. https://doi.org/10.1016/j.jece.2022.108582

Hu, X., Qiao, L., Hao, X., Lin, C., & Liu, T. (2022). Research on the impact of entry points on urban arterial roads in the framework of Kerner's three-phase traffic theory. *Physica A: Statistical Mechanics and Its Applications*, *605*, 127962. https://doi.org/10.1016/j.physa.2022.127962

IEA, 2023. (2023, May 29). https://www.iea.org/

Innovation and Business/Market Opportunities associated with Energy Transitions and a Cleaner Global Environment Issue Paper G20 Ministerial Meeting on Energy Transitions and Global Environment for Sustainable Growth | 2 Executive summary (2019).

Kahn, M. E. (2007). Gentrification trends in new transit-oriented communities: Evidence from 14 cities that expanded and built rail transit systems. *Real Estate Economics*, *35*(2), 155–182. https://doi.org/10.1111/j.1540-6229.2007.00186.x

Kraus, L., Wittowsky, D., & Proff, H. (2023). Multi-method analysis to identify criteria interrelations for sustainability assessment of urban transportation services. *Journal of Cleaner Production*, *412*. https://doi.org/10.1016/j.jclepro.2023.137416

Krause, R. M., Lane, B. W., Carley, S., & Graham, J. D. (2016). Assessing demand by urban consumers for plug-in electric vehicles under future cost and technological scenarios. *International Journal of Sustainable Transportation*, *10*(8), 742–751. https://doi.org/10.1080/15568318.2016.1148213

Litman, T. (2016). *Transportation Affordability: Evaluation and improvement strategies affordable-accessible housing in a dynamic city (www.vtpi.org/aff_acc_hou.pdf) View project investigating the process of traditional design principles formation in the Iranian-Kurdish urban quarters view project*. www.researchgate.net/publication/37183850

Pathak, D. K., Thakur, L. S., & Rahman, S. (2019). Performance evaluation framework for sustainable freight transportation systems. *International Journal of Production Research*, *57*(19), 6202–6222. https://doi.org/10.1080/00207543.2019.1602741

Perrotta, A. F. (2017). Transit fare affordability: Findings from a qualitative study. *Public Works Management & Policy*, *22*(3), 226–252. https://doi.org/10.1177/1087724X16650201

Plyushteva, A. (2023). Affording mobility: Attending to the socio-material affordances of transport un/affordability. *Journal of Transport Geography*, *108*, 103558. https://doi.org/10.1016/j.jtrangeo.2023.103558

Sheu, B.-H., Chiu, C.-C., Lu, W.-T., Lien, C.-C., Liu, T.-K., & Chen, W.-P. (2019). Dual-axis rotary platform with UAV image recognition and tracking. *Microelectronics Reliability*, *95*, 8–17. https://doi.org/10.1016/j.microrel.2019.02.005

Singh, S., Chakma, S., Alawa, B., Kalyanasundaram, M., & Diwan, V. (2023). Identification, characterization, and implications of microplastics in soil – a case study of Bhopal, Central India. *Journal of Hazardous Materials Advances*, *9*, 100225. https://doi.org/10.1016/j.hazadv.2022.100225

Stepanchuk, O., Bieliatynskyi, A., Pylypenko, O., & Stepanchuk, S. (2017). Surveying of traffic congestions on arterial roads of Kyiv City. *Procedia Engineering*, *187*, 14–21. https://doi.org/10.1016/j.proeng.2017.04.344

Tong, H., & Kang, J. (2021). Characteristics of noise complaints and the associations with urban morphology: A comparison across densities. *Environmental Research*, *197*, 111045. https://doi.org/10.1016/j.envres.2021.111045

Upadhyay, C. K., Vasantha, G. A., Tiwari, V., Tiwari, V., & Pandiya, B. (2020). Strategic upturn of reverse logistics with crowdshipping: Transportation explication for India. *Transportation Research Procedia*, *48*, 247–259. https://doi.org/10.1016/j.trpro.2020.08.019

Wallimann, H., Blättler, K., & von Arx, W. (2023). Do price reductions attract customers in urban public transport? A synthetic control approach. *Transportation Research Part A: Policy and Practice*, *173*, 103700. https://doi.org/10.1016/j.tra.2023.103700

Wang, L., & Liu, J. (2017). *Transportation cost index: A comprehensive performance measure for transportation and land use systems and it's application in OR, FL, and Utah*. https://doi.org/10.15760/trec.169

Washington, D. C., Carruthers, R., Dick, M., & Saurkar, A. (2005). *The World Bank Group affordability of public transport in developing countries*. www.worldbank.org/transport/

Wei, F., Walls, W. D., Zheng, X., & Li, G. (2023). Evaluating environmental benefits from driving electric vehicles: The case of Shanghai, China. *Transportation Research Part D: Transport and Environment, 119*. https://doi.org/10.1016/j.trd.2023.103749

Zaidi, Z., & Pelling, M. (2014). Vulnerability to drought and heatwave in London. In *Assessment of vulnerability to natural hazards* (pp. 125–148). Elsevier. https://doi.org/10.1016/B978-0-12-410528-7.00006-0

Zuk, M., Bierbaum, A. H., Chapple, K., Gorska, K., & Loukaitou-Sideris, A. (2018). Gentrification, displacement, and the role of public investment. *Journal of Planning Literature, 33*(1), 31–44. https://doi.org/10.1177/0885412217716439

Chapter 4

Accessible Sustainable Smart Charging Infrastructure Planning for Cities

Sanket Raval, Nilesh Patel, Thangadurai N., Hemant Raval, and Sanchari Deb

4.1 INTRODUCTION

4.1.1 Background

In the 21st century, demand for fossil fuel increases due to its usage in the power sector, transportation sector, and petroleum products. It is observed that pollution due to ICEs in the transportation sector is deteriorating the environment on the entire transportation route. To mitigate this problem, ICEs are replaced with EVs in developed and developing countries (Hasan and Mustafi, 2022). The governments of developing countries give subsidies to encourage citizens to buy EVs (IEA, 2021). But, citizens are still reluctant to adopt EVs due to the unavailability of charging infrastructure. In a country like India, where each state has a different EV policy, only major cities have charging infrastructure (Singh *et al.*, 2021). So, for intercity or interstate routes, it is very difficult to avoid range anxiety. Moreover, establishing charging infrastructure at randomly favorable locations also creates a problem of accessibility. Coordinated charging with a V2G facility and sustainable charging infrastructure will encourage faster adoption of EVs in urban and rural areas. So, it is necessary to have Accessible, Sustainable, Smart Charging Infrastructure (ASSI) planning for each city to solve the problem of the adoption of EVs.

4.1.2 Motivation

In the year 2021, there were 16.5 million EVs on the road, and the number of EV car models increased to 450. It is estimated that 2 million EV cars were sold in 2022 (first quarter), which is a 75% increase from the same time period in 2021 (IEA, 2022). Heavy-duty vehicles (HDV) are also promoted to achieve the zero-emission target. Due to the increase in EV sales, the stock performance of EVs has outperformed among automakers since 2019. To fuel EVs, 500,000 public charging stations were installed in 2021, and the total number of available public charging stations increased to 1.8 million (IEA, 2022). But, there is still a lack of planning for Accessible, Sustainable, Smart Charging Infrastructure in major

 DOI: 10.1201/9781003464556-4

cities worldwide. Accessible, Sustainable, Smart Charging Infrastructure planning includes a wide coverage of EV charging stations for all types of EVs, the inclusion of V2G technology, the integration of renewables in the charging station infrastructure, and battery swapping facilities in charging stations. It also includes an efficient financial model for EV sales and EV charging infrastructure. The ultimate goal of Accessible, Sustainable, Smart Charging Infrastructure planning is to set up charging infrastructure in such a way that it will promote EVs to achieve the zero-emission goal and will remove the hurdles for EV owners when EV charging.

4.1.3 Organization of the Chapter

In the following chapter, ASSI is discussed in detail. Charging infrastructure, the concept of smart charging, and the definition of ASSI in terms of a smart city are discussed in detail in section 2. In section 3, various latest charging standards are stated. The most important parts of this chapter, the global demonstration project on ASSI and case studies, are presented in sections 4 and 5. At the end of this chapter, in section 6, the conclusion of this chapter and the way forward are drawn for the inclusion of ASSI for cities.

4.2 OVERVIEW OF ASSI

4.2.1 Charging Infrastructure

Batteries of EVs are charged by charging stations, which work as an intermediator between EVs and the power network. Different types of charging infrastructures are available for EV charging, which are based on charging methods. Basically, charging methods are classified into three categories as stated:

1) Conductive method
2) Inductive method
3) Battery swapping method

4.2.1.1 Conductive Charging

In this charging method, the terminals of the battery are directly connected to the socket, as shown in Figure 4.1. AC and DC charging is possible depending on the availability of an on-board charger in the EV (Maheshwari *et al.*, 2014). If the EV has on-board charger, then AC charging can be done, and the charger converts AC to DC for battery charging. In DC charging, DC power is directly transferred to batteries in the EV, and it does not require an on-board charger (Knez *et al.*, 2019). The cost of DC charging is more than AC charging due to the converter.

There are three types of charging levels (Rahman *et al.*, 2016), as shown in Table 4.1. Level 1 charging is developed for domestic use, and it is slow charging.

Figure 4.1 Conductive Charging.

Table 4.1 Types of Conductive Charging

Level	*Voltage*	*Charging time*	*Use*
Level 1 (AC)	120 V	10–13 hr	Domestic
Level 2 (AC)	240 V	1–3 hr	Commercial
Level 3 (DC)	208 V	0.5–1.44 hr	Public

It requires a standard SAE J1772 connector for charging. Level 2 requires dedicated installation for usage, and it is developed for commercial or private utilization. Levels 1 and 2 use single-phase supply AC. Level 3 is developed for the public charging station, which has the lowest charging time. It uses DC power, and the cost of level 3 charging is around 30,000 to 160,000 USD. DC charging is possible with the pantograph, and it is used in suburban public transportation (Das *et al.*, 2019).

4.2.1.2 *Inductive Charging*

This type of charging is also known as wireless charging, as shown in Figure 4.2. It requires a greater number of auxiliary equipment like a high-frequency transformer, switchgear, cables, SCADA, etc., so the cost of inductive charging is higher (Ahmad *et al.*, 2017). Land requirements in inductive charging are less compared to conductive charging. It is severely affected by EMI (Electro-Magnetic Interference). The latest developments make the efficiency of this type of charging to 90%, which is generally one or two percent less than plug-in charging. It utilizes a primary pad, which is placed on the floor, and secondary pad, which is fitted at

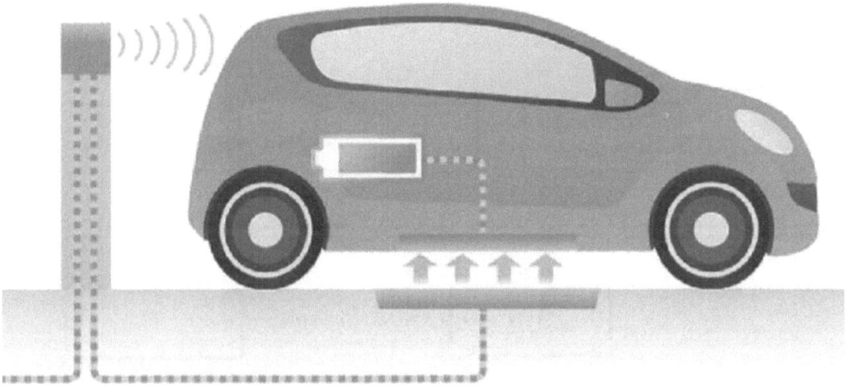

Figure 4.2 Inductive Charging.

the lower end of the EV as shown in Figure 4.2. The principle of electromagnetic induction is used for this type of charging, and energy from the primary pad is transferred to the secondary pad. The secondary pad is connected to the battery for charging.

Two types of inductive charging are possible: stationary charging and dynamic charging (Das *et al.*, 2019). In stationary charging, the EV is required to be in a stationary position for charging, and in dynamic charging, charging is possible while driving. In dynamic charging, the power source is arranged underneath the road surface, and it acts as a primary pad, as discussed previously.

4.2.1.3 Battery Swapping

The conductive method takes time to charge the battery, as shown in Table 4.1. In this method, the charged battery is swapped with the discharged battery at the battery charging station (Ahmad *et al.*, 2020). This method can be incorporated with a demand response program to support the voltage and stability of the system during peak demand on the power system. This method only takes time to swap the battery, which is much less compared to the other two methods. It offers the best alternative option for a heavy-duty EV to charge its battery. Currently, it faces major challenges in battery ownership, interchangeability, and feasibility. It is complex in nature and very expensive.

4.2.2 Charging Infrastructure with Renewable Energy

To feed power from the substation to charging stations requires the reconfiguration of the distribution system. To solve this problem, distributed energy sources like wind energy and solar energy can be used, but they are intermittent and do not provide power all day. Researchers offered solutions to this problem through

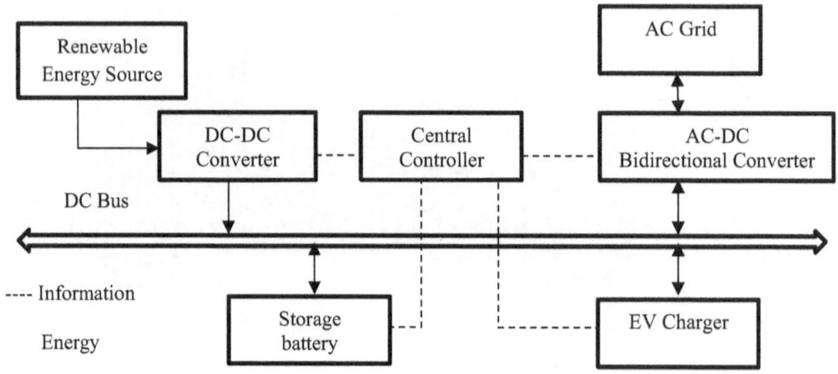

Figure 4.3 Typical Renewable-based Charging Infrastructure.

the optimal operation of renewable power, grid power, and developed renewable-based charging infrastructure (RCI) (Khalid *et al.*, 2019). The main advantage of RCI is that renewable energy sources offer DC link voltage, and it can be matched with EV charging demand. It requires converters, a storage battery, an energy management system, and a central controller. Figure 4.3 shows the typical renewable-based charging infrastructure. Its challenges include power quality, stability, power balance, and pricing.

4.2.3 Smart Charging

Uncoordinated charging can harm the electrical grid's ability to function properly and may cause serious concerns such as voltage instability, degradation of reliability indices, power losses, and harmonic distortions (Deb *et al.*, 2017a, 2017b, 2018, 2022; Wang *et al.*, 2021; Ahmad *et al.*, 2022; Rahman *et al.*, 2022; Ashfaq *et al.*, 2021). A considerable amount of control over the charging process can be provided via smart charging. Smart charging enables EV charging processes to adapt to the needs of both EV users and the power infrastructure. In Table 4.2, many smart charging techniques (IRENA, 2019) are described. There are many different smart charging schemes, including ON/OFF control, V1G, V2G, and V2B, as shown in Table 4.2. It has been noted that V1G, V2G, and V2B all exhibit high levels of maturity and acceptance for ON/OFF control.

4.2.4 Defining ASSI in the Context of a Smart City

The unavailability of easily accessible charging infrastructure is one of the challenges hindering the mass deployment of EVs. The establishment of Accessible, Sustainable, Smart Charging Infrastructure (ASSI) in urban areas is of utmost importance since cities will be the first to adopt electric vehicles. By *accessibility,* it means that the charging stations must be at a convenient distance from the

Table 4.2 Types of Smart Charging Strategies

Type	*Control Strategy*	*Uses*
Uncontrolled with TOU tariff	None	Long-term grid capacity management and peak shaving with implicit demand response
Basis Control	ON/OFF	Congestion management of the grid
Unidirectional V1G	Increase and decrease of real-time charging rate	Ancillary services and frequency regulation
Bidirectional V2G	Instant response to grid conditions with the ability to feed electricity back into the grid; most vehicles and EVSE hardware modifications required	Ancillary services, frequency regulation, voltage regulation, load following, reliability improvement, and short-term integration of renewable energy
Bidirectional V2X (V2H, V2B)	V2G and building/home management system integration	Microgrid optimization

charging demand points. By *sustainable*, it means that the charging infrastructure will meet the current and future charging needs in an economical way without compromising power grid security. By *smart,* it means that there will be some degree of control over the charging activities, taking into account EV charging needs and the load curve of the power grid.

4.3 CHARGING STANDARDS

To standardize the charging process in particular regions or countries, several organizations have established charging standards (Das *et al.*, 2020). IEC (International Electro-Technical Commission) is the Europe-based standards authority who developed IEC 61851-x Ed.x, IEC 62196-x Ed.x, IEC 61980-x Ed.x, etc. The United States-based professional association SAE (Society of Automotive Engineers) framed SAE J2293, SAE J1772, and SAE J1773. Japan has developed a charging standard named CHAdeMO, and China has developed the GB/T charging standard. UL (Underlined Laboratories) declared UL2202, UL2251, UL2594, UL2734, UL2871, UL9741, etc. (Note: x indicates multiple charging standards available in that series with serial number). Country-specific adopted standards (Sachan *et al.*, 2022) are mentioned in Table 4.3.

Other organizations like IEEE, ESO, NTCAS, JARI, JIS, and ARAI have also developed their own standards for EV charging (Deb and Sachan, 2022). The majority of standards are developed for conductive charging, and very few standards are developed for wireless charging, battery swapping (i.e., IEC/TS 62840–1), and V2G. IEC, SAE, and UL standards are summarized in the upcoming section.

Table 4.3 Country-specific Adopted Standard

Category	*Standard*	*United States*	*Europe*	*China*	*Japan*
AC charging and connectors	IEC 62196–2		*		
	IEC 60309–1	*	*	*	*
	IEC 60309–2	*	*	*	*
	SAE J1772	*			*
	GB/T 20234–2			*	
DC charging and connectors	IEC 62196–3	*	*		
	GB/T 20234.3			*	
V2G	IEC 15118	*	*		
	CHAdeMO V2X	*			*

4.3.1 IEC Charging Standard

Various IEC charging standards are summarized in Table 4.4 (IEC, 2022).

Table 4.4 IEC Charging Standards

Sr. No.	*Standard*	*Details*
1	IEC 61851–1 Ed 3.0	• The global requirement for on-board and off-board conductive charging (AC voltage up to 1000 V and DC voltage up to 1500 V). • Defines rating of supply voltage, charging modes, and other specific requirements. • The updated version of IEC 61851–1 Ed 2.0 and IEC 61851–1 Ed 1.0.
2	IEC 61851–21–1 Ed 1.0	• Specifies EMC requirements for on-board charging from AC and DC supply (only for conductive charging).
3	IEC 61851–21–2 Ed 1.0	• Specifies EMC requirements for off-board charging (for conductive and wireless charging).
4	IEC 61851–23-Ed 1.0 (combined with IEC 61851–1 Ed 2.0)	• Requirements of DC Electric Vehicle combined charging system and control communications. • It proves that DC charging is more efficient than AC charging.
5	IEC 62196–1 Ed 3.0	• Specifies plugs for industrial and multiphase applications. • It is an updated version of IEC 62196–1 Ed-2.0.
6	IEC 62196–2 Ed 1.0	• Refers to plugs and connectors with nominal rated operating voltage up to 500 V, 50–60 Hz, and maximum rated current rating limit up to 63 A (3-Phase AC) and 70 A (1-Phase AC). • Specifies connectors, i.e., J1772 connector (for 1-Phase AC), SCAME connector (1-Phase AC and 3-Phase AC), and CHAdeMO (for DC charging).

Sr. No.	Standard	Details
7	IEC 62196–3 Ed 1.0	• For vehicle coupler with pins and contact tubed of standardized configurations for AC and DC. • Includes CHAdeMO, J1772 (for combo charging station), and Tesla supercharger.
8	IEC 619801 Ed 1.0	• Defines wireless charging method from grid or rechargeable energy storage systems (RESS or Batteries).

4.3.2 SAE Charging Standard

Most popular SAE charging standards are briefly stated in Table 4.5 (SAE, 2022).

4.3.3 UL Charging Standard

UL charging standards are detailed in Table 4.6 (UL, 2022).

4.3.4 Comparison of IEC and SAE Charging Standard

In SAE, the power level is considered as Level, and in IEC, the level of power is considered as Mode. The comparison is shown in Table 4.7 (Arif *et al.*, 2021). The application of charging mode/level is stated in Table 4.8 (Arif *et al.*, 2021)

Table 4.5 SAE Charging Standards

Sr. No.	Standard	Details
1	SAE J1772	• It refers to conductive charging requirements in North America.
2	SAE J1773	• It gives minimum interfacing compatibility for inductive charging. • It transfers power at a higher frequency than power line frequency.

Table 4.6 UL Charging Standards

Sr. No.	Standard	Details
1	UL 2202	• It covers requirements for conductive charging.
2	UL 2251	• It refers to plugs, receptacles, vehicle inlets, and connectors for conductive charging.

Table 4.7 Comparison between IEC and SAE Charging Standards

Standard	*Mode/Level*	*Voltage*	*Current*	*Supply*
IEC 62196	Mode 1	120 V	16 A	AC (1-Phase)
	Mode 2	240 V	32 A	
	Mode 3	250 V	32–250 A	
	Mode 4	600 V	400 A	DC
IEC 61851	Mode 1	120 V	16 A	AC (1-Phase)
	Mode 2	240 V	80 A	
	Mode 4	200–450V	80 A	DC
SAE J1772	Level 1	120 V	16 A	AC (1-Phase)
	Level 2	240 V	32–80 A	
	Level 1	200–450V	80 A	DC
	Level 2	200–450V	200 A	

Table 4.8 Application of Mode/Level

Mode/level	*Application*
Mode 1/Level 1	Slow charging (home/office)
Model 2/Level 2 and Mode 3	Public and private charging
Mode 4	Fast charging

4.4 GLOBAL DEMONSTRATION PROJECT ON ASSI

The initial demonstration projects executed globally and whether they considered accessibility, sustainability, and smart charging are reported in Table 4.9 (Tdworld, 2022; Electrive, 2022; Ofgem, 2022; Arcadis, 2022; Bchydro, 2022; Arup, 2022a, 2022b, 2022c; Mahmood *et al.*, 2021; Speth and Funke, 2021; Borden and Boske, 2013).

4.5 CASE STUDIES

A number of case studies on different aspects of ASSI and their findings are reported in Table 4.10 (Deb *et al.*, 2018, 2019, 2021a, 2021b; Mortimer *et al.*, 2022).

4.6 CONCLUSIONS

In this chapter, the need for Accessible, Sustainable, Smart Charging Infrastructure for cities is discussed in terms of charging types, charging strategies, and charging standards. Three types of charging i.e. conductive, inductive and battery swapping, and charging infrastructure with renewable energy sources are

Table 4.9 Global Demonstration Projects

Project name	*Location*	*Accessibility*	*Sustainability*	*Smart Charging*	*Findings*
Electric Vehicle Charging Station Pilot	Maryland, USA	Yes	No	No	• A five-year pilot project, under which Potomac Edison will install more than 50 charging stations across Maryland • Rebates of up to 300 dollars to residential customers for the installation of EV charging stations at home • Range anxiety of EV drivers will be reduced by the installation of some public charge stations under this project
Plug and Charge Pilot Project	France	Yes	No	Yes	• Uses digital certificates for the communication between the vehicle and the grid per the ISO 15118 standard • Elimination of RFID cards and the smartphone app • Only the plug-in is needed for the charging to initiate; the encrypted data from the EV is automatically transferred to the grid while processing the payment in the background
CHARGE Pilot Project	Great Britain	Yes	Yes	No	• Strategic planning of the charging network considering power distribution as well as the road network • Enabling lower-cost connections of charging points for destination as well as on-street charging • Development of the ConnectMore software tool
Electric Vehicle Charging Infrastructure Planning for Cambridge Pilot Project	Cambridge, UK	Yes	No	No	• Arcadis is working with the city of Cambridge to develop a curbside and street light EV charging pilot • Identify the most suitable locations for charger placement • Work with stakeholders and discuss the constraints to be considered for curbside and street light charging
The British Columbia Electric Vehicle Smart Infrastructure Project	British Columbia, Canada	No	Yes	Yes	• Development and deployment of Level 2 chargers and DC fast current infrastructure • Smart charging trials
Highways England Rapid EV Charge point pilot project	UK	Yes	No	No	• 66 rapid charge points installed • 97.4% network coverage achieved • Develop EV-compliant procurement model

(Continued)

Table 4.9 Continued

Project name	*Location*	*Accessibility*	*Sustainability*	*Smart Charging*	*Findings*
Welsh Government EV charging strategy pilot project	Wales, UK	Yes	No	Yes	• Produced an inclusive strategy that responded to the unique needs across Wales • Develop bespoke charging behavior scenarios for different EV users based on demographic data
Ola Beyond Nagpur Pilot Project by OLA	Nagpur, India	Yes	Yes	No	• Established battery charging and battery swapping infrastructure in Nagpur • Investigated operational issues, i.e., the impact of temperature on charging and battery life • Investigation on the integration of renewables with the charging infrastructure, charging behavior, and EV performance
Hero Electric and EV Motors India Pilot for Rapid Charging E-Bikes Pilot Project	Across India	Yes	No	No	• It aims to provide a fast charging station network, 'Plug N go,' at strategic locations by EV Motors and to facilitate EV users at the Hero Dealerships • Guarantees an approximate range of 130–140 km daily with this fast-charging solution
HoLa	Autobahn, Germany	Yes	No	No	• Prototype pilot project for the electrification of long-haul freight transport
RouteCharge Pilot Project	Berlin, Germany	Yes	No	No	• Pilot project testing the feasibility of battery swapping • Three battery swap stations at the start-middle-end of the test track
Plug-In Everywhere Pilot Project	Austin, USA	Yes	Yes	No	• Pilot analyzing renewable-based EV charging in Austin
Plug-In Partners Pilot Project	Austin, USA	Yes	Yes	No	• Pilot project to promote home charging in Austin
Pecan Street Pilot Project	Austin, USA	Yes	Yes	No	• Pilot project to test and conduct various clean energy technologies in the Austin area

Table 4.10 Case Studies

Case study	*Description*	*Test system*	*Diligence*
1	Impact of charging load on power distribution network	33 bus distribution network	• The 33 bus test system is robust enough to withstand the placement of four charging stations at bus 2, representing the strongest bus of the system • The SAIDI jumped by 44% when five charging stations were added on bus 2, which is unacceptable • The voltage of the weakest bus decreased to 0.7351 per unit after installing a single charging station there • Instead of situating the charging stations at a single bus, it is advised to spread them out
2	Charging station placement for Guwahati city	Highway network of Guwahati, Assam, India	• A two-stage planning model for charger placement in the highway network of Guwahati • Location of chargers taking into account the cost, comfort of EV drivers, and grid security • Chargers at three nodes are sufficient to meet the net charging demand
3	Charging hotspot prediction for Helsinki	Helsinki, Finland	• Data-driven approach for finding the charging hotspots of Helsinki, Finland • Home charging, as well as charging at shopping complexes and office scenarios, was considered • High charging demand was observed during evening hours because of home charging
4	Charging station placement for Tianjin	Tianjin, China	• A two-stage planning model for charger placement in the road network of Tianjin, China • Multi-objective planning model considering cost, waiting time, accessibility, and security of power grid • Reduction in search space because of two-stage modelling further reduced computational complexity • Post-placement analysis revealed that operating parameters of the power grid, such as voltage profile, reliability indices, and power losses, are within safe limits

(Continued)

Table 4.10 Continued

Case study	*Description*	*Test system*	*Diligence*
5	EV public charging station planning utilizing real-time charging data from Germany	Urban and suburban areas of Germany	• Computation of charger utilization rate of existing chargers located in urban and suburban areas of Germany • There was no conclusive evidence of coordinated planning for additional charging stations within cities or other municipal districts
6	Distribution network planning considering charger placement	69 bus distribution network	• Three models for distribution network planning considering EV charger load • Model 1 considered voltage stability as an objective function, model 2 considered composite reliability indices as an objective function, and model 3 considered cost as an objective function • Comparison of the three models
7	Analytical strategy for deploying charging stations in the power distribution network of Allahabad, India	30 bus distribution network of Allahabad	• Charger placement in 30 bus distribution network of Allahabad considering voltage deviation and line losses • The network of distribution could withstand the placement of six charging stations
8	Big data analytics-based charging infrastructure planning	Mumbai, India	• Establish a correlation between charging demand and charger utilization rate for Mumbai, India
9	Charging station placement for Allahabad, India	Allahabad, India	• Single-objective charger placement scheme for Allahabad, India • Post-placement analysis confirmed that distribution network parameters are least affected, indicating the efficacy of the approach

explained. Due to the higher charging time of conductive and inductive charging, it is required to establish more battery swapping stations in cities. To coordinate operation with power grid, various types of charging strategies are detailed with their application. Popular IEC, SAE and UL charging standards for conductive and inductive charging are stated in terms of technical aspects.

Worldwide demonstrated EV projects are compared in terms of accessibility, sustainability, and smart charging and it can be observing that none of them have facilitated ASSI. Nine case studies are also discussed to investigate different aspects of ASSI in terms of effect on power system network. It is concluded that to promote EVs in all the categories of road transportation, it is required to have Accessible, Sustainable, Smart Charging Infrastructure so that EV owners get confidence of charging as fueling of ICE vehicles

REFERENCES

Ahmad, A., Alam, M. S., & Chabaan, R. (2017). A comprehensive review of wireless charging technologies for electric vehicles. IEEE Transactions on Transportation Electrification, 4(1), 38–63.

Ahmad, F., Alam, M. S., Alsaidan, I. S., & Shariff, S. M. (2020). Battery swapping station for electric vehicles: Opportunities and challenges. IET Smart Grid, 3(3), 280–286.

Ahmad, F., Iqbal, A., Ashraf, I., & Marzband, M. (2022). Optimal location of electric vehicle charging station and its impact on distribution network: A review. Energy Reports, 8, 2314–2333.

Arcadis (2022). https://www.arcadis.com/en/knowledge-hub/perspectives/global/2020/resilience-to-recovery/electric-vehicle-charging-infrastructure-planning-for-the-city-of-cambridge-massachusetts (Accessed 30th May 2022).

Arif, S. M., Lie, T. T., Seet, B. C., Ayyadi, S., & Jensen, K. (2021). Review of electric vehicle technologies, charging methods, standards and optimization techniques. Electronics, 10(16), 1910.

Arup (2022a). www.arup.com/projects/highways-england-rapid-electric-vehicle-charge-points-programme (Accessed 30th May 2022).

Arup (2022b). www.arup.com/projects/welsh-government-electric-vehicle-charging-strategy (Accessed 30th May 2022).

Arup (2022c). https://www.arup.com/projects/highways-england-rapid-electric-vehicle-charge-points-programme (Accessed 30th May 2022).

Ashfaq, M., Butt, O., Selvaraj, J., & Rahim, N. (2021). Assessment of electric vehicle charging infrastructure and its impact on the electric grid: A review. International Journal of Green Energy, 18(7), 657–686.

Bchydro: ev-project-reports.pdf (bchydro.com) (Accessed 30th May 2022)

Borden, E. J., & Boske, L. B. (2013). Electric vehicles and public charging infrastructure: Impediments and opportunities for success in the United States (No. SWUTC/13/600451-00064-1). Southwest Region University Transportation Center (US).

Das, H. S., Rahman, M. M., Li, S., & Tan, C. W. (2020). Electric vehicles standards, charging infrastructure, and impact on grid integration: A technological review. Renewable and Sustainable Energy Reviews, 120, 109618.

Das, S., Sasidharan, C., & Ray, A. (2019). Charging India's bus transport. Alliance for an Energy Efficient Economy. https://aeee.in/our-publications/charging-indias-bus-transport

Deb, S. (2021a). A Novel Data Driven Methodology for Charging Hotspots Computation for the City of Helsinki. In 2021 International Conference on Electrical, Communication, and Computer Engineering (ICECCE) (pp. 1–6). IEEE.

Deb, S., Kalita, K., & Mahanta, P. (2017a, December). Impact of Electric Vehicle Charging Stations on Reliability of Distribution Network. In 2017 International Conference on Technological Advancements in Power and Energy (TAP Energy) (pp. 1–6). IEEE.

Deb, S., Kalita, K., & Mahanta, P. (2017b, December). Review of Impact of Electric Vehicle Charging Station on the Power Grid. In 2017 International Conference on Technological Advancements in Power and Energy (TAP Energy) (pp. 1–6). IEEE.

Deb, S., & Sachan, S. (2022). Proposed Power Systems Planning in Indian Scenario for Integrating EV Charging Infrastructure. In Developing Charging Infrastructure and Technologies for Electric Vehicles (pp. 25–37). IGI Global.

Deb, S., Tammi, K., Gao, X. Z., Kalita, K., Mahanta, P., & Cross, S. (2021b). A robust two-stage planning model for the charging station placement problem considering road traffic uncertainty. IEEE Transactions on Intelligent Transportation Systems, 23(7), 6571–6585.

Deb, S., Tammi, K., Kalita, K., & Mahanta, P. (2018). Impact of electric vehicle charging station load on distribution network. Energies, 11(1), 178.

Deb, S., Tammi, K., Kalita, K., & Mahanta, P. (2019). Charging station placement for electric vehicles: A case study of Guwahati city, India. IEEE Access, 7, 100270–100282.

Electrive (2022). https://www.electrive.com/2020/07/08/pan-european-pilot-to-look-at-usability-of-charging-stations/ (Accessed 30th May 2022).

Hasan, M. R., & Mustafi, N. N. (2022). Environmental impacts of the use of electric vehicles. Environmental Claims Journal, 34(1), 56–79.

IEA (2021). Global EV Outlook 2021. IEA. www.iea.org/reports/global-ev-outlook-2021 (Accessed 1st June 2022).

IEA (2022). Global EV Outlook 2022. IEA. www.iea.org/reports/global-ev-outlook-2022 (Accessed 1st June 2022).

IEC (2022). Electric vehicles | IEC. www.iec.ch/transportation/electric-vehicles (Accessed 10th June 2022).

IRENA (2019). https://www.irena.org/publications/2019/May/Innovation-Outlook-Smart-Charging (Accessed 24th November 2020).

Khalid, M. R., Alam, M. S., Sarwar, A., & Asghar, M. J. (2019). A comprehensive review on electric vehicles charging infrastructures and their impacts on power-quality of the utility grid. ETransportation, 1, 100006.

Knez, M., Zevnik, G. K., & Obrecht, M. (2019). A review of available chargers for electric vehicles: United States of America, European Union, and Asia. Renewable and Sustainable Energy Reviews, 109, 284–293.

Maheshwari, P., Tambawala, Y., Nunna, H. K., & Doolla, S. (2014). A Review on Plug-in Electric Vehicles Charging: Standards and Impact on Distribution System. In IEEE International Conference on Power Electronics, Drives and Energy Systems (PEDES) (pp. 1–6). IEEE.

Mahmood, M., Alam, M. S., Deb, S., & Rafat, Y. (2021). Review of Charging Infrastructure Pilot Projects: Global Scenario and Lessons Learnt. In ISUW 2021: Proceedings of the 7th International Conference and Exhibition on Smart Energy and Smart Mobility for Smart Cities (p. 239). Springer Nature.

Mortimer, B. J., Hecht, C., Goldbeck, R., Sauer, D. U., & De Doncker, R. W. (2022). Electric vehicle public charging infrastructure planning using real-world charging data. World Electric Vehicle Journal, 13(6), 94.
Ofgem (2022). Project Summary. ofgem.gov.uk (Accessed 30th May 2022).
Rahman, I., Vasant, P., Singh, B., Wadud, M., & Adnan, N. (2016). Review of recent trends in optimization techniques for plug-in hybrid, and electric vehicle charging infrastructures. Renewable and Sustainable Energy Reviews, 58, 1039–1047.
Rahman, S., Khan, I. A., Khan, A. A., Mallik, A., & Nadeem, M. F. (2022). Comprehensive review & impact analysis of integrating projected electric vehicle charging load to the existing low voltage distribution system. Renewable and Sustainable Energy Reviews, 153, 111756.
Sachan, S., Deb, S., Singh, P. P., Alam, M. S., & Shariff, S. M. (2022). A comprehensive review of standards and best practices for utility grid integration with electric vehicle charging stations. Wiley Interdisciplinary Reviews: Energy and Environment, 11(3), e424.
SAE (2022). www.sae.org/standards/ (Accessed 10th June 2022).
Singh, V., Singh, V., & Vaibhav, S. (2021). Analysis of electric vehicle trends, development and policies in India. Case Studies on Transport Policy, 9(3), 1180–1197.
Speth, D., & Funke, S. Á. (2021). Comparing options to electrify heavy-duty vehicles: Findings of german pilot projects. World Electric Vehicle Journal, 12(2), 67.
Tdworld (2022). www.tdworld.com/electrification/article/20972298/electric-vehicle-charging-station-pilot-project-to-be-launched-in-maryland (Accessed 30th May 2022).
UL (2022). www.ul.com/ (Accessed 10th June 2022).
Wang, L., Qin, Z., Slangen, T., Bauer, P., & van Wijk, T. (2021). Grid impact of electric vehicle fast charging stations: Trends, standards, issues and mitigation measures-an overview. IEEE Open Journal of Power Electronics, 2, 56–74.

Chapter 5

Sustainable Transportation

Policy, Planning, and Implementation

Rajesh Kumar Upadhyay, Varun Pratap Singh, Ashwani Kumar

5.1 INTRODUCTION TO SUSTAINABLE TRANSPORTATION

The notion of sustainable transportation has garnered considerable interest in recent decades as global societies confront the environmental, social, and economic consequences associated with conventional transportation systems [1]. This chapter explores the fundamental principles, importance, and historical backdrop of sustainable transportation, providing insights into the policies, planning, and strategies for implementation that can facilitate a more sustainable trajectory for the future.

5.1.1 Defining Sustainable Transportation

Sustainable transportation encompasses a holistic strategy that seeks to fulfil the present transportation requirements of individuals and societies while mitigating detrimental effects on the environment and future generations. The concept encompasses a diverse range of transportation modes, such as road, rail, air, water, and non-motorised options like walking and cycling. The principal objective is to achieve a state of equilibrium that encompasses economic advancement, social fairness, and ecological conservation [2].

There are several fundamental elements that are integral to the concept of sustainable transportation, as outlined by various sources [3].

- **Energy efficiency:** One of the key objectives is to enhance energy efficiency by advocating for the optimal utilisation of energy resources and facilitating the shift towards sustainable and renewable energy sources for the purpose of powering transportation systems [4].
- **Emissions Reduction:** The objective is to mitigate the release of greenhouse gas emissions, air pollutants, and noise pollution linked to transportation activities [5].
- Accessibility and equity: The principle of accessibility and equity pertains to the imperative of providing transport options that are equally

 DOI: 10.1201/9781003464556-5

available to individuals, irrespective of their socioeconomic status or physical capabilities.
- **Integration of Land Use:** Promoting land use planning strategies that prioritise the development of compact, mixed-use areas in order to mitigate the necessity for extensive travel distances.
- **Ensuring Safety and Health:** Placing utmost importance on implementing safety measures for all individuals utilising transportation systems while advocating for the adoption of active transportation methods to enhance public health outcomes.

5.1.2 Importance of Sustainable Transportation

The importance of sustainable transportation extends beyond environmental considerations. It exerts influence on multiple facets of contemporary society, encompassing [6]:

- **Environmental preservation**: It is a significant concern in relation to traditional transportation methods, as they are heavily dependent on fossil fuels. This reliance contributes to the escalation of greenhouse gas emissions and the subsequent deterioration of air quality. The adoption of sustainable transportation is of utmost importance in addressing the challenges posed by climate change, preserving finite natural resources, and safeguarding delicate ecosystems [7].
- **Public Health**: The implementation of sustainable transport options, such as walking and cycling, has been shown to have a positive impact on public health. These options not only contribute to the reduction of air pollution but also encourage physical activity, resulting in improved public health outcomes and a decrease in the prevalence of chronic diseases.
- **Economic Benefit:** One of the advantages of adopting sustainable transport is the potential for economic benefits, particularly in terms of long-term cost savings. Through the reduction of fuel consumption and maintenance costs, both individuals and businesses are able to reallocate resources towards more productive endeavours [8].
- **Social equity:** It is enhanced through the implementation of sustainable transportation systems, as they contribute to increased accessibility, particularly in areas that are typically underserved. This improved accessibility leads to better opportunities for education, employment, and community engagement.
- **Enhancing Urban Liveability:** The promotion of public transportation and non-motorised modes of transportation has the potential to mitigate traffic congestion, minimise noise pollution, and alleviate the demand for excessive parking infrastructure, thereby enhancing the overall liveability and attractiveness of urban areas.

5.1.3 Historical Overview of Sustainable Transportation

The origins of sustainable transport can be attributed to the early 20th century, during which urban planners began to contemplate the consequences of rapid urbanisation and the proliferation of motor vehicles. Nevertheless, it garnered considerable interest during the latter half of the 20th century [9].

The decades of the 1970s and 1980s witnessed a significant transformation in transport policies as a direct response to the oil crises and an increasing recognition of environmental concerns [10]. The emergence of concepts such as transit-oriented development, carpooling, and the expansion of public transportation has been observed. During the 1990s, the concept of "sustainable transportation" emerged as a prominent topic of discussion in international conferences and academic research. The incorporation of sustainability, efficiency, and equity has emerged as essential components in the field of transportation planning [11].

Since the beginning of the 21st century, there has been a significant global push towards the development and implementation of sustainable transport policies. Illustrative instances encompass investments in public transport, advocacy for the adoption of electric vehicles, and the development of infrastructure conducive to cycling. Despite the advancements made, there are still enduring challenges in the worldwide execution. This chapter examines potential solutions for a transport system that effectively addresses the requirements of both current and future generations. The comprehension of the fundamental nature and historical background of sustainable transportation enables policymakers and stakeholders to develop a system that yields societal, environmental, and economic benefits [12].

5.2 POLICY FRAMEWORK FOR SUSTAINABLE TRANSPORTATION

5.2.1 Developing Nations Sustainable Transportation Policies Framework Features

Developing countries encounter distinct obstacles when it comes to the implementation of sustainable transport policies, primarily stemming from the swift urbanisation process, constrained resources, and diverse socioeconomic circumstances [13]. Notwithstanding these obstacles, numerous developing nations have made noteworthy advancements in the promotion of sustainable transportation [14]. The policy frameworks of the organisation exhibit several notable characteristics, as illustrated in Figure 5.1.

a. **Integrated Transport Planning:** The adoption of integrated transport planning approaches, which take into account various modes of transportation, such as buses, trains, cycling, and pedestrian pathways, is becoming more prevalent in developing nations. This approach aims to establish a comprehensive and efficient network.

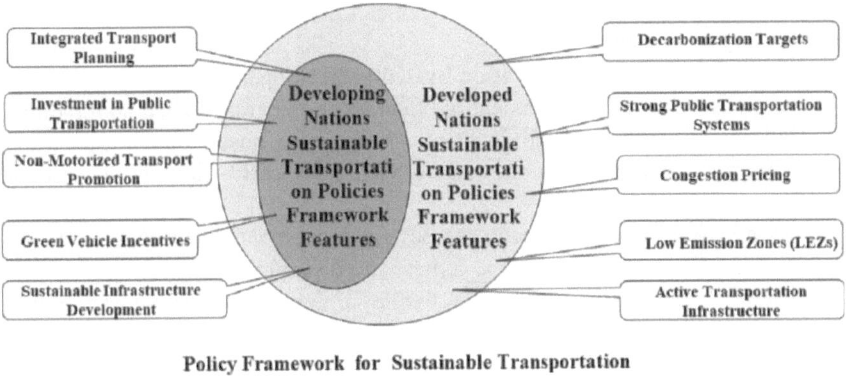

Figure 5.1 Policy Framework for Sustainable Transportation.

b. **Investment in Public Transport:** Governments are allocating resources towards the expansion and enhancement of public transport infrastructures with the aim of offering affordable and easily accessible alternatives for commuters, consequently mitigating reliance on privately owned vehicles.
c. **Promotion of Non-Motorised Transport:** The promotion of non-motorised transport, including activities such as cycling and walking, is a prevalent component of policy frameworks implemented in developing countries. In addition to mitigating emissions, this measure also enhances public health and promotes greater accessibility.
d. **Green Vehicle Incentive:** In order to promote the adoption of eco-friendly vehicles, several developing nations provide economic incentives, tax exemptions, or subsidies for electric vehicles or vehicles powered by alternative fuels.
e. **Sustainable Infrastructure Development:** A focus on the development of sustainable infrastructure, which encompasses the establishment of environmentally-friendly buildings, energy-efficient transportation centres, and integrated transit-oriented developments [15].

5.2.2 Developed Nations Sustainable Transportation Policies Framework Features

Developed nations have taken a leading role in implementing sustainable transport initiatives and have established comprehensive policy frameworks to effectively tackle environmental issues and mitigate carbon emissions. Several notable characteristics of their policies are outlined as follows [16]:

a. **Decarbonisation Targets:** Developed nations have established ambitious decarbonisation objectives for their transportation sectors, with the intention of mitigating greenhouse gas emissions through a range of strategies, including the adoption of electric vehicles and the promotion of renewable energy resources [17].

b. **Robust Public Transportation Systems:** Developed countries place a high emphasis on the establishment of effective and interconnected public transportation systems, which frequently incorporate buses, trains, trams, and subways. These systems aim to offer viable alternatives to the utilisation of private vehicles.
c. **Congestion pricing:** These schemes are implemented in urban centres of certain developed countries as a means to discourage the utilisation of private vehicles during peak hours, thereby mitigating traffic congestion and emissions.
d. **Low Emission Zones (LEZs):** Refers to specific geographical areas that have been designated for the exclusive use of low-emission vehicles. The primary objective of LEZs is to encourage the adoption of cleaner vehicles and subsequently mitigate air pollution levels.
e. **Active transportation infrastructure**: It is a key focus for developed nations, as they allocate resources towards the creation of secure and comprehensive infrastructure that promotes walking and cycling. This strategic investment aims to incentivise residents to embrace active modes of transportation.

5.3 INTERNATIONAL AGREEMENTS AND INITIATIVES

In order to effectively address sustainability issues related to transportation, it is imperative to foster international cooperation and establish agreements, given the global nature of these challenges [18]. Numerous international agreements and initiatives have been implemented with the aim of fostering sustainable transport, as exemplified by [19]:

a. **Paris Agreement:** Within the United Nations Framework Convention on Climate Change (UNFCCC), the Paris Agreement aims to limit global warming and mitigate the impact of climate change. Countries are committed to reducing their greenhouse gas emissions, including those from the transportation sector [20].
b. **Sustainable Development Goals (SDGs):** Adopted by UN member states, the SDGs include Goal 11, which focuses on making cities and human settlements inclusive, safe, resilient, and sustainable. Transportation policies play a crucial role in achieving this goal [21].
c. **United Nations Decade of Action for Road Safety:** This initiative aims to reduce road traffic deaths and injuries globally. Sustainable transportation policies often prioritise road safety measures [22].
d. **World Urban Forum:** Organised by UN-Habitat, this forum brings together policymakers, experts, and stakeholders to discuss sustainable urbanisation, including transportation-related topics [23].

5.4 LEGISLATIVE AND REGULATORY MEASURES

Effective, sustainable transportation policies often require legislative and regulatory measures to enforce compliance and foster change [24]. Some legislative and regulatory approaches include [25]:

a. **Emission Standards:** Setting mandatory emission standards for vehicles to limit the amount of pollutants they release into the atmosphere.
b. **Fuel Efficiency Regulations:** Implementing regulations that mandate minimum fuel efficiency standards for vehicles, encouraging manufacturers to produce eco-friendly vehicles.
c. **Land Use Zoning:** Zoning regulations that promote mixed-use development and prioritise compact urban growth to reduce the need for long-distance commuting.
d. **Green Procurement:** Governments adopt green procurement policies, which favor environmentally friendly vehicles and infrastructure in their procurement processes.
e. **Vehicle Access Restrictions:** Implementing restrictions on the access of certain vehicles, especially in urban areas with high pollution, to promote cleaner transportation options.

5.5 FUNDING AND FINANCIAL INCENTIVES

Financial incentives and funding mechanisms play a crucial role in supporting the implementation of sustainable transportation policies. Some common approaches include [26]:

a. **Grants and Subsidies:** Governments provide grants and subsidies to support the development of public transportation infrastructure and promote the adoption of cleaner vehicles.
b. **Tax Incentives:** Offering tax credits or deductions to individuals and businesses that invest in sustainable transportation options or use eco-friendly vehicles.
c. **Congestion Charge Revenue:** Reinvesting revenue generated from congestion charges into sustainable transportation projects and initiatives.
d. **Public-Private Partnerships (PPPs):** Collaborating with private sector entities to fund and develop sustainable transportation infrastructure projects.
e. **Green Bonds:** Green bonds are issued to raise funds specifically for sustainable transportation projects.

Through the implementation of comprehensive policy frameworks, nations, regardless of their level of development, can expedite their shift towards sustainable transport systems. This transition not only aids in safeguarding the environment but also enhances the overall well-being and quality of life of their respective populations. International agreements and regulatory measures serve as a comprehensive framework for addressing transport challenges on a global scale, fostering collective efforts [27]. Additionally, funding and financial incentives play a crucial role in providing the required support and driving the successful implementation of these measures. The user did not provide any text to rewrite.

5.6 PLANNING FOR SUSTAINABLE TRANSPORTATION

The attainment of sustainable transport objectives necessitates the implementation of efficient planning strategies. The process encompasses the implementation of various strategies aimed at effectively managing transportation demand, integrating land use and transportation planning, advocating for the use of multiple modes of transportation, and ensuring the proper maintenance of infrastructure. Every facet of transport sustainability has a significant impact on shaping its future [28].

5.6.1 Transportation Demand Management (TDM)

TDM strategies are increasingly being adopted on a global scale as urban areas confront the challenges posed by traffic congestion and environmental issues. Several strategies, such as carpooling, ride-sharing, congestion pricing, and flexible work hours, are being employed in order to mitigate the usage of single-occupancy vehicles. The future prospects of Transportation Demand Management (TDM) appear optimistic due to the continuous advancements in technology, which facilitate enhanced data collection and analysis. These developments hold the potential to yield more efficient and tailored transportation solutions. The potential transformation of transport demand management (TDM) implementation may occur through the integration of smart mobility platforms and autonomous vehicles [29].

The achievement of effective TDM implementation is contingent upon the cultivation of public awareness, government endorsement, and stakeholder collaboration. Furthermore, the presence of dependable and effective public transportation options has a significant influence on the implementation of Transportation Demand Management (TDM) strategies [30].

5.6.2 Integrated Land Use and Transportation Planning

There is a growing trend among numerous cities to adopt principles of integrated land use and transportation planning. This entails the development of urban areas characterised by the proximity of residential areas, commercial establishments, and vital amenities, thereby mitigating the necessity for extensive commuting. The potential for creating more sustainable and liveable urban environments is significantly enhanced by the future of integrated planning. The increasing urbanisation of cities is expected to result in a higher demand for meticulously designed mixed-use developments [31].

The attainment of successful integration necessitates the presence of political will, community engagement, and collaboration among urban planners, policy-makers, and developers. In addition, the establishment of affordable housing in close proximity to employment hubs is a crucial determinant [32].

5.6.3 Multi-Modal Transportation Planning

The adoption of multi-modal transportation planning is increasingly prevalent in urban regions. Municipalities are allocating resources towards the enhancement of cycling infrastructure, pedestrian walkways, and the enhancement of public transportation networks as a means to promote the adoption of alternative transportation methods [33]. The potential for multi-modal transportation to thrive in the coming years appears optimistic, given the increasing emphasis placed by cities on enhancing accessibility, safety, and convenience for all individuals utilising various modes of transportation. The integration between different transportation modes may be further enhanced through technological advancements [34].

The success of multi-modal transportation planning is contingent upon several crucial factors, namely sufficient financial resources, effective coordination among diverse transportation agencies, and a steadfast commitment to prioritising the safety of vulnerable road users.

5.6.4 Infrastructure Development and Maintenance

Numerous geographical areas are currently allocating resources towards the development of sustainable transport infrastructure, encompassing various initiatives such as the establishment of electric vehicle charging stations, the implementation of bike-sharing facilities, and the enhancement of environmentally friendly public transport systems. The future outlook for infrastructure development and maintenance appears promising, as governmental entities acknowledge the significance of enhancing current infrastructure and constructing new sustainable facilities. The availability of funding, government policies, and the capacity to adapt to evolving technology and user demands have a substantial impact on the development and maintenance of infrastructure [35].

The process of planning for sustainable transportation necessitates the adoption of a comprehensive methodology that encompasses various factors, including transportation demand management, integrated land use and transportation planning, the incorporation of multiple modes of transportation, and the establishment and upkeep of infrastructure. The efficacy of these strategies is contingent upon a multitude of factors, encompassing technological progress, societal endorsement, governmental regulations, and the accessibility of financial resources. As urban areas and countries persist in their efforts to prioritise sustainable transport, the forthcoming years hold the potential for the emergence of inventive strategies aimed at tackling transport obstacles and establishing more ecologically conscious and streamlined mobility systems.

5.7 SUSTAINABLE TRANSPORTATION TECHNOLOGIES

The implementation of sustainable transportation technologies is of utmost importance in mitigating the environmental consequences associated with the transportation industry. The aforementioned technologies have the primary objective of reducing the release of greenhouse gases, enhancing energy efficiency, and encouraging the adoption of environmentally sustainable transportation alternatives [10]. Several notable sustainable transportation technologies encompass electric vehicles (EVs) and associated charging infrastructure, alternative fuel technologies, intelligent transportation systems (ITS), and active transportation modes such as cycling and walking. The adoption of these technologies has the potential to contribute to a more environmentally sustainable transportation landscape [36].

5.8 ELECTRIC VEHICLES AND CHARGING INFRASTRUCTURE

Electric vehicles (EVs) have gained significant popularity due to their low greenhouse gas emissions and potential to reduce dependence on fossil fuels. The global EV market has been rapidly growing, with sales reaching over 3 million units in 2020. Countries around the world are implementing policies to support EV adoption, such as offering subsidies, tax incentives, and expanding charging infrastructure [37]. The development of charging infrastructure is essential to overcome range anxiety and encourage more people to switch to electric vehicles. As of 2021, there were over 200,000 public charging stations globally, with both governments and private companies investing in expanding this infrastructure further [36].

5.9 ALTERNATIVE FUEL TECHNOLOGIES

Alternative fuel technologies provide feasible alternatives to traditional fossil fuels, thereby aiding in the mitigation of carbon emissions. Examples of alternative fuels include biofuels, hydrogen, and natural gas. Biofuels are produced through the utilisation of sustainable resources, including crops and waste materials, which are renewable in nature. According to recent projections, the global biofuels market is anticipated to exhibit a consistent growth trajectory, characterised by a compound annual growth rate (CAGR) of 3.8% over the period spanning from

2020 to 2027 [5]. The utilisation of hydrogen fuel cell technology is experiencing a notable surge, particularly within the transportation industry, with projected investments set to reach $1.5 billion per year by the year 2030. Furthermore, in different geographical areas, compressed natural gas (CNG) and liquefied natural gas (LNG) are being employed as alternative sources of fuel for buses and commercial vehicles [38].

5.10 INTELLIGENT TRANSPORTATION SYSTEMS

Intelligent Transportation Systems (ITS) utilise sophisticated communication and information technologies in order to enhance the effectiveness and security of transportation systems. These systems include a range of applications, including real-time traffic management, electronic toll collection, and smart parking [39]. The projected value of the global Intelligent Transportation Systems (ITS) market is estimated to reach $51.09 billion by the year 2027. This growth is primarily attributed to the rising trend of urbanisation and the consequential demand for effective traffic management solutions. The implementation of Intelligent Transportation Systems (ITS) has demonstrated encouraging outcomes, with observed reductions in travel times of up to 25% and a corresponding decrease in traffic congestion of up to 30%. Governments and transportation agencies are making investments in Intelligent Transportation Systems (ITS) with the aim of improving mobility, reducing emissions, and enhancing overall efficiency in transportation [40].

5.11 ACTIVE TRANSPORTATION MODES (CYCLING AND WALKING)

Active transportation modes, such as cycling and walking, are increasingly being recognised as sustainable and health-promoting alternatives to motorised forms of transportation. The incorporation of cycling lanes and pedestrian-friendly infrastructure in numerous urban areas has effectively incentivised individuals to opt for these modes of transport for short-distance journeys. The anticipated trajectory of cycling infrastructure investment indicates a forthcoming expansion, as the global market is forecasted to attain a value of $1.6 billion by the year 2027. Walking is a widely adopted means of transportation, as evidenced by the fact that approximately 33% of global trips are conducted on foot. The promotion of active transport can yield a multitude of advantages, such as the mitigation of emissions, the enhancement of public health, and the improvement of urban liveability [41].

The ongoing development and increasing adoption of sustainable transport technologies offer substantial prospects for mitigating environmental concerns and establishing a transport system that is both sustainable and resilient in the long term. The support and adoption of these technologies by governments, industries, and individuals are all essential in facilitating the development of a more environmentally friendly and efficient transportation infrastructure [42]. Table 5.1 provides a parametric analysis of various sustainable transportation technologies.

Table 5.1 Parametric Comparison between Different Sustainable Transportation Technologies

Parameters	*Electric Vehicles and Charging Infrastructure*	*Alternative Fuel Technologies*	*Intelligent Transportation Systems*	*Active Transportation Modes (Cycling and Walking)*
Environmental Impact	Low greenhouse gas emissions	Varied depending on fuel type	Reduce congestion and emissions	Zero emissions, eco-friendly
Energy Source	Electricity	Biofuels, hydrogen, natural gas	Electricity and communication systems	Human-powered
Infrastructure Requirements	Charging stations and grid upgrades	Fuelling stations and supply chain	Communication networks and sensors	Pathways, bike racks, pedestrian crossings
Technological Maturity	Advanced, rapidly evolving	Developing, gaining popularity	Advanced, implemented in some areas	Traditional, but improving safety and comfort
Range and Performance	Improving steadily	Varies by technology	NA	Dependent on human ability
Adoption Challenges	High initial cost, range anxiety	Limited fuel availability	Initial investment, privacy concerns	Safety concerns, weather limitations
Government Incentives	Subsidies, tax credits, and rebates	Subsidies and tax incentives	Funding and supportive policies	Infrastructure investment, bike-friendly cities
Consumer Acceptance	Growing with increasing awareness	Varies by region and user	Acceptance increasing over time	Already widely accepted in many areas
Fuelling/ Charging Time	Charging time varies (rapid chargers exist)	Varies depending on fuel type	NA	Instantaneous for walking, charging time for cycling
Energy Efficiency	High efficiency in electric powertrains	Varies by fuel and technology	More efficient traffic management	Highly energy-efficient

5.12 IMPLEMENTATION STRATEGIES FOR SUSTAINABLE TRANSPORTATION

The successful implementation of sustainable transportation initiatives necessitates a comprehensive strategy that encompasses various elements, including collaborative efforts, innovative approaches, partnerships between public and private entities, active involvement of stakeholders, drawing lessons from successful precedents, and continuous evaluation and monitoring. Through the strategic utilisation of the unique capabilities possessed by various stakeholders and the adoption of novel methodologies, societies can progress towards a transportation future characterised by enhanced sustainability, efficiency, and environmental consciousness. This section examines various strategies that can effectively support the implementation of sustainable transport projects. These strategies encompass the involvement of public-private partnerships and the assessment of the outcomes and effects of such initiatives. The user's text does not contain any information to be rewritten [43].

5.12.1 Public-Private Partnerships (PPP)

Public-private partnerships (PPPs) are of paramount importance in facilitating the mobilisation of resources, expertise, and innovation for sustainable transportation projects. These collaborative efforts facilitate the integration of the respective advantages of the public and private sectors, thereby enhancing the efficiency and efficacy of implementation processes. Private enterprises have the capacity to offer financial resources, technological advancements, and operational proficiency, whereas the government plays a role in providing regulatory assistance and representing the interests of the general public. Public-private partnerships (PPPs) have played a crucial role in the advancement of sustainable transportation infrastructure, the facilitation of electric vehicle uptake, and the enlargement of public transportation systems. Governments can expedite sustainable transportation advancements by cultivating transparent and well-organised partnerships that harness the capabilities of the private sector.

5.12.2 Stakeholder Engagement and Collaboration

The attainment of support, resolution of concerns, and achievement of success in sustainable transportation initiatives heavily rely on the implementation of efficient stakeholder engagement and collaboration. Stakeholders encompass a range of entities, such as government agencies, businesses, community organisations, environmental advocates, and the public. Incorporating stakeholder involvement throughout the stages of planning and implementation facilitates a more comprehensive comprehension of a wide range of needs and perspectives. The utilisation of collaborative decision-making processes has the potential to facilitate consensus-building and foster the development of transportation solutions that are more sustainable, socially equitable, and environmentally responsible. Moreover, the inclusion of the general public in the decision-making process cultivates a feeling of ownership and generates support for the executed initiatives.

5.12.3 Case Studies of Successful Implementations

The analysis of case studies pertaining to the successful implementation of sustainable transport initiatives offers valuable insights and exemplary approaches that can inform future projects. Illustrating concrete instances of successful initiatives in the real world can function as exemplars for emulation by others. Case studies could encompass various initiatives aimed at enhancing urban transport systems. These may involve the implementation of bike-friendly infrastructure on a citywide scale, the integration of intelligent transport systems within urban areas, or the undertaking of projects focused on electrifying public transport. The examination of these achievements and the comprehension of the obstacles encountered can provide valuable insights for decision-making and serve as a roadmap for future implementations. Gaining knowledge from previous experiences has the potential to enhance resource allocation efficiency and policy effectiveness.

5.12.4 Evaluation and Monitoring of Sustainable Transportation Initiatives

Continuous evaluation and monitoring are crucial for ensuring the effectiveness and impact of sustainable transport initiatives. It is imperative for governments and organisations to establish unambiguous performance metrics in order to effectively evaluate and monitor advancements made towards sustainability objectives. Metrics encompass a range of indicators, such as the mitigation of greenhouse gas emissions, enhancements in air quality, augmented utilisation of public transportation, or elevated rates of active transportation modes. The practice of regular monitoring facilitates the timely implementation of adjustments and enhancements, thereby ensuring the continued alignment of initiatives with their intended objectives. In addition, the practice of providing clear and comprehensive disclosure of findings promotes responsibility and cultivates public confidence in the endeavours aimed at establishing a transport system that is more environmentally and socially sustainable.

5.13 CHALLENGES AND SOLUTIONS IN SUSTAINABLE TRANSPORTATION

The endeavour to achieve sustainable transportation encounters substantial obstacles that necessitate the development of inventive resolutions. The implementation of public transport networks and electric vehicle charging infrastructure presents significant challenges in terms of financing and cost considerations. These projects necessitate substantial resources in order to be successfully executed. Nevertheless, the implementation of financial incentives and the establishment of public-private partnerships present viable approaches to surmounting these challenges. The advancement of societal objectives is impeded by political and institutional obstacles, thereby requiring the presence of effective leadership and cooperative efforts. The successful adoption of any initiative necessitates the crucial factors of public acceptance and behavioural change. Achieving these outcomes requires the implementation of effective communication strategies and the provision of targeted incentives. The development of sustainable infrastructure, such as the establishment of electric vehicle charging networks and the implementation of bike lanes, continues to present a significant obstacle, particularly in rural regions. By employing data-driven methodologies to tackle these obstacles, the implementation of sustainable transport can propel us towards a future characterised by enhanced cleanliness and efficiency [44].

5.13.1 Financing and Cost Considerations

Acquiring the requisite financial resources for the effective execution of sustainable transport initiatives is a pivotal element. The successful implementation of environmentally sustainable transport systems necessitates substantial financial commitments towards the development of infrastructure, technological advancements, and research endeavours. Various stakeholders, including governmental bodies, private entities, and international organisations, are confronted with the challenge of managing limited financial resources while simultaneously considering the trade-off between immediate expenditures and future gains. This section examines the complex obstacles associated with funding sustainable transport initiatives and suggests possible remedies, including the utilisation of public-private partnerships and the implementation of innovative financing models, to surmount these barriers.

5.13.2 Political and Institutional Barriers

The transition of transport systems towards sustainability encounters various political and institutional challenges. The implementation of policies often faces opposition from influential interest groups, intricate bureaucratic processes, and the resistance posed by long-standing transportation practices. In order to surmount these obstacles, it is imperative to possess robust political determination and exhibit effective leadership. Furthermore, the promotion of interagency collaboration and the implementation of institutional reforms can effectively enhance the incorporation of sustainability principles into transportation decision-making processes, thereby catalysing favourable transformations within the industry.

5.13.3 Public Acceptance and Behavioural Change

The induction of behavioural change among individuals is crucial for the public acceptance and adoption of sustainable transportation practices. The success of eco-friendly mobility initiatives is significantly influenced by individuals' travel preferences, habits, and attitudes. It is imperative to comprehend the psychological determinants that impact travel decision-making, including the appeal of owning a private vehicle and the reluctance to adopt unfamiliar modes of transportation. This section explores various communication strategies, public awareness campaigns, and incentive-based approaches that can effectively promote public acceptance and drive behavioural changes towards sustainable transportation options.

5.13.4 Overcoming Infrastructure Limitations

The promotion of environmentally friendly mobility solutions relies heavily on the establishment and advancement of sustainable transportation infrastructure. Nevertheless, there are several obstacles that emerge when it comes to establishing and enhancing such infrastructure [45]. In order to establish electric vehicle charging networks and implement bike lanes and pedestrian-friendly designs, meticulous planning and specialised technical knowledge are imperative [46]. This section examines potential strategies to address the existing infrastructure limitations and establish smooth intermodal connectivity. Furthermore, this study acknowledges the distinct obstacles encountered in rural and remote regions, highlighting the importance of implementing adaptable and situation-specific strategies to ensure the availability of enduring transportation alternatives.

5.14 FUTURE TRENDS AND INNOVATIONS IN SUSTAINABLE TRANSPORTATION

The potential of sustainable transport in the coming years is considerable, owing to technological advancements and the emergence of inventive approaches that facilitate environmentally conscious mobility. One of the prominent trends in the transportation sector pertains to the swift advancement of autonomous vehicles and the emergence of Mobility as a Service (MaaS) as a novel concept which facilitates the provision of seamless and efficient transportation alternatives. Smart cities are utilising data and connectivity to incorporate different forms of sustainable transport, thereby improving urban mobility and mitigating environmental consequences [47]. The adoption of greener transportation systems is being propelled by the emergence of advanced batteries and the integration of renewable energy. The implementation of forward-thinking policies and the development of strategic planning are of utmost importance in order to facilitate the widespread acceptance and utilisation of sustainable transportation practices. By adopting and incorporating these forthcoming trends and innovations, it is possible to construct a transportation ecosystem that is environmentally friendly, highly productive, and capable of being maintained for future generations [48].

5.14.1 Autonomous Vehicles and Mobility as a Service (MaaS)

The emergence of autonomous vehicles and the concept of Mobility as a Service (MaaS) is significantly influencing the trajectory of sustainable transportation in the future. The utilisation of autonomous vehicles, propelled by sophisticated artificial intelligence and sensor technologies, holds the promise of fundamentally transforming transportation systems. This transformation can be achieved through the alleviation of traffic congestion, optimisation of fuel consumption, and augmentation of overall road safety. On the contrary, MaaS aims to achieve a cohesive integration of diverse transportation modes, including ride-sharing, public transit, and micro-mobility, within a unified platform that prioritises the needs and preferences of the user. This section delves into the potential benefits and obstacles associated with autonomous vehicles and Mobility as a Service (MaaS). It examines their potential environmental implications, the transformation of transportation business models, and the significance of regulations in guaranteeing responsible implementation.

5.14.2 Smart Cities and Sustainable Transportation Integration

By leveraging data and technology to improve urban mobility, smart cities are taking the lead in the continuing movement towards more environmentally responsible modes of urban transportation. Incorporating environmentally friendly modes of transportation into the overall planning and development of smart cities is an important step towards reducing emissions, improving air quality, and expanding opportunities for urban habitation [17]. This section examines the role of intelligent transportation systems, real-time data analytics, and urban planning methods in the process of developing ecologically sustainable transportation networks that are networked with one another. In addition, we explore the impact that intelligent infrastructure, such as charging stations for electric vehicles and intelligent traffic control systems, can have on the path that sustainable urban transportation takes.

5.14.3 Emerging Technologies and Their Impact

There is a significant possibility that developing technology may significantly modify environmentally friendly modes of transportation in the foreseeable future. This section examines the transformative effects of innovative technologies, such as advancements in battery technology that increase the range and charging speed of electric vehicles, as well as the integration of renewable energy sources into transportation systems, an example of which is the use of solar-powered charging stations. Other examples of transformative technologies include advancements in battery technology that improve the safety of electric vehicles, in addition, we perform an investigation into the potential influence that Hyperloop networks, flying taxis, and other novel transportation concepts could have on the transformation of long-distance and urban mobility. The purpose of this investigation is to determine whether or not they are able to shorten travel times and cut down on carbon emissions [49]. A complete examination of developing technologies and their potential to drastically transform sustainable transport in the future is shown in Table 5.2.

Table 5.2 Emerging Technologies and Their Potential to Reshape Sustainable Transportation in the Future

Emerging Technology	*Description*	*Potential Impact on Sustainable Transportation*
Advanced Battery Tech	Next-gen batteries with higher energy density and faster charging	Extended range and faster charging for electric vehicles
		Reduced reliance on fossil fuels and lower greenhouse gas emissions
Renewable Energy	Integration of renewable energy sources like solar, wind, and hydro	Enhanced performance and scalability for electric mobility
		Powering electric vehicle charging stations and other transportation infrastructure with clean energy sources
		Reducing carbon emissions and environmental impact
Hyperloop	High-speed transportation system using maglev technology in low-pressure tubes	Sustainable energy supply for transportation electrification
		Ultra-fast and efficient long-distance travel
		Lower energy consumption and reduced carbon footprint
Autonomous Vehicles	Self-driving vehicles equipped with AI and sensors for navigation	Potentially relieving congestion on roadways and reducing air travel demand
		Improved road safety and reduced accidents
		Greater fuel efficiency through optimised driving patterns
Mobility as a Service (MaaS)	Integrated mobility platforms that offer seamless access to various	Increased accessibility, especially for elderly and disabled individuals
		Streamlined transportation options, making it easier for users to plan and pay for multi-modal journeys
		Reduced individual car ownership and increased use of shared mobility

Emerging Technology	*Description*	*Potential Impact on Sustainable Transportation*
Flying Vehicles	Electric vertical take-off and landing (eVTOL) aircraft for urban air mobility	Enhanced urban mobility and efficient use of transportation resources Overcoming traffic congestion and reducing travel times in urban areas Lowering carbon emissions by replacing short-haul flights and urban commuting with electric flying vehicles
Sustainable Materials	Usage of eco-friendly and lightweight materials in vehicle manufacturing	Potential for transformative impact on urban transportation and logistics Reduced vehicle weight leading to improved fuel efficiency Lower environmental impact from vehicle production and end-of-life disposal

5.15 POLICY AND PLANNING IMPLICATIONS FOR THE FUTURE

When it comes to effectively capitalising on the benefits that come with sustainable mobility technologies and trends, comprehensive policies and strategic planning play a significant role. In this section, we will discuss the policy consequences that are related to the shift towards environmentally friendly transportation systems. It investigates the role that governments play in establishing rules for emissions, encouraging the development of environmentally friendly technologies through research and development initiatives, and enabling the use of ecologically friendly modes of transportation. In addition, we investigate the value of multi-stakeholder collaboration, public involvement, and international cooperation in the context of addressing issues pertaining to global transportation and developing a sustainable future for future generations.

In conclusion, a large amount of collaboration across a wide variety of stakeholders is required because of the significance of environmentally friendly transportation solutions in the fight against environmental concerns. The advancement of sustainable transportation relies heavily on the success of this cooperative effort. It is possible to nurture a transportation system that is both cleaner and more efficient if one addresses the issues connected to financing, political impediments, public acceptance, and constraints in infrastructure. The implementation of emerging technologies and the advancement of smart city development present a big opportunity to generate an effect that is both sustainable and helpful to human and environmental welfare. This is a significant opportunity that should not be missed. By employing methodologies grounded in data analysis and fostering a collective dedication to the principles of sustainability, humankind has the potential to forge a path towards a more promising and environmentally conscious future.

REFERENCES

[1] P. Newman and J. Kenworthy, *Sustainability and cities: overcoming automobile dependence*. Island Press, 1999.

[2] P. L. Schiller and J. Kenworthy, *An introduction to sustainable transportation: Policy, planning and implementation*. Routledge, 2017.

[3] K. Martens, *Transport justice: Designing fair transportation systems*. Routledge, 2016.

[4] A. Kundu, A. Kumar, V. P. Singh, C. S. Meena, and N. Dutt, "Chapter 1: Introduction to thermal energy resources and their smart applications," in *Thermal energy systems: Design, computational techniques, and applications*. A. Kumar, V. P. Singh, C. S. Meena, and N. Dutt, Eds. CRC Press, 2023, pp. 1–15. https://doi.org/10.1201/9781003395768

[5] S. Jain, N. Kumar, V. P. Singh, S. Mishra, and N. K. Sharma, "Transesterification of algae oil and little amount of waste cooking oil blend at low temperature in the presence of NaOH," *Energies*, vol. 16, no. 1, pp. 1–12, 2023, doi: 10.3390/en16031293.

[6] J. Faulin, S. E. Grasman, A. A. Juan, and P. Hirsch, "Sustainable transportation: Concepts and current practices," in *Sustainable transportation and smart logistics*, Elsevier, 2019, pp. 3–23.

[7] K. Chauhan and V. P. Singh, "Prospect of biomass to bioenergy in India : An overview," *Mater. Today Proc.*, 2023, doi: 10.1016/j.matpr.2023.01.419.

[8] M. Nayal, A. K. Sharma, S. Jain, and V. P. Singh, "Chapter 11 Green hydrogen production methods, designs and applications," in *Highly efficient thermal renewable energy systems design, optimization and applications*, 1st ed., R. W. V. Verma, S. Thangavel, N. Dutt, A. Kumar, Eds. CRC Press, 2024, pp. 178–193.

[9] E. Weiner, *Urban transportation planning in the United States: An historical overview*. US Department of Transportation, 1997.

[10] A. R. Singh, S. K. Singh, and V. P. Singh, "Process parameters optimization of carbon nano tube based catalytic transesterification of algal oil," *Mater. Today Proc.*, 2023, doi: 10.1016/j.matpr.2023.01.418.

[11] A. K. Sharma, M. Nayal, S. Jain, and V. P. Singh, "Chapter 07 Optimization Techniques of Solar Thermal and Hybrid Energy Systems," in Highly Efficient Thermal Renewable Energy Systems Design, Optimization and Applications, 1st ed., R. W. V. Verma; S Thangavel, N Dutt, A Kumar, Ed. CRC Press, 2024, pp. 1–14.

[12] V. P. Singh and S. Jain, "Economic analysis of a large scale solar updraft tower power plant," *Sustain. Energy Technol. Assess.*, vol. 58, p. 103325, 2023, doi: https://doi.org/10.1016/j.seta.2023.103325.

[13] M. Hasselwander, J. F. Bigotte, A. P. Antunes, and R. G. Sigua, "Towards sustainable transport in developing countries: Preliminary findings on the demand for mobility-as-a-service (MaaS) in metro Manila," *Transp. Res. Part A Policy Pract.*, vol. 155, pp. 501–518, 2022.

[14] M. Nayal, A. K. Sharma, S. Jain, and V. P. Singh, "Chapter 06 Design and Modelling of Solar, Geothermal and Hybrid Energy Systems.pdf," in Highly Efficient Thermal Renewable Energy Systems Design, Optimization and Applications, 1st ed., R. W. V. Verma; S Thangavel, N Dutt, A Kumar, Ed. CRC Press, 2024, pp. 1–15.

[15] A. Kundu, A. Kumar, N. Dutt, V. P. Singh, and C. S. Meena, "Chapter 7: Modelling and simulation of thermal energy system for design optimization," in *Thermal energy*

systems: Design, computational techniques, and applications, A. Kumar, N. Dutt, V. Singh, and C. Meena, Eds. CRC Press, 2023, pp. 103–137.
[16] H. Suzuki, R. Cervero, and K. Iuchi, *Transforming cities with Transit: Transit and land-use integration for sustainable urban development.* World Bank Publications, 2013.
[17] R. K. Shubham Srivastava, D. Verma, S. Thusoo, A. Kumar, and V. P. Singh, "Nanomanufacturing for energy conversion and storage devices," in *Nanomanufacturing and nanomaterials design: Principles and applications.* S. Singh, S. K. Behura, A. Kumar, and K. Verma, Eds. CRC Press, 2022, pp. 165–174. https://doi.org/10.1201/9781003220602
[18] T. Goldman and R. Gorham, "Sustainable urban transport: Four innovative directions," *Technol. Soc.*, vol. 28, no. 1–2, pp. 261–273, 2006.
[19] V. P. Singh, A. Karn, G. Dwivedi, T. Alam, and A. Kumar, "Experimental Assessment of Variation in Open Area Ratio on Thermohydraulic Performance of Parallel Flow Solar Air Heater," Arab. J. Sci. Eng., vol. 41, no. 12, pp. 1–17, 2022, doi: 10.1007/s13369-022-07525-7.
[20] Y. Gauri, A. Kumar, A. K. S. Gangwar, A. Kumar, C. S. Meena, V. P, Singh, N. Dutt, and A. Prasad, "Biomedical study of femur bone fracture and healing," in Advanced Materials for Biomedical Applications, 2022, pp. 283–298.
[21] V. Mohamad Taghvaee, A. A. Arani, M. Nodehi, J. K. Shirazi, L. Agheli, H. M. N. Ghojogh, N. Salehnia, A. Mirzaee, S. Taheri, R. M. Saber, and H. Faramarzi, "Sustainable development goals: Transportation, health and public policy," *Rev. Econ. Polit. Sci.*, vol. 8, no. 2, pp. 134–161, 2023.
[22] V. P. Singh, S. Jain, A. Karn, A. Kumar, G. Dwivedi, C. S. Meena, and R. Cozzolino, "Mathematical modeling of efficiency evaluation of double-pass parallel flow solar air heater," Sustainability, vol. 14, no. 17, p. 10535, 2022, doi: 10.3390/su141710535.
[23] F. Rozon, C. Mcgregor, and M. Owen, "Towards a transparent and dynamic forecasting of renewable energy technologies capacity and cost to 2050," *SSRN Electron. J.*, vol. June, pp. 1–16, 2023, doi: https://dx.doi.org/10.2139/ssrn.4472244.
[24] J. Zietsman and L. R. Rilett, *Sustainable transportation: Conceptualization and performance measures.* Southwest University Transportation Center, Texas Transportation Institute, 2002.
[25] V.P. Singh, S. Jain, A. Karn, G. Dwivedi, A. Kumar, S. Mishra, N.K. Sharma, M.Bajaj, H.M. Zawbaa and S. Kamel, "Heat transfer and friction factor correlations development for double pass solar air heater artificially roughened with perforated multi-V ribs," Case Stud. Therm. Eng., vol. 39, no. September, p. 102461, 2022, doi: https://doi.org/10.1016/j.csite.2022.102461.
[26] J. Zhou, "Sustainable transportation in the US: A review of proposals, policies, and programs since 2000," *Front. Archit. Res.*, vol. 1, no. 2, pp. 150–165, 2012.
[27] G. Singh, S. Rathor, A. Karn, and V. P. Singh, "Recent progress in solar dryers using phase changing material: A review," *Int. J. Energy Resour. Appl.*, vol. 2, no. 1, pp. 57–77, 2023, doi: 10.56896/IJERA.2023.2.1.005.
[28] M. Saini, A. Sharma, V. P. Singh, G. Dwivedi, and S. Jain, "Solar thermal receivers – a review," *Adv. Mater. Manuf. Energy Eng.*, vol. II, pp. 310–325, 2022, doi: 10.1007/978-981-16-8341-1.
[29] E. Ferguson, *Travel demand management and public policy.* Routledge, 2018.

[30] A. Bond and R. Steiner, "Sustainable campus transportation through transit partnership and transportation demand management: A case study from the University of Florida," *Berkeley Plan. J.*, vol. 19, no. 1, 2006.

[31] P. Waddell, "Integrated land use and transportation planning and modelling: Addressing challenges in research and practice," in *Transport models in urban planning practices*. Routledge, 2016, pp. 71–92.

[32] N. Dutt, A. Binjola, A. J. Hedau, A. Kumar, V. P. Singh, and C. S. Meena, "Comparison of CFD results of smooth air duct with experimental and available equations in literature," *Int. J. Energy Resour. Appl.*, vol. 1, no. 1, pp. 40–47, 2022, doi: 10.56896/IJERA.2022.1.1.006.

[33] R. Tayal, R. K. Upadhyay, M. Yadav, and R. Singh, "The development and validation of the organizational effectiveness scale using confirmatory factor analysis," *Glob. Bus. Rev.*, vol. 22, no. 3, pp. 821–843, 2019, doi: 10.1177/0972150918811720.

[34] T. Litman, *Introduction to multi-modal transportation planning*. Victoria Transport Policy Institute Canada, 2017.

[35] D. Pojani and D. Stead, "Sustainable urban transport in the developing world: Beyond megacities," *Sustainability*, vol. 7, no. 6, pp. 7784–7805, 2015.

[36] T. Litman and D. Burwell, "Issues in sustainable transportation," *Int. J. Glob. Environ. Issues*, vol. 6, no. 4, pp. 331–347, 2006, doi: 10.1504/IJGENVI.2006.010889.

[37] U. Eberle and R. Von Helmolt, "Sustainable transportation based on electric vehicle concepts: a brief overview," *Energy Environ. Sci.*, vol. 3, no. 6, pp. 689–699, 2010.

[38] M. Raymond Byrne and M. J. Polonsky, "Impediments to consumer adoption of sustainable transportation: Alternative fuel vehicles," *Int. J. Oper. Prod. Manag.*, vol. 21, no. 12, pp. 1521–1538, 2001, doi: 10.1108/EUM0000000006293.

[39] M. Chugh, N. Chanderwal, R. Upadhyay, and D. K. Punia, "Effect of knowledge management on software product experience with mediating effect of perceived software process improvement: An empirical study for Indian software industry," *J. Inf. Sci.*, vol. 46, no. 2, pp. 258–272, 2019, doi: 10.1177/0165551519833610.

[40] A. Balasubramaniam, A. Paul, W.-H. Hong, H. Seo, and J. H. Kim, "Comparative analysis of intelligent transportation systems for sustainable environment in smart cities," *Sustainability*, vol. 9, no. 7. 2017, doi: 10.3390/su9071120.

[41] L. Zhao and Y. Jia, "Intelligent transportation system for sustainable environment in smart cities," *Int. J. Electr. Eng. Educ.*, p. 0020720920983503, 2021, doi: 10.1177/0020720920983503.

[42] D. R. Jordan and T. A. Horan, "Intelligent transportation systems and sustainable communities: Findings of a national study," *Transp. Res. Rec.*, vol. 1588, no. 1, pp. 70–76, 1997.

[43] V. T. Cheang, R. A. Hedderwick, and C. McGregor, "Benchmarking supercritical carbon dioxide cycles against steam Rankine cycles for Concentrated Solar Power," *Sol. Energy*, vol. 113, pp. 199–211, 2015, doi: https://doi.org/10.1016/j.solener.2014.12.016.

[44] C. Macharis and B. Kin, "The 4 A's of sustainable city distribution: Innovative solutions and challenges ahead," *Int. J. Sustain. Transp.*, vol. 11, no. 2, pp. 59–71, 2017.

[45] V. P. Singh, C. S. Meena, A. Kumar, and N. Dutt, "Double pass solar air heater: A review," *Int. J. Energy Resour. Appl.*, vol. 1, no. 2, pp. 22–43, 2022, doi: 10.56896/IJERA.2022.1.2.009.

[46] V. P. Singh and G. Dwivedi, "Technical analysis of a large-scale solar updraft tower power plant," *Energies*, vol. 16, no. 1, p. 103325, 2023, doi: 10.3390/en16010494.
[47] V. P. Singh, A. Dwivedi, A. Karn, A. Kumar, S. Singh, S. Srivastava, and K. Srivastava, "Nanomanufacturing and design of high-performance piezoelectric nanogenerator for energy harvesting," in *Nanomanufacturing and nanomaterials design: Principles and applications.* S. Singh, S. K. Behura, A. Kumar, and K. Verma, Eds. CRC Press, 2022, pp. 241–272. https://doi.org/10.1201/9781003220602
[48] M. Janic, "Sustainable transport in the European Union: A review of the past research and future ideas," *Transp. Rev.*, vol. 26, no. 1, pp. 81–104, 2006.
[49] W.-M. Wey and J.-Y. Huang, "Urban sustainable transportation planning strategies for livable City's quality of life," *Habitat Int.*, vol. 82, pp. 9–27, 2018.

Chapter 6

Transitions from IC engine to EV and HEV

Francis Luther King M., John Baruch, E. Elayarani

6.1 INTRODUCTION

The common perception that oil is a finite resource that will be completely depleted within the next half-century is driving up oil prices every day. Nevertheless, the total number of automobiles is projected to rise from 700 million to 2.5 billion throughout the same period [1, 2]. Different approaches are required and being offered. Mainstream automakers' proposals for EVs and HEVs signal a shift in the way cities approach transportation. Additionally, several groups and energy specialists are proposing new legislation to boost study. Electric cars are a viable option for transportation since they are both affordable and environmentally benign, but their high purchase prices may discourage some people from making the switch. Marketed electric vehicles, like the Tesla Roadster, need massive battery packs to store the energy required to power the vehicle. The lithium-ion batteries used in these vehicles typically cost approximately $100,000. Attempts to lower the price of electric cars are a priority for everyone. In order to do this, it may be worthwhile to think about having an ICE-powered car converted into a brand-new electric car. There are kits available that may be used to do this conversion, but along with the cost of batteries, they can total over $10,000.

In the context of EV research, the term "revolutionary" is often used to describe cutting-edge automobiles. While expensive custom-built solutions are possible, low-cost alternatives employing commercially available, high-quality components are also possible [3]. For this reason, converting ICE cars into EVs is a desirable option during the transitional era of cohabitation. Electric vehicle conversion describes this procedure [4, 5]. It is a difficult technical technique that requires familiarity with mechanics and electronics. The electric motor and its controller, as well as a battery pack and its charging infrastructure, are the primary systems that must be coordinated. There are a number of potential methods for storing energy, but batteries seem to be the most popular [2, 6, 7]. As a result, a battery charging method is essential.

 DOI: 10.1201/9781003464556-6

6.1.1 How EVs Work

Wires need to be installed in lieu of the pipes and hoses in order to complete the transformation from an internal combustion engine to an electric motor. It is necessary to dismantle the internal combustion engine, the transmission, the gas tank, and the exhaust system before proceeding with the installation of an electric motor, a controller, and a battery for an electric car. In order to be successful in this line of work, you need to have previous experience in the domains of both mechanical and electrical engineering. The great majority of people do not possess these skills, which is why they either hire a company to make the conversion for them or purchase a kit that has all of the essential parts and walks them through the procedure.

6.1.2 Benefits

In 2015, the average amount spent on gasoline over the course of a year was $1,681.50, as reported by the American Automobile Association (AAA). The unpredictability of gas costs is compounded by the difficulty of doing routine maintenance on an internal combustion engine, which requires regular oil changes. Electric cars are more reliable and have a greater resale value than gas-powered vehicles. This is because electric vehicles have fewer moving parts than gas-powered vehicles. The vast majority of electrical components are completely motionless and do not include any moving elements. The life expectancy of the electric car's motor is limitless, and its component components are likely to outlast the chassis of the vehicle. The only substantial expenditures associated with electric cars are the cost of the electricity they use and the cost of replacing the batteries every three to four years. Your converted vehicle needs to have a range of 60 to 80 miles before requiring recharging, a top speed of 50 to 90 miles per hour, and a good rate of acceleration. It will take anything from six to 12 hours to fully charge the car to its maximum capacity.

6.1.3 Cost

The cost of the components needed to convert to an electric vehicle is around $6,000, and the cost of the batteries and installation ranges from approximately $1,000 to $3000. A more costly refit might end up costing you more than $20,000 in total. However, in exchange for all of this, you will have a car that has minimal emissions and operating expenses that are just a few pennies per mile. In addition, the reliability of your electric vehicle will be far higher, and it will call for significantly less maintenance than a conventional vehicle.

6.1.4 Which Can Be Converted

The perfect vehicle would have a modest curb weight (between 2,000 and 3,000 pounds) and a manual gearbox. Heavy cars may reduce the amount of distance that can be travelled by an electric motor, and automatic gearboxes use an excessive amount of power due to the fact that they need the engine to be continually idling. You will need a body design that is capable of holding all of the batteries that you will be installing, so look for something with enough space. A vehicle that is not only light but also spacious, such as a Volkswagen Rabbit, Honda Civic, Nissan Sentra, Ford Escort, or light pickup truck, come highly recommended by automotive experts. The perfect donor vehicle will have a functional gearbox and a functioning body and interior, but the engine will be completely dead. It is possible to breathe fresh life into the body of an older vehicle by converting it from a gas-powered to an electric vehicle.

6.2 AUTOMOTIVE ECOSYSTEM

In the same general timeframe of the introduction of the internal combustion engine automobile in 1870, the first operational electric vehicle made its debut. Around the year 1900, the first hybrid automobiles, which were powered by motors in the wheel hubs, were developed. Despite the fact that electric vehicles came very close to breaching the ice, cars powered by internal combustion engines (ICE) have proven to be more successful than their battery-powered counterparts. At the turn of the 21st century, there was a resurgence in car buyers' interest in electric and/or alternative fuel vehicles. This interest was sparked by concerns regarding the surrounding damage caused by the exhaust of hydrocarbon-fuelled vehicles, the sustainability of the current infrastructure for transportation based on hydrocarbons, and advancements in electric vehicle technology. Elon Musk's Tesla Motors has been a sensation in the field of electric autos ever since the debut of the Model 3, which catapulted the company to the forefront of the industry's conversation. The world's largest original equipment manufacturers (OEMs) have all stated that they want to build electric vehicles (EVs) in the near future.

6.3 EVOLUTION

It has been close to a century and a half since the world saw the transition of cars from primitive, high-performance vehicles to contemporary, high-tech autos. Significant headway has been achieved by automotive stakeholders in the development of the automotive ecosystem, beginning with the production of vehicles and continuing through the management of supply chains and the customer experience (both pre- and post-sales). The Original Equipment Manufacturers (OEMs) have established their own approved car sales and servicing network. Post-sale support for automobiles is something that OEMs and consumers alike anticipate receiving, and this includes things like vehicle services and the availability of replacement parts for vehicle maintenance. Based on the experiences of the Indian market, the

OEM's built network is performing well in terms of sales and servicing up to the point when the cars' warranties expire. To contemplate a parallel service network for their automobiles is something that vehicle owners find appealing for a variety of reasons. The OEM network was evolving at the same time as the expanded automotive ecosystem, which was taking into consideration the ease of use, cost, and availability of more individualised services for the car. The parallel ecosystem gained popularity at the same rate over time, and as time went on, this network grew and contributed to the expansion and development of the vehicle industry. The ICE powertrain parallel ecology is likewise similarly powerful and capable at the current time.

6.4 EV TYPES

Electric vehicles may either operate purely on electric power or with the assistance of an internal combustion engine (ICE). The most basic relies only on battery power, however, there are several variants that may make use of various alternate energy delivery systems. It is possible to refer to them as hybrid electric vehicles (HEVs). Technical Committee 69 (Electric Road Vehicles) of the International Electro technical Commission has recommended that cars with two or more forms of power source, storage, or converters may be considered HEVs so long as at least one of them provides electrical energy [8]. Many hybrid vehicle configurations, including internal combustion engine (ICE) and battery, battery and flywheel, battery and capacitor, battery and fuel cell, etc., are all within the definition's purview. So, both laypeople and experts began referring to ICE and electric motor hybrids as HEVs, ultra-capacitor-assisted EVs, and battery and fuel cell hybrids as FCEVs [9, 10]. Using these now-standard terms, we may classify EVs as follows:

- Battery Electric Vehicle (BEV)
- Hybrid Electric Vehicle (HEV)
- Plug-in Hybrid Electric Vehicle (PHEV)
- Fuel Cell Electric Vehicle (FCEV)

6.4.1 Battery Electric Vehicle (BEV)

Battery electric vehicles (BEVs) are electric vehicles (EVs) in which the drive train is powered only by batteries. Since battery electric vehicles (BEVs) can only run on the power stored in their batteries, the range of BEVs is directly proportional to the size of their batteries. Most can go between 100 and 250 kilometres on a single charge [11], with the most advanced going even further—between 300 and 500 kilometres. These distances are affected by a wide variety of variables, including the driver's physical condition and driving style, the vehicle's design, the road conditions, the weather, the battery's age, and so on. When the battery

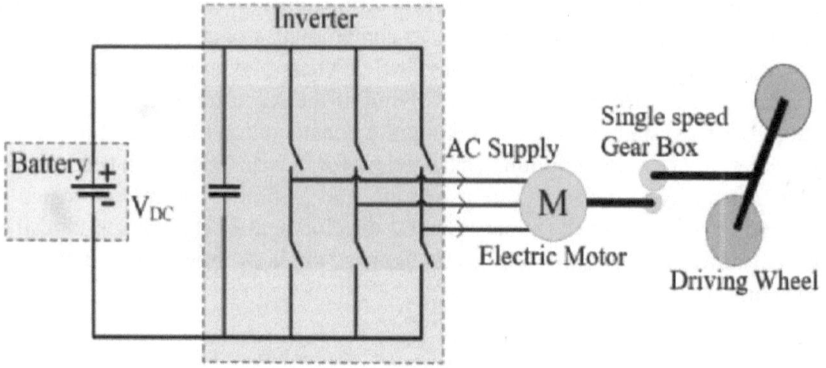

Figure 6.1 Battery Electric Vehicle.

Source: Adapted from Un-Noor et al. [118]

life is over, it takes a lot longer than it would to refill a regular internal combustion engine car. In certain cases, recharging the batteries may take as long as 36 hours [12, 13]. While there are faster and slower methods, none of them come close to the speed with which a petrol tank can be replenished. The charger's setup, infrastructure, and operational power all affect how long it takes to charge a device. Simple design, straightforward operation, and uncomplicated design are all benefits of BEVs. These are good for the environment since they do not contribute to global warming, they do not generate unwanted noise, and they are not a source of GHGs. High torques may be generated instantly by electric propulsion, even at slow speeds. For these reasons, and because of their limited range, they are the ideal vehicle for usage in congested metropolitan environments. Figure 6.1 depicts the battery electric vehicle.

6.4.2 Hybrid Electric Vehicle (HEV)

Some hybrid electric vehicles are able to convert to electric propulsion in low-energy situations. It is useful in urban areas with heavy traffic and stop-and-go conditions, and it helps you conserve gas by shutting off the engine. These enhancements help cut down on greenhouse gas production as well. The HEV automatically changes to the ICE when more speed is required. When used in tandem, the two drivetrains may significantly boost efficiency. Many turbocharged vehicles, like the Acura NSX, utilise hybrid power systems to mitigate or eliminate turbo lag. Additionally, it improves functionality by facilitating speed increases and bridging the time gap between gear changes. Not only may HEVs use their internal combustion engines to replenish their power supplies, but they can also use their regenerative braking systems to recover energy. Therefore, HEVs are typically internal combustion engine (ICE) vehicles that use an electrical drive

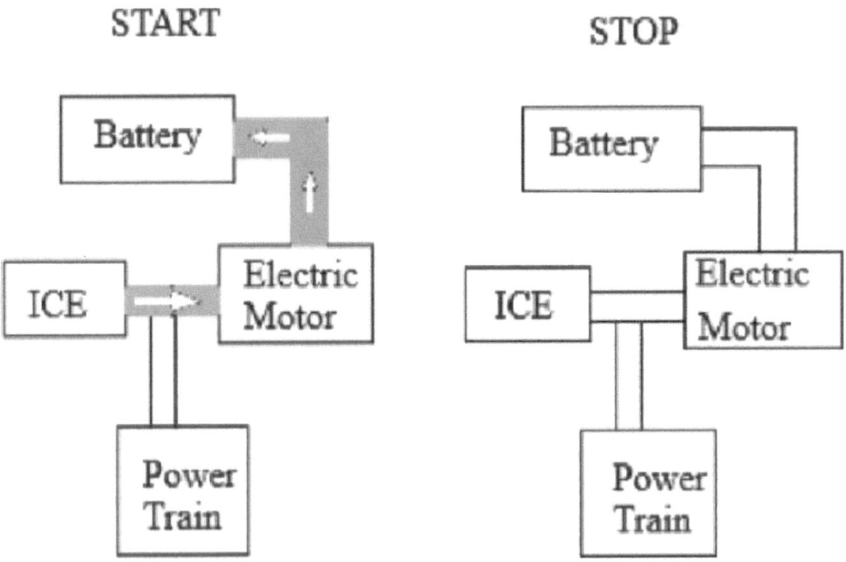

Figure 6.2 Hybrid Electric Vehicle.

Source: Adapted from Un-Noor et al. [118]

system to increase fuel economy or boost performance. Automakers are increasingly turning to HEV layouts to provide these amenities. The energy fluxes in a simple HEV are shown in Figure 6.2. The internal combustion engine (ICE) may use the vehicle's motor as a generator to create electricity and charge the battery before driving off. Since increased velocity is required for passing, the power train is driven by both the ICE and the motor. The powertrain uses regenerative braking, in which the engine is used as a generator when braking. While idling, ICE operates the car and the engine as a generator, topping up the battery. As soon as the vehicle comes to a halt, the flow of electricity is cut off. Energy management systems are common in HEVs. In order to maximise fuel economy, the one shown in Figure 6.2 divides power between the internal combustion engine (ICE) and the electric motor (EM) based on factors such as vehicle speed, driver input, battery state of charge (SOC), and motor speed.

6.4.3 Plug-in Hybrid Electric Vehicle (PHEV)

The PHEV idea was developed to increase the all-electric driving range of HEVs [14–19]. PHEV relies primarily on electric propulsion, necessitating a larger battery capacity. PHEVs begin operation in "all electric" mode, using only electrical power, and only switch to the internal combustion engine (ICE) for additional acceleration or battery recharging when the electric motors have reached their limit. In this case, the ICE is employed to increase mileage. PHEVs are equipped to get a direct charge from the power grid, and they may also benefit from regenerative braking. When compared to HEVs, PHEVs have a lesser carbon footprint since they can operate on electricity alone for the vast majority of their driving time. Also, they save money on gas since they have lower fuel requirements. Sales of the Chevrolet Volt and Toyota Prius are evidence of the growing popularity of such vehicles.

6.4.4 Fuel Cell Electric Vehicle (FCEV)

Vehicles that run on fuel cells (FCEVs) are also sometimes referred to as FCVs. It is named as such because chemical processes inside fuel cells provide power for the vehicle [20]. FCEVs are sometimes referred to as "hydrogen fuel cell vehicles" since hydrogen is used as the fuel to power this reaction. FCEVs transport the hydrogen in specialised high-pressure tanks, and the oxygen needed for the power generation process is drawn from the surrounding air. The fuel cells power an electric motor, which in turn propels the vehicle. Batteries and supercapacitors are examples of storage technologies that may be used to temporarily hold large amounts of energy [9, 10, 21, 22]. Fuel cell electric vehicles on the market today, such as the Toyota Mirai and Honda Clarity, rely on batteries for this function. Fuel cell electric vehicles (FCEVs) shown in Figure 6.3 only emit water as a by-product during the power generation process. They have the potential to minimise their carbon footprint even further than other EVs since they can generate their own power in a carbon-free manner. Perhaps the most immediately relevant benefit is that refuelling one of these cars takes about as long as filling a regular vehicle at a petrol station. Since this is the case, it is expected that these cars will become more widely used in the near future [8–10, 22]. Hydrogen fueling station scarcity is a huge problem right now, but the lack of these stations was also a problem when BEV and PHEV charging stations were first introduced a few years ago. The high

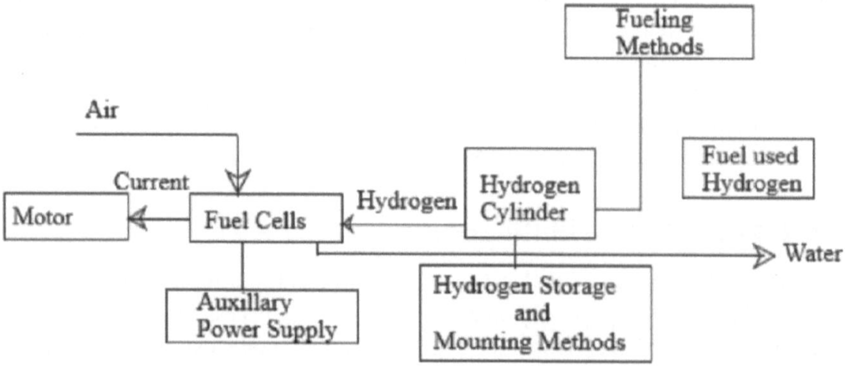

Figure 6.3 Fuel Cell Electric Vehicle.

Source: adapted from Un-Noor et al. [118]

cost of fuel cells, which is above $200 per kW according to a study by the U.S. Department of Energy (DOE), is another drawback. ICE costs less than $50 per kW. The potential for explosive hydrogen to escape from the storage tanks is also a cause for worry. If these problems are resolved, FCEVs would be a serious contender for the future of transportation. However, BEVs only seem to be superior in the areas of needed wind energy and gasoline cost per kilometre. One of the biggest downsides of FCEVs seems to be the expense of fuel since there is now no affordable, sustainable, and environmentally acceptable method of manufacturing hydrogen, and the refuelling infrastructure lags behind that of BEVs; however, these issues may soon no longer exist.

6.5 CONFIGURATION

As opposed to its internal combustion engine predecessors, electric vehicles may be easily adapted to a variety of situations [8]. This is because conventional vehicles need complex mechanical systems that electric vehicles do not. Only the motor really moves in an EV. Various methods of regulation and control systems may be used to regulate it. An assortment of power sources may be used to drive the motor. These two parts need only be physically linked with cables in order to function, therefore, they may be located anywhere in the car. As a result of this adaptability, a variety of configurations have arisen and are used for various vehicles. When broken down into its component parts, electric vehicles (EVs) are really a system [8] that consists of an energy source, a propulsion system, and an auxiliary system. The energy source, its refuelling system, and the energy management system together make up the energy source subsystem. An electric motor, power converter, controller, transmission, and driving wheels make up the propulsion subsystem. Auxiliary subsystem components include a generator, climate control, and power steering. Figure 6.4 shows the EV subsystem.

Regenerative operations, such as regenerative braking, may provide a reverse flow of power. To store the energy returned by regenerative processes, the energy source must be amenable to such feedback. Capacitors/flywheels (CFs) and the vast majority of EV batteries work together well for this purpose [8].

6.6 EV SETUP

As seen in [8], EVs may take on a variety of shapes and sizes. Figure 6.5a depicts a conventional front-engine front-wheel drive car where ICE was replaced by an electric motor. The gearbox and clutch combine to provide maximum torque at low RPM and minimum torque at high RPM. In addition, a differential lets the wheels spin at different rates. If you look at Figure 6.5b, you'll see a setup where the clutch is not included. Since it lacks a gearbox and instead uses a fixed gear, it is unable to achieve the ideal torque-speed characteristics. The

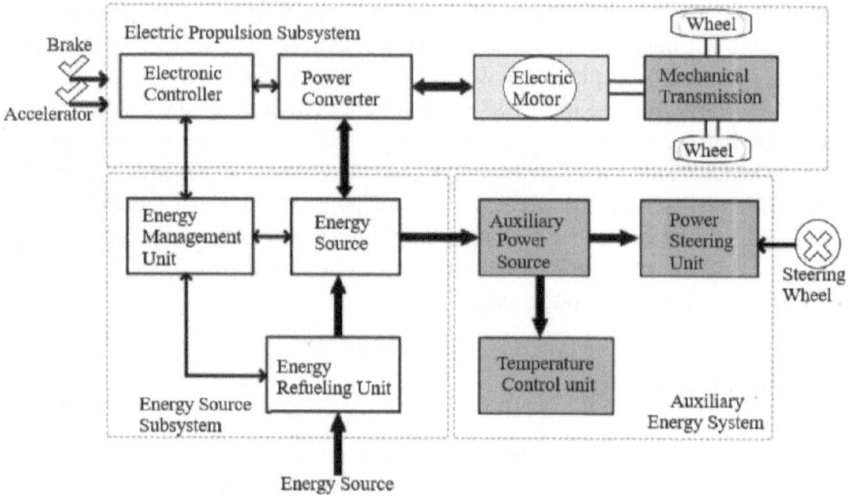

Figure 6.4 Electrical Vehicle Subsystems.

Source: Adapted from Un-Noor et al. [118]

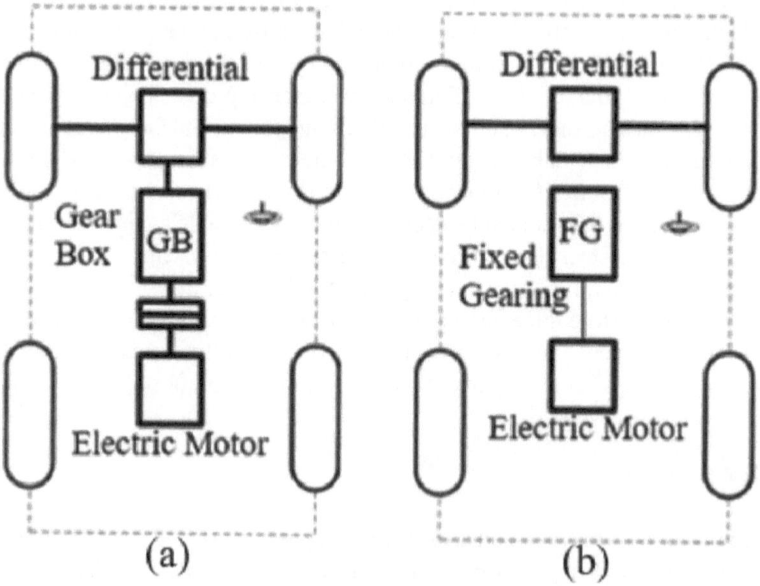

Figure 6.5 Variations on EVs with Front-wheel Drive. (a) Car having front-wheel drive and an electric motor instead of an internal combustion engine (ICE); (b) a car without a clutch; (c) setup where the motor, gearbox, and differential all work together to power the front wheels; (d) setup where separate motors drive the front wheels using fixed-gear ratios to provide differential action.

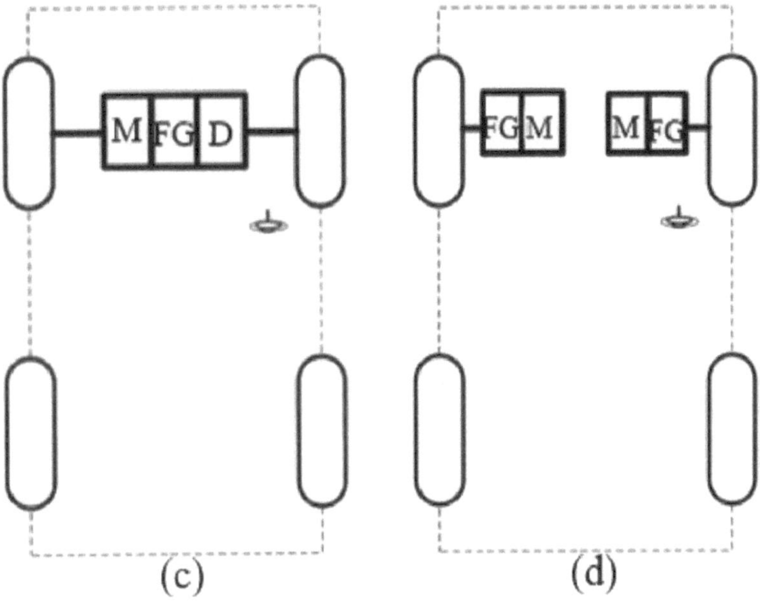

Figure 6.5 (Continued)

engine, gear, and differential in Figure 6.5c are all integrated into a single unit that powers both wheels. Both the Nissan Leaf and the Chevrolet Spark have front-mounted electric motors that power the front wheels. Differential action might be obtained by utilising two motors for the two wheels, as illustrated in Figure 6.5d. Motors installed in the wheels, creating what is known as "in-wheel drive," further eliminate mechanical contact. We chose a planetary gear system because of its benefits.

Electric vehicles are not limited to a front-wheel drive arrangement. This setup is used for the Model S with a single motor. Nissan's Blade Glider is an all-electric vehicle (EV) featuring an in-wheel motor layout for the rear wheels. Using in-wheel motors improves handling since they can send varied torques to the two rear wheels. All-wheel drive (AWD) arrangements may also be employed to increase control and power, albeit they come with additional costs, weight, and complexity. In this instance, the front and rear axles may be driven independently by separate motors. Better cornering performance and handling may be achieved with AWD setups thanks to torque vectoring, which is especially helpful in slick circumstances. With an in-wheel motor setup, you can have an all-wheel drive. The Hiriko Fold is a compact city automobile that benefits greatly from this since each wheel has a motor, suspension, brakes, and steering actuator. Such setups may facilitate easy parking and turning, in addition to providing all-wheel drive and all-wheel steering.

By eliminating the central motor, gearbox, differential, universal joints, and drive shaft, in-wheel motor layouts lighten the driving train [23]. They also provide greater control, better turning, and more room for batteries, fuel cells, or cargo, but the motor is coupled to the power and control systems via cables that may break in hard environments, vibration, and acceleration. Sato et al. suggested a wireless in-wheel motor system (W-IWM) in [24]. Two power-transferring coils replace the cables. Road vibrations may misalign the motor and vehicle, affecting secondary side voltage. Magnetic resonance coupling is favoured for wireless power transmission in such settings [25] because it can overcome misalignments [26] suggested using a hysteresis comparator and supplying secondary inverter power to a controller to offset secondary voltage changes. Wireless power transfer (WPT) using magnetic resonance coupling in a series-parallel setup can transmit 90% in both directions at 2 kW [27]. W-IWM supports regenerative braking.

6.7 HEV SETUP

Hybrid vehicles rely on both an ICE and an electric motor. The combinations that result from the numerous ways these two may be set up to turn the wheels fall into four broad classes [4]:

- Series hybrid
- Parallel hybrid
- Series-parallel hybrid
- Complex hybrid

6.7.1 Series Hybrid

This is the quickest and easiest way to construct a HEV. Here, just the motor, not the wheels, is coupled to the drivetrain since the engine is utilised to power a generator. It is equivalent to an EV with a conventional internal combustion engine (ICE) generator [8].

Advantages

A highly effective and streamlined power plant
Flexibility in Power Plant Design
Enhanced transmission line
Fast 'black box' service swaps might be possible.
Infinitely durable
A technological system that has reached maturity
Lightning-fast reaction
The ability to produce no emissions

Limitations

Massive traction drive system
Need for reliable algorithms
Several intermediate forms of energy transformation

6.7.2 Parallel Hybrid

Both the internal combustion engine and the motor are linked to the wheels in this setup. They may work together or independently to provide the energy. It is like an internal combustion engine car but with added electricity [8]. The electric motor may recharge the energy storage during braking by transferring kinetic energy back into the system, or the internal combustion engine (ICE) can do so when it generates more power than is needed to move the wheels.

Advantages

Positioned to achieve the complete absence of emissions
Advantage to the economy
Modularity increased

Limitations

Expensive
Handling complexity
Need for reliable algorithms
The need for high voltage to guarantee effectiveness

6.7.3 Series-Parallel Hybrid

It is a hybrid system that offers benefits from both but at a higher price and with greater complexity. The addition of a planetary gear unit adds considerable complexity to the drive train [28]. A planetary gear set consists of a sun gear related to the generator, a ring gear attached to the motor's output shaft, an internal combustion engine (ICE) coupled to the planetary carrier, and pinion gears holding everything together. A transmotor, an electric machine with a floating stator, provides a simpler alternative to this concept. The engine is linked to the stator, while the rotor remains coupled to the wheel of the drive train through the gearing. Adjusting the engine speed for a given vehicle speed is done by manipulating the motor speed, which is the relative speed between the rotor and the stator [28].

6.7.4 Complex Hybrid

The main difference between this and the series-parallel system is that the former allows for bidirectional power flow and the latter only for unidirectional. However, in the jargon of the modern business world, this setup is also called a series-parallel system. However, several vehicles have embraced this technology despite its high complexity and expense in order to utilise dual-axle propulsion [8]. The wheels may be driven by one of several different power sources in a sophisticated hybrid system with the help of a continuously variable transmission (CVT). Electric arrangements, termed e-CVT after its development and introduction by Toyota Motor Co., may be employed for such procedures. CVTs may be built in a number of different ways, including hydraulic, mechanical, hydro-mechanically, or electrically [29], demonstrating two power splitting techniques, input splitting and complicated splitting. The term "input splitting" was coined because of the presence of a power-splitting mechanism at the transmission's input. Particular Toyotas and Fords employ this method [29]. There are 24 distinct possible configurations for the electric machines, wheels, engines, and planetary gear (PG) used in power-split HEVs. The figure increases by more than a thousand if another PG is used. In a best-case scenario that uses only one, PG is presented. Two-motor hybrid designs are advantageous for 4WD setups because they eliminate the requirement for a power transmission system to the rear wheels (because they receive their own motor) and provide the benefit of energy reproduction through regenerative braking [30].

6.8 ENERGY SOURCES

EVs may acquire their power from several sources. High energy density and power density are two of the parameters such sources must meet [8]. Fast charging, extended service and cycle life, low cost, and upkeep are additional qualities desired in an ideal energy source. High specific energy and power are needed to boost acceleration and driving range, respectively. Due to the many properties needed for the optimal source, several sources or energy storage systems (ESS) are discussed and employed in various combinations to meet power and energy needs [8].

6.8.1 Battery

EVs have relied on batteries for a long time, but new battery technologies are being developed to meet performance targets. Lead-acid batteries, Ni-Cd, Ni-Zn, Zn/air, Ni-MH, Na/S, Lipolymer, and Li-ion batteries are popular. Yong et al. also demonstrated a graphene-based EV battery [31]. Consider the benefits and drawbacks of each battery type before buying. EV battery packs are composed of many battery cells. The 85 kWh pack of the Tesla Model S comprises 7104 Li-Ion cells. To avoid early end of life (EOL), all these cells should have the same SOC at all times [32]. A power electronic control device called a cell voltage equaliser may do this by actively equalising each cell's SOC and voltage. Equalisers are classified by construction and operation. Resistive equalisers burn additional power at higher-voltage cells to maintain all cells at the same voltage. However, capacitive equalisers swap capacitors to transfer energy from high-energy cells to low-energy ones. Inductive capacitors may be simple, Cuk, or single or multiple transformer-based, but they all use inductors to transmit energy from higher energy cells to lower energy cells [32–35]. Each setup has advantages and disadvantages. Lithium-ion batteries are ubiquitous. It replaced lead-acid batteries and matured. Popular EVs like the Nissan Leaf and Tesla Model S employ these batteries [36]. Lithium batteries also have room for improvement [36]. Lithium batteries will dominate EVs for a long time since better battery technologies have been developed but are too expensive to manufacture.

6.8.2 Ultracapacitors (UCs)

In UCs, two electrodes are separated by an ion-enriched liquid dielectric. The negative electrode collects positive ions, and the positive electrode collects negative ions when a voltage is applied. As a result, the power density is significantly increased, and the charges are physically held on electrodes. Ultracapacitors typically have a high cycle life because of the lack of chemical reactions at the electrodes, however, this also results in a poor energy density. Due to its low internal resistance, it is also very efficient, although charging it from a very low state of charge will result in a large output current [38, 39]. Since the terminal voltage of a UC is proportional to its SOC, the UC may function over its whole voltage range. Especially in metropolitan environments, EVs are subjected to frequent start/stop circumstances. This results in a battery drain rate that is very variable. Normally, batteries do not need to provide a lot of juice, but during extreme situations like acceleration or hill climbing, a lot of juice very quickly is needed [8]. In high-performance electric vehicles, the peak power need may be up to sixteen times the average power [8]. Since UCs can provide sustained high power for brief periods, they are well suited to this kind of application. The process of absorbing energy from regenerative braking occurs rapidly [9]. Batteries and UC systems work together to overcome their individual weaknesses and provide a robust and dependable power supply. Additionally, UCs are a great choice because of their inexpensive price, load-leveling capabilities, temperature adaptability, and extended service life [8, 28].

6.8.3 Fuel Cell

Electricity is produced by fuel cells via an electrochemical process. A fuel cell (FC) consists of an anode (A), a cathode (C), and an electrolyte (E). When fuel is put into the anode, it undergoes oxidation; the resulting ions then flow through the electrolyte to the cathode, where they react with the other reactant. Power is generated when electrons, created by oxidation at the anode, are used to charge a nearby conductor. Because of its high energy density, non-polluting nature (it produces only water as exhaust), and natural abundance in the form of various chemicals such as hydrocarbons [8], hydrogen is employed in FCEVs.

Fuel cell electric vehicles (FCEVs) now on the market, such as Toyota's Mirai, employ cylinders to store hydrogen for use in their propulsion systems. Fuel cells may be broken down into subcategories based on the components they utilise. Using fuel cells instead of conventional generators to power EVs has several benefits, including quiet operation, quick recharging, zero or low emissions, long lifespan, and high-density current output [39]. The expensive cost of this technology is a major downside. Larger fuel tanks are needed for FCEVs since hydrogen has a lower energy density compared to fuels generated from petroleum. These tanks must also be able to adequately contain the hydrogen and reduce the danger of explosion in the event of an accident. FC efficiency is power-dependent; as power consumption rises, FC performance often degrades. Most losses are caused by voltage drops in internal resistances. FCs have a quicker response time than UCs or batteries, on the whole. As a result, FCs are often used in conjunction with other forms of storage, such as batteries or UCs, to provide energy for the system. The FC is used to charge the battery pack that the Toyota Mirai relies on to power the vehicle's powertrain. The energy produced by regenerative braking is also sent to the batteries. With this setup, refuelling the FC is faster than charging the batteries, and there is greater freedom to move about [37].

6.8.4 Flywheel

The energy is utilised to spin the flywheel, and the flywheel continues to spin due to inertia, making it an effective energy storage device. While in storage, the flywheel serves as a motor. The kinetic energy stored in the flywheel may be utilised to spin a generator and create electricity when it is required. The rotors of modern flywheels may be composed of carbon composites and are hung in a vacuum chamber on magnetic bearings. Flywheels have several benefits over alternative storage types for EVs due to their low weight, high speed and efficiency at absorbing power from regenerative braking, and quickness at providing a large amount of power in a short period during rapid acceleration. They are particularly popular for hybrid race cars because to their ability to withstand the increased g-forces experienced during hard braking and acceleration. The energy generated by regenerative braking is not effectively captured by storage devices like batteries or UCs under these conditions. However, flywheels, because of their rapid reaction, are more efficient in such situations because they make greater use of regenerative braking, which in turn lessens the strain on the brake pads. This innovation was used in the hybrid Porsche 911GT3R. Various materials have their advantages and disadvantages when used to construct flywheels.

6.8.5 Motor Used

The electric motor is at the centre of the EV's propulsion system [40–45]. The motor draws power from the battery and transforms it into mechanical energy to propel the vehicle forward. When in regenerative mode, it operates as a generator, returning power to the mains. The number of motors in an EV may range from one, as in the Toyota Prius, to three, as in the Acura NSX; this varies depending on the vehicle's purpose. The specifications for an electric vehicle motor are laid forth in references [8] and include high power, high torque, broad speed, dependability, robustness, cost, less noise, and compact size. While direct current (DC) motor drives exhibit certain desirable qualities, such as those needed for EV application, their inefficiency, bulky construction, lack of dependability due to the commutator or brushes contained in them, and related maintenance demand have made them less appealing [8, 28]. Different motor types arose to satisfy the demands of the automotive industry with the development of power electronics and control systems, with induction and permanent magnet (PM) motors being the most popular [28, 46]

6.8.6 Brushed DC Motor

The stator of these motors is made of permanent magnets (PM), while the rotor uses brushes to feed electricity to the stator. One potential benefit is that these motors can provide maximum torque at low speeds. However, it has a number of drawbacks, including a cumbersome design, poor performance, excessive heat from the brushes, and a consequent reduction in efficiency. Since the heat is produced at the rotor's epicentre, it is also hard to dissipate. Brushed DC motors are no longer utilised in EVs for these reasons [46].

6.8.7 Permanent Magnet Brushless DC Motor (BLDC)

These motors use a permanent magnet (PM) rotor (often NdFeB [8]) and an inverter to give alternating current (AC) to the stator. Since the rotor has no windings, it uses less copper and is thus more efficient than induction motors. Additionally, they are more dependable, have a higher torque density, and produce more specific power [8]. It is also smaller in size, generates heat in the stator, and so can be dissipated more efficiently. Nonetheless, the consistent power range is restricted. This is because its field-weakening capacity is severely limited. Increasing the speed causes the stator windings to produce more back EMF, which in turn reduces the torque. It is also more expensive due to the incorporation of PM [28, 46]. However, more field windings may improve overall efficiency and expand the usable speed range [8, 47]. Due to the existence of both permanent magnet and field windings, such configurations are sometimes referred to as PM hybrid motors. However, the intricacy of such setups also limits their usefulness; the speed ratio is insufficient to fulfil the demands of EV usage, especially in off-roaders [28]. By combining a reluctance motor with a permanent magnet motor, a PM hybrid motor may be created. Modulating the power converter's conduction angle may increase the efficiency of PM BLDCs and allow them to operate at speeds up to four times the base speed [8], with the possible caveat that efficiency may drop at very high speeds due to demagnetisation of PM. PM BLDCs may be built with buried magnets, which can increase the air gap flux density, or with surface magnets, and they need less magnet material. There is a practical use for BLDCs in tiny automobiles that need no more than 60 kW of electricity [48].

6.8.8 Permanent Magnet Synchronous Motor (PMSM)

These machines are among the most sophisticated available because of their versatility in speed adjustment without the use of gears. This improvement allows these motors to use less power while taking up less space. This setup is ideal for in-wheel applications because of its ability to provide substantial torque even at low velocities. It is also feasible to build PMSMs with an outside rotor that does not need rotor bearings. However, these machines' one major drawback occurs during in-wheel operations, when massive iron loss is encountered at high speeds, rendering the system unstable [49]. Due to their high energy density, NdFeB permanent magnets (PMs) are utilised in PMSMs. Due to the sinusoidal nature of the air-flux gap's couplings, these motors may be controlled using sinusoidal voltage supply and vector control [46]. More than two-thirds (at least 26 models) of presently available BEVs employ permanent magnet synchronous motors (PMSM) [11].

6.8.9 Induction Motor (IM)

There are induction motors in both vintage electric vehicles like the GM EV1 and cutting-edge ones like Teslas [50]. This way of driving commutatorless motors is the most developed of the few available [9]. The capacity of such a system to reduce loss under arbitrary load conditions is illustrated in [51]. By isolating the IM's field control from its torque control, field orientation control may make an IM behave like an independently excited DC motor. By reducing the flux, we can increase the range of speeds over the basic speed while maintaining the same power [30], and by manipulating the IM's field orientation, we can increase the range by a factor of three to five above the base speed [52].

6.8.10 Switched Reluctance Motor (SRM)

SRMs are synchronous motors powered by unipolar inverter-generated current, also called doubly salient motors (DSMs), due to the presence of salient poles in both the stator and the rotor. They have several desirable characteristics for usage in electric vehicles, including cheap cost, fast speed, minimal risk of dangers, an inherently large constant power range, and a high power density. Due to the elimination of the need for preventative maintenance, the fault tolerance and dependability of such motors may be greatly improved. However, in comparison to permanent magnet (PM) machines, they are bulkier, heavier, and produce a lot more noise due to their variable torque nature. Due to the fringe effect of slots and poles and the high saturation of the pole-tips, designing and controlling such machines is challenging despite their straightforward architecture [8, 28, 46]. The expensive price of the rare-earth elements used in PM machines has led to a growing interest in SRMs. Nidec's 2012 demonstration of an advanced SRM revealed near-IPM-like performance at a reasonable cost. Research on SRMs primarily focuses on methods to reduce noise and torque ripple. A twin stator arrangement emerged from these studies; it has lower inertia and noise, higher torque density, and a wider speed range than standard SRMs [53, 54]. Finite element analysis in the design phase can help cut down on overall loss [55].

6.8.11 Synchronous Reluctance Motor (SynRM)

The Reluctance of Time, in Synchrony Synchronous speed motor combines the best features of permanent magnet and induction motors. They share the advantages of IMs in being reliable and fault-tolerant while also possessing the advantages of PM motors in being compact and powerful. They share a control technique with PM motors. Controllability, manufacturing, and low power factor are three areas that need improvement before SynRM can be used in EVs. Nonetheless, studies have been conducted, and some progress has been achieved as well, with the rotor design being the primary area of concern. Increasing the saliency, which results in a greater power factor, is one technique to enhance this motor's performance.

6.8.12 PM Assisted Synchronous Reluctance Motor

Power Modules (PMs) integrated inside the rotor of a Synchronous Reluctance Motor (SynRM) allow for higher power factors in SynRMs. A smaller number of PMs are employed, and hence, fewer flux connections are established, making it comparable to an IPM but not identical to one. Properly dosed PMs applied to the rotor's core boost efficiency while causing very small adjustments to the stator and almost no back EMF.

With this idea, you will not have to worry about demagnetisation as you do with IPMs when you overuse them or heat them up too much. This motor's performance can match that of IPM motors with the use of an efficient optimisation method. BRUSA Elektronik AG has shown off a SynRM with PM support that works well in EVs (Sennwald, Switzerland). Better design methods, control systems, and state-of-the-art production technologies may also boost the quality of PM-assisted SynRMs.

6.8.13 Axial Flux Ironless Permanent Magnet Motor

It features a rotor on the outside that does not have a slot, and it also avoids using iron. There is no stator core either, therefore, the machine may be lighter. Due to the radial nature of the field, the power density in this gap is increased. Like most modern motors, this one can run at varying speeds. The rotors of this machine may be mounted on the outside edges of the wheels, and the stator windings can be located in the middle of the axle, which is a significant improvement over conventional designs. Because there are no slots to block current, efficiency is increased, and copper loss is reduced [46]

6.9 POWER CONVERSION TECHNOLOGY

Batteries and ultra-capacitors (UC) store DC energy. They get electricity from grid-connected AC lines, either wired or wirelessly. The motors need this energy transferred back. Regenerative braking and V2G also feed power back to the batteries [56]. Power converters, which exist in many types, are needed to fill EV energy storage and push the car [48, 57] details power electronics converters. [58] DC-DC boost converters increase battery voltage to operate DC motors [48]. DC-DC converters combine power and complementary energy [59]. A universal DC-DC converter enhances battery-to-DC link power flow and buck reverse power flow. DC-DC boost converters may employ DSPs [60]. Dual inverter is the latest technology to drive AC motors like PMSMs [48].

6.10 CHARGING EVs

Electric vehicles may be charged using either direct current (DC) or alternating current (AC) systems. Power supplies may be charged at a variety of levels, each corresponding to a unique combination of current and voltage. The amount of time needed for a complete charge is proportional to the power setting. Similarly, the safety and effectiveness of wireless charging have been studied and tested for years.

6.11 CHALLENGES

Since EVs are better for the environment, several nations are passing legislation to encourage their use as a replacement for conventional vehicles [61]. However, without the challenges, a whole transition to EVs would be impossible. What we see in the present market illustrates this. The general public's opinion of EVs is crucial to their success on the market and their eventual widespread adoption [62]. The consumer attitude is a major factor in the widespread use of EVs. Equally important obstacles preventing widespread EV adoption include a lack of charging infrastructure, limited driving range, lengthy charging times, a lack of government assistance, high costs, and a lack of customer awareness [63]. The current state of battery technology is a major hurdle that increases the price, decreases the range, and lengthens the charging time of EVs [64]. Both consumers and producers have benefited from the efforts of policymakers and authorities to reduce the barrier effects via the development of suitable regulations and incentives. These could have eased the market entry of EVs, albeit they would not have guaranteed their longevity [65]. Therefore, to further enhance EV spread, a comprehensive study and analysis of current impediments is required.

6.11.1 Technology Limitations

Battery range, safety aspect, component requirements, and power generation are the biggest challenges to EV adoption. The capacity of the battery, weight of the vehicle, and aerodynamic design drag determine EV range [66]. EV batteries are bigger and less powerful than CV batteries; hence, more batteries are required for the same range [67]. Charger availability affects EV range. Weight affects an electric vehicle's performance and marketability. Energy storage costs increase with electric car size, deterring purchasers. Increase driving range via lightweight materials, high-energy-density batteries, and RES charging [68]. This method has greater installation and maintenance costs, diurnal charging, and lower battery life in freestanding photovoltaic cell-based EVs [69]. Despite the promise of new batteries, battery ingredients remain scarce, like fossil fuels. EVs' safety also drives their popularity. EV batteries have overheating, fire dangers, thermal runaway, and excessive energy storage [70].

Electric vehicle cyber security vulnerabilities include unprotected BMS communication [71]. However, incorrect battery disposal is a major environmental concern. Battery recycling technology has lagged behind promises of a greener future. The cost of EV manufacturing is greatly affected by the absence of standards. EV manufacturers may use induction motors, permanent magnet motors, and permanent magnet-assisted synchronous motors, which impact standardisation [72]. Hydrogen (H_2) fuel cell electric vehicles (FCEVs) are "clean cars," however, their methane-to-H_2 extraction process emits CO2. Wind, tidal, hydro, and solar can supply the huge amounts of electricity required for hydrogen extraction, but they are not generally used for this purpose. H_2's volatility makes FCEVs unsafe.

A strong, durable charging infrastructure is needed before imagining an EV for a particular vehicle market. Range concern is the biggest EV purchase barrier. The lack of on-road charging stations in most EV-using locations causes it [73]. EV-populated areas must invest in a coherent network of many linked charging/swapping infrastructure to solve this problem. Addressing charging infrastructure deficiency requires examining charging station specifications and safety requirements [74]. Inductive charging shields the user/EV from the actual charging circuitry underneath the road surface. This technology's charging efficiency is limited to 85% by the EV's distance from the charging emitter. After 250 mm, charge effectiveness drops dramatically [75, 76]. "Fast Charge" circuitry that may charge an EV in ten minutes needs safety circuits and components at a charging station or at home [77, 78]. To guarantee safe and efficient charging, global EV charging standards regulate Level 1, Level 2, and Level 3 charging [79]. Level 1 (8–16 h) employs 120 V outlets/connectors without further installations. Cable size and thermal limits limit current rating and charging time. Level 2 (4–8 h) uses 240 V with specific connectors/adaptors owing to rated voltage and power management requirements.

Level 3 technology utilises 480 V and takes 10–15 min for 80% SOC [80, 81]. For domestic usage, this technology is pricey. It needs three-phase alternating current (AC)/direct current (DC) connections above 22 kW and a minimum current of 32 Amps for AC and 25 Amps for DC. Ratings of this size cause safety problems for a residence and necessitate costlier stronger insulated cabling with shorter charger-vehicle lines. Multiple procedures and complicated BMS are needed to limit battery deterioration and protect consumers [82]. EV charging's power transfer and storage impacts must be addressed by consumers and producers. The government's power grid management depends on electric vehicle grid integration (EVGI). Electric vehicle charging increases grid loads, creating voltage instability and peak load demands. Due to EV charging's nonlinear loading, power reserve margins and power quality suffer. Increased hot spot temperatures reduce distribution transformer lifespan [83]. For 42 million EVs, excess power demand might reach 92%, emphasising the need for power management systems. Existing uncoordinated EV charging systems produce disordered peak load demand and system harmonics. Harmonic distortions (20%–45%) lower grid power quality and reduce transformer longevity [84]. Harmonic distortion is maximum with Level 3 charging. For a 69 kV network to maintain power quality, total harmonic distortion (THD) must be less than 5% [85, 86]. EV charging generally generates THD larger than 10%. With more EVs, standard rate charging may potentially have power quality issues. Swappable battery packs are being introduced by committees worldwide to address charging infrastructure issues [87]. Instead of technical charging standards, this infrastructure is positioned for storage. However, huge capital expenditure, EV layout change costs, and land area requirements hinder it. In addition, batteries must be homogeneous in a market, unlike previous EVs [88–90]. Similar to charging infrastructure, uncoordinated charging in the battery swapping station (BSS) system causes congestion [91–93]. Countries must coordinate EV battery charging/swapping and power management systems to ensure sustainable charging systems with good power quality.

6.11.2 Battery Difficulties

The need for advanced battery technology prevents EVs from becoming mainstream. Range, cost, charging time, safety, operating temperature range, materials utilised, recycling potential, and disposal capabilities affect EV battery performance. Due to their high energy density, power density, weight, cost-effectiveness, non-toxicity, and quick charging, lithium-ion batteries dominate the industry

[93, 94]. Li-ion batteries are fire dangers. These batteries' strong thermal runaway potential creates extreme temperatures, making them explosive [74, 95, 96]. Overcharging and improper insulation of lithium-ion batteries may cause fires and endanger passengers. Any battery technology's capacity, which determines EV range, is a downside.

Heavier, high-voltage batteries boost range but take up more space than a CV gasoline tank [97, 98], impacting vehicle handling and performance. Even with quick charging, these batteries take longer to charge than CVs to refill [81], which hinders EV adoption. Battery makers must maintain ideal battery temperatures regardless of climate and geography. Battery heat management systems increase system complexity. Battery materials and production difficulties limit EV adoption. Materials determine battery performance. EV cathodes are LiCoO, $LiMn_2O_4$, and $LiFePO_4$, whereas anodes are graphite. Because of their high price, toxicity, bad conductivity, and poor safety, battery makers are always looking for new alternatives [99].

Composite materials like micro- or nano-scaled particles may prevent substantial volume changes from treating these materials [100]. Li-ion batteries for EVs need a lot of industrial fossil fuels. The lack of lithium deposits to provide 100 million 40 kWh Li-based batteries for EVs is a hidden issue. Li-ion battery cost reduction will also be difficult [101]. Because of their enormous quantity, complexity, and material diversity, EV batteries become a waste management issue at end-of-life [102, 103]. Recycling might reduce hazardous waste and management costs [104–106]. Battery recycling requires grouping, shipping, storage, dismantling, residual energy, and pollution, which raises disposal costs [103]. Novel battery materials would raise prices, reducing EV adoption.

6.11.3 Consumer Purchase Capability and Behaviour

Consumer behaviour is influenced by attitudes, perceptions, experiences, and society [107]. Young and middle-aged people, particularly technical workers, prefer EVs [108, 109].

However, hurdles hinder EV uptake. EV range, charging problems, and high initial cost affect customer behaviour [110, 111]. Media coverage of EV fires raises public awareness of EV safety. This contradicts research that suggests EVs are not more flammable than CVs [112]. EV purchasing is further hindered by public ignorance of EV environmental advantages [113]. Consumers choose EV performance above environmental benefits [29]. Battery life drives EV purchases [114]. The lack of EV versions discourages purchasers. Thus, EV range, style, look, and usefulness might attract more buyers [115, 116]. Government regulations and incentives with financial rewards also affect EV sales. Financial incentives, tax refunds, free parking, etc., should encourage EV purchases [91]. Monetary incentives will help EVs enter the market, but they will not last. Understanding customer mindsets and attitudes will improve EV dissemination policies [65].

6.12 CONCLUSION

Electric vehicles (EVs) have the potential to become the transportation standard of the future while also preventing the worst effects of climate change. They provide a practical replacement for cars powered only by fossil fuels, which are in increasingly short supply [117]. This chapter provided an overview of electric vehicle (EV) categories, configurations, energy sources, motors, power conversion, and charging methods. Each section's defining technologies have been summarised and characterised. We've spoken about how EVs have an effect on many industries, as well as the immense potential they have to advance a better and greener energy system via synergy with smart grids and the facilitation of the integration of renewable sources. Current electric vehicle (EV) limitations and potential improvements have been outlined. We have also included the most up-to-date algorithms and methods for optimisation and control. The current state of the electric vehicle market has been briefly discussed. The article concludes with a summary of its findings, which provide a clear picture of the state of the industry and the gaps in knowledge that still need to be filled.

REFERENCES

1. Dixon, J. Energy storage for electric vehicles. International Conference on Industrial Technology, pp. 20–26, 2010.
2. Khaligh, A., Li, Z. Battery, Ultracapacitor, Fuel Cell, and Hybrid Energy Storage Systems for Electric, Hybrid Electric, Fuel Cell, and Plug-in Hybrid Electric Vehicles: State of the Art. IEEE Trans. Veh. Technol. 2010, 59, 2806–2814.
3. Keoun, B.C. Designing an electric vehicle conversion. Conference Record Southcon/95, pp. 303–308, March 1995.
4. Bakar, S.A.A.; Masuda, R.; Hashimoto, H.; Inaba, T.; Jamaluddin, H.; Rahman, R.A.; Samin, P.M. Ride comfort performance of electric vehicle conversion with active suspension system. Proceedings of SICE Annual Conference (SICE), pp. 1980–1985, August 2012.
5. Asimakopoulos, P.; Boumis, T.; Patsias, C.E.; Safacas, A.; Mitronikas, E. Experience derived from the conversion of a conventional car to a hybrid electric vehicle -analysis of the powertrain. International Symposium on Power Electronics Electrical Drives Automation and Motion (SPEEDAM), pp. 1040–1045, June 2010.
6. Andrew, M.; Weinlein, C. The Lead-Acid Battery—Demonstrating the Systems Design Approach to a Practical Electric Vehicle Power Source. IEEE Trans. Veh. Technol. 1983, 32, 21–25.
7. Gao, Y.; Ehsani, M. Investigation of Battery Technologies for the Army's Hybrid Vehicle Application. Proc. Veh. Technol. Conf. 2002, 3, 1505–1509.
8. Chan, C.C. The State of the Art of Electric and Hybrid Vehicles. Proc. IEEE 2002, 90, 247–275.
9. Camacho, O.M.F.; Nørgård, P.B.; Rao, N.; Mihet-Popa, L. Electrical Vehicle Batteries Testing in a Distribution Network using Sustainable Energy. IEEE Trans. Smart Grid. 2014, 5, 1033–1042.

10. Camacho, O.M.F.; Mihet-Popa, L. Fast Charging and Smart Charging Tests for Electric Vehicles Batteries using Renewable Energy. Oil Gas Sci. Technol. 2016, 71, 13–25.
11. Grunditz, E.A.; Thiringer, T. Performance Analysis of Current BEVs Based on a Comprehensive Review of Specifications. IEEE Trans. Transp. Electr. 2016, 2, 270–289.
12. SAE International. SAE Electric Vehicle and Plug-in Hybrid Electric Vehicle Conductive Charge Coupler. In SAE Standard J1772; Society of Automotive Engineers (SAE), Warrendale, PA, 2010.
13. Yilmaz, M.; Krein, P.T. Review of Battery Charger Topologies, Charging Power Levels, and Infrastructure for Plug-In Electric and Hybrid Vehicles. IEEE Trans. Power Electr. 2013, 28, 2151–2169.
14. Marchesoni, M.; Vacca, C. New DC–DC Converter for Energy Storage System Interfacing in Fuel Cell Hybrid Electric Vehicles. IEEE Trans. Power Electr. 2007, 22, 301–308.
15. Schaltz, E.; Khaligh, A.; Rasmussen, P.O. Influence of Battery/Ultracapacitor Energy-Storage Sizing on Battery Lifetime in a Fuel Cell Hybrid Electric Vehicle. IEEE Trans. Veh. Technol. 2009, 58, 3882–3891.
16. Kramer, B.; Chakraborty, S.; Kroposki, B. A review of plug-in vehicles and vehicle-to-grid capability. Proceedings of the 34th IEEE Industrial Electronics Annual Conference, Orlando, FL, pp. 2278–2283, 10–13 November 2008.
17. Williamson, S.S. Electric drive train efficiency analysis based on varied energy storage system usage for plug-in hybrid electric vehicle applications. Proceedings of the IEEE Power Electronics Specialists Conference, Orlando, FL, pp. 1515–1520, 17–21 June 2007.
18. Wirasingha, S.G.; Schofield, N.; Emadi, A. Plug-in hybrid electric vehicle developments in the US: Trends, barriers, and economic feasibility. Proceedings of the IEEE Vehicle Power and Propulsion Conference, Harbin, pp. 1–8, 3–5 September 2008.
19. Gao, Y.; Ehsani, M. Design and Control Methodology of Plug-In Hybrid Electric Vehicles. IEEE Trans. Ind. Electron. 2010, 57, 633–640.
20. EG&G Technical Services, Inc. The Fuel Cell Handbook, 6th ed. U.S. Department of Energy: Morgantown, WV, 2002.
21. Thounthong, P.; Raël, S.; Davat, B. Utilizing fuel cell and supercapacitors for automotive hybrid electrical system. Proceedings of the Applied Power Electronics Conference and Exposition, Austin, TX, Volume 1, pp. 90–96, 6–10 March 2005.
22. Why the automotive future will be dominated by fuel cells. IEEE Spectrum. Available online: http://spectrum.ieee.org/green-tech/fuel-cells/why-the-automotive-future-will-be-dominated-by-fuel-cells (accessed on 8 May 2017).
23. Tahami, F.; Kazemi, R.; Farhanghi, S. A Novel Driver Assist Stability System for All-Wheel-Drive Electric Vehicles. IEEE Trans. Veh. Technol. 2003, 52, 683–692.
24. Sato, M.; Yamamoto, G.; Gunji, D.; Imura, T.; Fujimoto, H. Development of Wireless In-Wheel Motor Using Magnetic Resonance Coupling. IEEE Trans. Power Electron. 2016, 31, 5270–5278.
25. Kurs, A.; Karalis, A.; Moffatt, R.; Joannopoulos, J.D.; Fisher, P.; Soljačić, M. Wireless Power Transfer Via Strongly Coupled Magnetic Resonances. Science 2007, 317, 83–86.
26. Imura, I.; Uchida, T.; Hori, Y. Flexibility of Contactless Power Transfer Using Magnetic Resonance Coupling to Air Gap and Misalignment for EV. World Electr. Veh. J. 2009, 3, 24–34.

27. Nakadachi, S.; Mochizuki, S.; Sakaino, S.; Kaneko, Y.; Abe, S.; Yasuda, T. Bidirectional contactless power transfer system expandable from unidirectional system. Proceedings of the 2013 IEEE Energy Conversion Congress and Exposition, Denver, CO, pp. 3651–3657, 15–19 September 2013.
28. Gao, Y.; Ehsani, M.; Miller, J.M. Hybrid electric vehicle: Overview and state of the art. Proceedings of the IEEE International Symposium on Industrial Electronics, Dubrovnik, pp. 307–316, 20–23 June 2005.
29. Kim, H.; Kum, D. Comprehensive Design Methodology of Input- and Output-Split Hybrid Electric Vehicles: In Search of Optimal Configuration. IEEE/ASME Trans. Mechatron. 2016, 21, 2912–2923.
30. Kim, D.; Hwang, S.; Kim, H. Vehicle Stability Enhancement of Four-Wheel-Drive Hybrid Electric Vehicle Using Rear Motor Control. IEEE Trans. Veh. Technol. 2008, 57, 727–735.
31. Li, Y.; Yang, J.; Song, J. Nano Energy System Model and Nanoscale Effect of Graphene Battery in Renewable Energy Electric Vehicle. Renew. Sustain. Energy Rev. 2017, 69, 652–663.
32. Williamson, S.S.; Rathore, A.K.; Musavi, F. Industrial Electronics for Electric Transportation: Current State of-the-Art and Future Challenges. IEEE Trans. Ind. Electron. 2015, 62, 3021–3032.
33. Cassani, P.A.; Williamson, S.S. Feasibility Analysis of a Novel Cell Equalizer Topology for Plug-In Hybrid Electric Vehicle Energy-Storage Systems. IEEE Trans. Veh. Technol. 2009, 58, 3938–3946.
34. Cassani, P.A.; Williamson, S.S. Design, Testing, and Validation of a Simplified Control Scheme for a Novel Plug-In Hybrid Electric Vehicle Battery Cell Equalizer. IEEE Trans. Ind. Electron. 2010, 57, 3956–3962.
35. Lee, Y.S.; Cheng, M.W. Intelligent Control Battery Equalization for Series Connected Lithium-Ion Battery Strings. IEEE Trans. Ind. Electron. 2005, 52, 1297–1307.
36. Why we still don't have better batteries. MIT Technology Review. Available online: www.technologyreview.com/s/602245/why-we-still-dont-have-better-batteries/ (accessed on 15 January 2023).
37. Ribeiro, P.F.; Johnson, B.K.; Crow, M.L.; Arsoy, A.; Liu, Y. Energy storage systems for advanced power applications. Proc. IEEE 2001, 89, 1744–1756.
38. Bartley, T. Ultracapacitors and batteries for energy storage in heavy-duty hybrid-electric vehicles. Proceedings of the 22nd International Battery Seminar & Exhibit, Fort Lauderdale, FL, 14–17 March 2005.
39. Chiu, H.J.; Lin, L.W. A Bidirectional DC-DC Converter for Fuel Cell Electric Vehicle Driving System. IEEE Trans. Power Electron. 2006, 21, 950–958.
40. Chan, C.C.; Chau, K.T. An Overview of Power Electronics in Electric Vehicles. IEEE Trans. Ind. Electron. 1997, 44, 3–13.
41. Chan, C.C.; Chau, K.T.; Jiang, J.Z.; Xia, W.A.X.W.; Zhu, M.; Zhang, R. Novel Permanent Magnet Motor Drives for Electric Vehicles. IEEE Trans. Ind. Electron. 1996, 43, 331–339.
42. Chan, C.C.; Chau, K.T.; Yao, J. Soft-switching vector control for resonant snubber based inverters. Proceedings of the IEEE International Conference Industrial Electronics, New Orleans, LA, pp. 605–610, 14 November 1997.
43. Chan, C.C.; Jiang, J.Z.; Chen, G.H.; Chau, K.T. Computer Simulation and Analysis of a New Polyphase Multipole Motor Drive. IEEE Trans. Ind. Electron. 1993, 40, 570–576.

44. Chan, C.C.; Jiang, J.Z.; Chen, G.H.; Wang, X.Y.; Chau, K.T. A Novel Polyphase Multipole Square-Wave Permanent Magnet Motor Drive for Electric Vehicles. IEEE Trans. Ind. Appl. 1994, 30, 1258–1266.
45. Chan, C.C.; Jiang, J.Z.; Xia, W.; Chan, K.T. Novel Wide Range Speed Control of Permanent Magnet Brushless Motor Drives. IEEE Trans. Power Electron. 1995, 10, 539–546.
46. Jose, C.P.; Meikandasivam, S. A Review on the Trends and Developments in Hybrid Electric Vehicles. In Innovative Design and Development Practices in Aerospace and Automotive Engineering. Springer: Singapore, 2017; pp. 211–229.
47. Chan, C.C.; Chau, K.T. Morden Elcetric Vehicle Technology. Oxford University Press: New York, 2001; pp. 122–133.
48. Lulhe, A.M.; Date, T.N. A technology review paper for drives used in electrical vehicle (EV) & hybrid electrical vehicles (HEV). Proceedings of the 2015 International Conference on Control, Instrumentation, Communication and Computational Technologies (ICCICCT), Kumaracoil, 18–19 December 2015.
49. Magnussen, F. On design and analysis of synchronous permanent magnet for field—Weakening operation. Ph.D. Thesis, KTH Royal Institute of Technology, 2004.
50. Model X Specifications. Tesla. Available online: www.tesla.com/support/model-x-specifications
51. Yamada, K.; Watanabe, K.; Kodama, T.; Matsuda, I.; Kobayashi, T. An efficiency maximizing induction motor drive system for transmissionless electric vehicle. Proceedings of the 13th International Electric Vehicle Symposium, Osaka, Volume II, pp. 529–536, 13–16 October 1996.
52. Boglietti, A.; Ferraris, P.; Lazzari, M.; Profumo, F. A New Design Criteria for Spindles Induction Motors Controlled by Field Oriented Technique. Electr. Mach. Power Syst. 1993, 21, 171–182.
53. Abbasian, M.; Moallem, M.; Fahimi, B. Double-Stator Switched Reluctance Machines (DSSRM): Fundamentals and Magnetic Force Analysis. IEEE Trans. Energy Convers. 2010, 25, 589–597.
54. Cameron, D.E.; Lang, J.H.; Umans, S.D. The Origin and Reduction of Acoustic Noise in Doubly Salient Variable-Reluctance Motors. IEEE Trans. Ind. Appl. 1992, 28, 1250–1255.
55. Chan, C.C.; Jiang, Q.; Zhan, Y.J.; Chau, K.T. A high-performance switched reluctance drive for P-star EV project. Proceedings of the 13th International Electric Vehicle Symposium, Osaka, Volume II, pp. 78–83, 13–16 Octobor 1996.
56. Onar, O.C.; Kobayashi, J.; Khaligh, A. A Fully Directional Universal Power Electronic Interface for EV, HEV, and PHEV Applications. IEEE Trans. Power Electron. 2013, 28, 5489–5498.
57. Bose, B.K. Power Electronics—A Technology Review. Proc. IEEE 1992, 80, 1303–1334.
58. Yaramasu, V.; Wu, B.; Sen, P.C.; Kouro, S.; Narimani, M. High-Power Wind Energy Conversion Systems: State-of-the-Art and Emerging Technologies. Proc. IEEE 2015, 103, 740–788.
59. Kok, D.; Morris, A.; Knowles, M. Novel EV drive train topology—a review of the current topologies and proposal for a model for improved drivability. Proceedings of the 2013 15th European Conference on Power Electronics and Applications (EPE), Lille, 2–6 September 2013.

60. Hegazy, O.; Van Mierlo, J.; Lataire, P. Analysis, control and comparison of DC/DC boost converter topologies for fuel cell hybrid electric vehicle applications. Proceedings of the 14th European Conference on Power Electronics and Applications (EPE 2011), Birmingham, 30 August–1 September 2011.
61. Hong, J.; Lee, H.; Nam, K. Charging Method for the Secondary Battery in Dual-Inverter Drive Systems for Electric Vehicles. IEEE Trans. Power Electron. 2015, 30, 909–921.
61. Egbue, O.; Long, S. Barriers to Widespread Adoption of Electric Vehicles: An Analysis of Consumer Attitudes and Perceptions. Energy Policy 2012, 48, 717–729.
62. Jreige, M.; Abou-Zeid, M.; Kaysi, I. Consumer Preferences for Hybrid and Electric Vehicles and Deployment of the Charging Infrastructure: A Case Study of Lebanon. Case Stud. Transp. Policy, 2021, 9(2), 466–476.
63. He, X.; Zhan, W.; Hu, Y. Consumer Purchase Intention of Electric Vehicles in China: The Roles of Perception and Personality. J. Clean. Prod. 2018, 204, 1060–1069.
64. Hannan, M.A.; Azidin, F.A.; Mohamed, A. Hybrid Electric Vehicles and Their Challenges: A Review. Renew. Sustain. Energy Rev. 2014, 29, 135–150.
65. Langbroek, J.H.; Franklin, J.P.; Susilo, Y.O. The Effect of Policy Incentives on Electric Vehicle Adoption. Energy Policy, 2016, 94, 94–103.
66. Mruzek, M.; Gajdáč, I.; L'Kučera,; Barta, D. Analysis of Parameters Influencing Electric Vehicle Range. Procedia Eng. 216, 134, 165–174.
67. Frieske, B.; Kloetzke, M.; Mauser, F. Trends in vehicle concept and key technology development for hybrid and battery electric vehicles. 2013 World Electric Vehicle Symposium and Exhibition (EVS27). IEEE, pp. 1–12.
68. Joost, W.J. Reducing Vehicle Weight and Improving US Energy Efficiency Using Integrated Computational Materials Engineering. Jom 2012, 64(9), 1032–1038.
69. Ghosh, A. Possibilities and Challenges for the Inclusion of the Electric Vehicle (EV) to Reduce the Carbon Footprint in the Transport Sector: A Review. Energies 2020, 13(10), 2602.
70. Zhang, J.; Zhang, L.; Sun, F.; Wang, Z. An Overview on Thermal Safety Issues of Lithium-Ion Batteries for Electric Vehicle Application. IEEE Access 2018, 6, 23848–23863.
71. Khalid, A.; Hernandez, A.; Sundararajan A.; Sarwat, A.I. Simulation-based analysis of equalization algorithms on active balancing battery topologies for electric vehicles. Proceedings of the Future Technologies Conference. Springer, Cham, 2019; pp. 708–728.
72. Bălţăţanu, A.; Florea, L.M. Comparison of electric motors used for electric vehicles propulsion. Proceeding of International Conference of Scientific Paper AFASES, 2013.
73. Andwari, A.M.; Pesiridis, A.; Rajoo, S.; Martinez-Botas, R.; Esfahanian, V. A Review of Battery Electric Vehicle Technology and Readiness Levels. Renew. Sustain. Energy Rev. 2017, 78, 414–430.
74. Liu, H.; Wei, Z.; He, W.; Zhao, J. Thermal Issues About Li-Ion Batteries and Recent Progress in Battery Thermal Management Systems: A Review. Energy Convers. Manag. 2017, 150, 304–330.
75. Franke, T.; Neumann, I.; Bühler, F.; Cocron, P.; Krems, J.F. Experiencing Range in an Electric Vehicle: Understanding Psychological Barriers. Appl. Psychol. 2012, 61(3), 368–391.

76. Yuan, X.; Li, L.; Gou, H.; Dong, T. Energy and Environmental Impact of Battery Electric Vehicle Range in China. Appl. Energy 2015, 157, 75–84.
77. Lin, Z. Optimizing and Diversifying Electric Vehicle Driving Range for US Drivers. Transp. Sci. 2014, 48(4), 635–650.
78. German, R.; Shili, S.; Desreveaux, A.; Sari, A.; Venet, P.; Bouscayrol, A. Dynamical Coupling of a Battery Electro-Thermal Model and the Traction Model of an EV for Driving Range Simulation. IEEE Trans. Veh. Technol. 2019, 69(1), 328–337.
79. Rahman, I.; Vasant, P.M.; Singh, B.S.M.; Abdullah-Al-Wadud, M.; Adnan, N. Review of Recent Trends in Optimization Techniques for Plug-In Hybrid, and Electric Vehicle Charging Infrastructures. Renew. Sustain. Energy Rev. 2016, 58, 1039–1047.
80. Eltoumi, F.M.; Becherif, M.; Djerdir, A.; Ramadan, H.S. The Key Issues of Electric Vehicle Charging Via Hybrid Power Sources: Techno-Economic Viability, Analysis, and Recommendations. Renew. Sustain. Energy Rev. 2020, 138, 110534. https://doi.org/10.1016/j.rser.2020.110534
81. Shareef, H.; Islam, M.M.; Mohamed, A. A Review of the Stage-of the-Art Charging Technologies, Placement Methodologies, and Impacts of Electric Vehicles. Renew. Sustain. Energy Rev. 2016, 64, 403–420.
82. Falvo, M.C.; Sbordone, D.; Bayram, I.S.; Devetsikiotis, M. EV charging stations and modes: International standards. 2014 International Symposium on Power Electronics, Electrical Drives, Automation and Motion. IEEE, pp. 1134–1139, 2014.
83. Deb, S.; Kalita, K.; Mahanta, P.; Impact of electric vehicle charging stations on reliability of distribution network. 2017 International Conference on Technological Advancements in Power and Energy (TAP Energy). IEEE, pp. 1–6, 2017.
84. Xiao, H.; Huimei, Y.; Chen, W.; Hongjun, L. A survey of influence of electrics vehicle charging on power grid. 2014 9th IEEE Conference on Industrial Electronics and Applications. IEEE, pp. 121–126, 2014.
85. Karmaker, A.K.; Roy, S.; Ahmed, M.R. Analysis of the impact of electric vehicle charging station on power quality issues. 2019 International Conference on Electrical, Computer and Communication Engineering (ECCE). IEEE, pp. 1–6, 2019.
86. Moses, P.S.; Deilami, S.; Masoum, A.S.; Masoum, M.A. Power quality of smart grids with plug-in electric vehicles considering battery charging profile. 2010 IEEE PES Innovative Smart Grid Technologies Conference Europe (ISGT Europe). IEEE, pp. 1–7, 2010.
87. https://Cordis.Europa.Eu/Project/Id/265684/Reporting (Accessed 11 April 2021).
88. Di Paolo, M. Analysis of harmonic impact of electric vehicle charging on the electric power grid, based on smart grid regional demonstration project—Los Angeles. 2017 IEEE Green Energy and Smart Systems Conference (IGESSC). IEEE, pp. 1–5, 2017.
89. Zhang, W.; Spence, K.; Shao, R.; Chang, L. Grid powersmoothing performance improvement for pv and electric vehicle (ev) systems. 2018 IEEE Energy Conversion Congress and Exposition (ECCE). IEEE, pp. 1051–1057, 2018.
90. Yang, S.; Yao, J.; Kang, T.; Zhu, X. Dynamic Operation Model of the Battery Swapping Station for EV (Electric Vehicle) in Electricity Market. Energy 2014, 65, 544–549.
91. Zeng, M.; Pan, Y.; Zhang, D.; Lu, Z.; Li, Y. Data-Driven Location Selection for Battery Swapping Stations. IEEE Access 2019, 7, 133760–133771.
92. Wang, G.; Zhang, F.; Sun, H.; Wang, Y.; Zhang, D. Understanding the Long-Term Evolution of Electric Taxi Networks: A Longitudinal Measurement Study on

Mobility and Charging Patterns. ACM Trans. Intell. Syst. Technol. (TIST) 2020, 11(4), 1–27.

93. Johannesson, L.; Egardt, B. A Novel Algorithm for Predictive Control of Parallel Hybrid Powertrains Based on Dynamic Programming. IFAC Proc. 2007, 40(10), 343–350.
94. Tian, Y.; Liu, J.; Yao, Q.; Liu, K. Optimal Control Strategy for Parallel Plug-In Hybrid Electric Vehicles Based on Dynamic Programming. World Electr Veh. J. 2021, 12(2), 85.
95. De Gennaro, M.; Paffumi, E.; Martini, G.; Giallonardo, A.; Pedroso, S.; Loiselle-Lapointe, A. A Case Study to Predict the Capacity Fade of the Battery of Electrified Vehicles in Real-World Use Conditions. Case Stud. Transp. Policy 2020, 8(2), 517–534.
96. Lopez, I.; Ibarra, E.; Matallana, A.; Andreu, J.; Kortabarria, I. Next Generation Electric Drives for HEV/EV Propulsion Systems: Technology, Trends and Challenges. Renew. Sustain. Energy Rev. 2019, 114, 109336.
97. Kamal, E.; Adouane, L.; Abdrakhmanov, R.; Ouddah, N. Hierarchical and Adaptive Neuro-Fuzzy Control for Intelligent Energy Management in Hybrid Electric Vehicles. IFAC-PapersOnLine 2017, 50(1), 3014–3021.
98. Mohammadian, S.K.; Zhang, Y. Thermal Management Optimization of an Air-Cooled Li-Ion Battery Module Using Pin-Fin Heat Sinks for Hybrid Electric Vehicles. J. Power Sources 2015, 273, 431–439.
99. Hanipah, M.R.; Mikalsen, R.; Roskilly, A.P. Recent Commercial Free-Piston Engine Developments for Automotive Applications. Appl. Therm. Eng. 2015, 75, 493–503.
100. Manthiram, A. Materials Challenges and Opportunities of Lithium Ion Batteries. J. Phys. Chem. Lett. 2011, 2(3), 176–184.
101. Li, J.; Daniel, C.; Wood, D. Materials Processing for Lithium-Ion Batteries. J. Power Sources 2011, 196(5), 2452–2460.
102. Deng, D. Li-Ion Batteries: Basics, Progress, and Challenges. Energy Sci. Eng. 2015, 3(5), 385–418.
103. Richa, K.; Babbitt, C.W.; Gaustad, G.; Wang, X. A Future Perspective on Lithium-Ion Battery Waste Flows from Electric Vehicles. Resour. Conserv. Recycl. 2014, 83, 63–76.
104. Tang, Y.; Zhang, Q.; Li, Y.; Li, H.; Pan, X.; McLellan, B. The Social-Economic-Environmental Impacts of Recycling Retired EV Batteries Under Reward-Penalty Mechanism. Appl. Energy 2019; 251: 113313.
105. Zhang, Y.; Li, Y.; Tao, Y.; Ye, J.; Pan, A.; Li, X.; Liao, Q.; Wang, Z. Performance Assessment of Retired EV Battery Modules for Echelon Use. Energy 2020, 193, 116555.
106. Ahmad, A.; Khan, Z.A.; Saad Alam, M.; Khateeb, S. A Review of the Electric Vehicle Charging Techniques, Standards, Progression and Evolution of EV Technologies in Germany. Smart Sci. 2018, 6(1), 36–53.
107. Ferguson, M.; Mohamed, M.; Higgins, C.D.; Abotalebi, E.; Kanaroglou, P. How Open Are Canadian Households to Electric Vehicles? A National Latent Class Choice Analysis with Willingness-to-Pay and Metropolitan Characterization. Transp. Res. Part D: Transp. Environ. 2018, 58, 208–224.
108. Li, W.; Long, R.; Chen, H.; Geng, J. A Review of Factors Influencing Consumer Intentions to Adopt Battery Electric Vehicles. Renew. Sustain. Energy Rev. 2017, 78, 318–328.

109. Prakash, N.; Kapoor, R.; Kapoor, A.; Malik, Y. Gender Preferences for Alternative Energy Transport with Focus on Electric Vehicle. J. Soc. Sci. 2014, 10, 114–122.
110. Bjerkan, K.Y.; Nørbech, T.E.; Nordtømme, M.E. Incentives for Promoting Battery Electric Vehicle (BEV) Adoption in Norway. Transp. Res. Part D: Transp. Environ. 2016, 43, 169–180.
111. Junquera, B.; Moreno, B.; Alvarez, R. Analyzing Consumer Attitudes Towards Electric Vehicle Purchasing Intentions in Spain: Technological Limitations and Vehicle Confidence. Technol. Forecast. Soc. Chang. 2016, 109, 6–14.
112. Bouguerra, S.; Layeb, S.B. Determining Optimal Deployment of Electric Vehicles Charging Stations: Case of Tunis City, Tunisia. Case Stud. Transp. Policy 219, 7(3), 628–642.
113. Haddadian, G.; Khodayar, M.; Shahidehpour, M. Accelerating the Global Adoption of Electric Vehicles: Barriers and Drivers. Electr. J. 2015, 28(10), 53–68.
114. Noppers, E.H.; Keizer, K.; Bolderdijk, J.W.; Steg, L. The Adoption of Sustainable Innovations: Driven by Symbolic and Environmental Motives. Glob. Environ. Change 2014, 25, 52–62.
115. Li, X.; Clark, C.D.; Jensen, K.L.; Yen, S.T.; English, B.C. Consumer Purchase Intentions for Flexible-Fuel and Hybrid-Electric Vehicles. Transp. Res. Part D: Transp. Environ. 2013, 18, 9–15.
116. Higgins, C.D.; Mohamed, M.; Ferguson, M.R. Size Matters: How Vehicle Body Type Affects Consumer Preferences for Electric Vehicles. Transp. Res. Part A Policy Pract. 2017, 100, 182–201.
117. Buhler, F.; Cocron, P.; Neumann, I.; Franke, T.; Krems, J.F. Is EV Experience Related to EV Acceptance? Results from a German field Study. Transport. Res. F Traffic Psychol. Behav. 2014, 25, 34–49.
118. Un-Noor, F.; Padmanaban, S.; Mihet-Popa, L.; Mollah, M.; Hossain, E. A Comprehensive Study of Key Electric Vehicle (EV) Components, Technologies, Challenges, Impacts, and Future Direction of Development. Energies 2017, 10(8), 1217.

Chapter 7

Efficient Design, Materials and Specifications of Electric Motors Used in Electric Vehicle Challenges

Manish D. Mahale, Bipasa Bimalendu Patra

7.1 INTRODUCTION

An electric vehicle (EV) is a vehicle that is fuelled by at least one electric engine, utilizing power put away in battery-powered batteries. In contrast to customary gas-powered motor (ICE) vehicles that depend on petroleum derivatives, EVs use power as their essential wellspring of energy, bringing about lower or zero tailpipe emanations, decreased reliance on petroleum derivatives, and expected natural advantages [1–3].

EVs can be classified into different types based on their power train configurations:

- **Battery Electric Vehicles (BEVs):** These vehicles are fueled exclusively by electric engines and depend on battery-powered batteries to store energy. BEVs do not have a gas-powered motor and produce zero tailpipe emanations. They are commonly charged by connecting to an outer power source, for example, a charging station or a home outlet.
- **Module Mixture (Plug-in) Electric Vehicles (PHEVs):** PHEVs consolidate electric engines with a gas-powered motor. They have a restricted electric reach and can work in all-electric mode for more limited distances. When the electric reach is exhausted, the gas-powered motor kicks in to give extra driving reach. PHEVs can be charged through outer power sources or can create power through regenerative slowing down.
- **Half and Half Electric Vehicles (HEVs):** HEVs utilize both a gas-powered motor and an electric engine. Nonetheless, the electric engine in HEVs helps the motor as opposed to being the essential wellspring of drive. The electric motor is primarily used for power assist during acceleration and regenerative braking. HEVs do not need to be plugged in for charging as the batteries are charged through the internal combustion engine and regenerative braking.

DOI: 10.1201/9781003464556-7

7.2 ADVANTAGE OF EVs OVER CONVENTIONAL FOSSIL FUEL-BASED VEHICLES

- **Environmental benefits:** EVs produce lower or zero tailpipe discharges, contingent upon the wellspring of power utilized for charging. They add to diminished air contamination, ozone-harming substance outflows, and reliance on petroleum derivatives.
- **Energy efficiency:** Electric engines are more energy-productive than gas-powered motors, bringing about better energy usage and decreased energy squandering during activity.
- **Lower operating costs:** Electric vehicles have lower working expenses than ICE vehicles. Power is by and large less expensive than gas or diesel fuel, and the support expenses of EVs can be lower because of fewer moving parts and diminished wear on parts like brakes and fumes frameworks.
- **Potential for renewable energy integration:** EVs can be charged utilizing power created from environmentally friendly power sources, for example, sunlight-based or wind power. This empowers the mix of environmentally friendly power into the transportation area, adding to a cleaner and more manageable energy framework.

7.2.1 Challenges Associated with Electric Vehicles

While electric vehicles (EVs) offer various advantages, there are, as yet, a few provocations that should be tended to for their far reaching reception [4–6]. Here are a few key difficulties related to electric vehicles:

- **Driving range and charging infrastructure:** Range tension is a worry for some potential EV proprietors. The restricted driving scope of some EV models, particularly contrasted with customary fuel-controlled vehicles, can be a hindrance to reception. Moreover, the accessibility and openness of charging foundations, including public charging stations and quick charging organizations, should be extended to help really long travel and lighten worries about charging accessibility.
- **Charging time:** Charging an electric vehicle commonly takes more time than refueling a traditional vehicle. While home charging should be possible short term, quick charging stations are expected to give speedier charging choices to those in a hurry. Improving charging innovation and framework to empower quicker charging times will be pivotal to upgrading accommodation and addressing shopper concerns.
- **Battery cost and technology:** The expense of electric vehicle batteries continues to be the main variable influencing the general expense of EVs. While battery costs have been diminishing, further headways in

battery innovation and economies of scale are expected to make electric vehicles more reasonable and cutthroat with gas-powered motor vehicles. Innovative work endeavors are centered around further developing energy thickness, life expectancy, and lessening the expense of batteries.

- **Battery range and durability:** Despite the fact that battery innovation has progressed fundamentally, there is still an opportunity to get better as far as expanding the scope of electric vehicles per charge. Upgrading battery energy thickness and sturdiness will take into consideration longer driving reaches and further developed battery life expectancy, tending to worries about battery execution and substitution costs.
- **Charging compatibility and standards:** There is a need for standardization and compatibility across different charging systems and connectors. Currently, various charging standards exist, which can create confusion and limit interoperability between different EV models and charging infrastructure. Establishing universal standards and promoting interoperability will help streamline the charging experience for EV owners.
- **Supply chain and raw materials:** The creation of electric vehicles depends on basic unrefined substances like lithium, cobalt, and nickel. Guaranteeing an economical and mindful inventory network for these materials, tending to potential stockpile limitations, and limiting the natural effect of mining and handling are significant contemplations for the drawn-out supportability of electric vehicle creation.
- **Second-life battery utilization and recycling:** Electric vehicle batteries typically have a lifespan of 8–15 years before they need replacement. Developing strategies for the repurposing and recycling of used EV batteries will be crucial to minimize waste and maximize resource utilization. Second-life applications, such as energy storage systems, can provide additional value and extend the lifespan of batteries beyond their use in vehicles.
- **Consumer awareness and education:** Many consumers still have limited knowledge and misconceptions about electric vehicles. Educating the public about the benefits, technology, and ownership experience of EVs is essential to increase awareness and promote their adoption. Providing clear and accurate information about EVs, including their range, charging options, and incentives, can help address misconceptions and improve consumer acceptance.
- **Policy and incentives:** Government strategies and impetuses assume a critical part in advancing the reception of electric vehicles. Steady approaches, for example, buy motivations, tax breaks, and interest in charging framework, can assist with speeding up the change to electric portability. Proceeding with strategy, support, and long haul arranging are important to establish an ideal market climate for EVs.

Addressing these challenges requires collaboration between industry stakeholders, policymakers, researchers, and infrastructure providers. Continued technological advancements, investment in charging infrastructure, and supportive policies will be crucial to overcoming these challenges and facilitating the widespread adoption of electric vehicles as a sustainable transportation solution. Investments in vehicle infrastructure, technical development, and policy support are necessary to make the transition from fossil fuels-based internal combustion engines [7] to cleaner and renewable energy sources. This transformation requires technical, economic, and policy framing with a main focus on research and development. The emerging areas like solar-based charging stations, nano techniques for energy conversion, and storage devices with special emphasis on high-performance piezoelectric nanogenerators for energy harvesting should be explored in the present decade [8–15]. Advanced modeling and simulation techniques based on AI and ML can be used for Electric vehicles [16, 17].

7.3 EFFICIENT DESIGN OF ELECTRIC VEHICLES

7.3.1 Design Process

Figures 7.1 and 7.2 show the design criteria and step involved in the electric vehicles design.

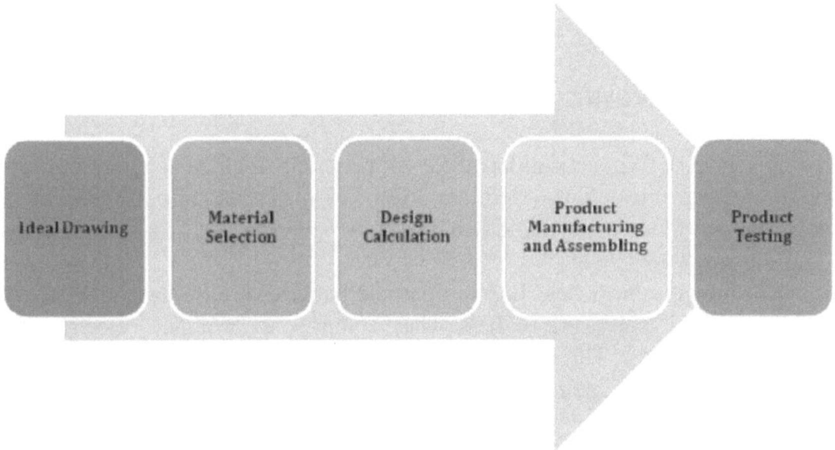

Figure 7.1 Design Procedure of EV.

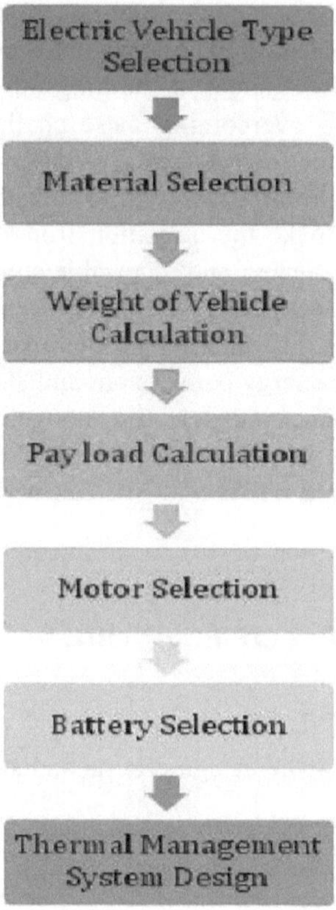

Figure 7.2 Parameter Calculation Process.

7.4 ELECTRIC VEHICLE MATERIAL

Electric vehicles (EVs) consolidate different materials to guarantee their productive and manageable activity. Here are a few key materials normally utilized in electric vehicles four wheeler and two wheeler (Figures 7.3 and 7.4):

- **Lithium ion batteries:** Lithium-particle batteries are the essential energy stockpiling framework in EVs. They comprise of terminals, electrolytes, and a separator. The terminals regularly contain lithium compounds, like lithium cobalt oxide ($LiCoO_2$), lithium iron phosphate ($LiFePO_4$), or lithium nickel manganese cobalt oxide (NMC). The electrolyte is normally a lithium salt disintegrated in a natural dissolvable (Figure 7.5).
- **Electric motors:** Electric motors in EVs often employ permanent magnets made of materials like neodymium, iron, and boron (NdFeB). These magnets provide high magnetic strength, enabling efficient power conversion.

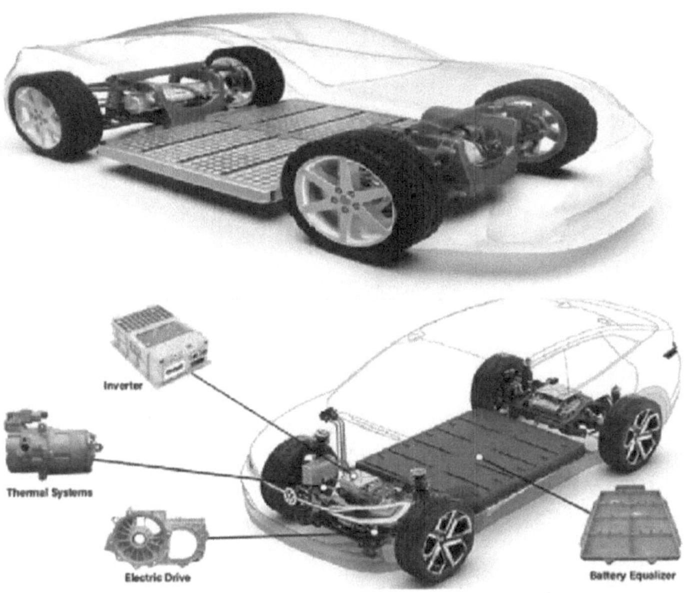

Figure 7.3 Generalized Four-Wheeler EV Model.

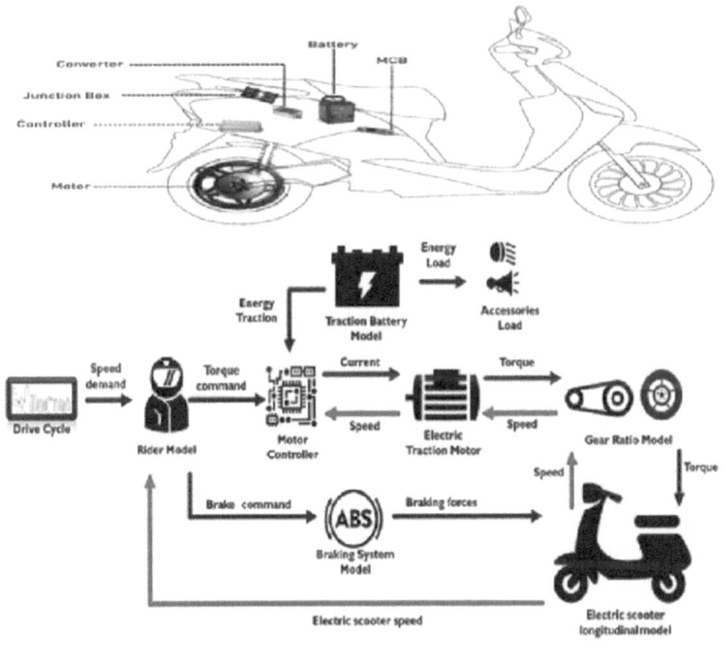

Figure 7.4 Generalized Two-Wheeler EV Model.

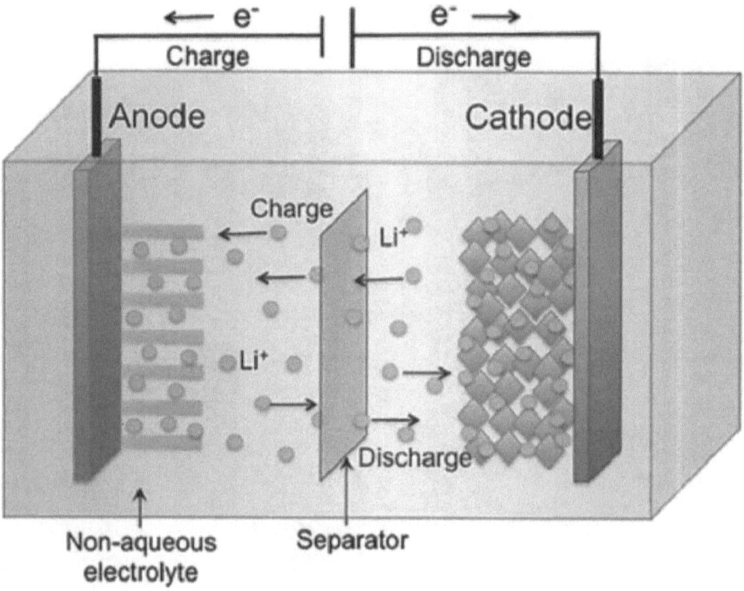

Figure 7.5 Electrodes of Lithium Ion Battery.

- **High-strength steel:** Electric vehicles incorporate lightweight but strong materials for their structure. High-strength steel alloys, such as advanced high-strength steel (AHSS), are used to reduce weight while maintaining structural integrity and safety.
- **Aluminum:** Aluminum is employed in various components of electric vehicles, including the chassis, body panels, and battery enclosures. It helps reduce overall vehicle weight, which improves energy efficiency and extends battery range.
- **Copper:** Copper is extensively used in electric vehicles due to its excellent electrical conductivity. It is utilized in wiring, cables, windings of motors, and other electrical components.
- **Carbon Fiber Reinforced Polymer (CFRP):** CFRP composites are lightweight yet solid materials that are utilized to build specific parts in EVs, for example, body boards, edges, and inside parts. CFRP decreases weight while keeping up with its primary unbending nature.
- **Thermal management materials:** Electric vehicles require efficient thermal management systems to maintain optimal operating temperatures for the battery, motor, and other components. Materials like thermally conductive adhesives, heat sinks, and phase-change materials are employed to dissipate heat effectively.

- **Semiconductors:** EVs heavily rely on power electronics for motor control, battery management, and other functions. Silicon-based semiconductors, such as silicon carbide (SiC) and gallium nitride (GaN), are increasingly used in power electronics for their high efficiency and superior performance.
- **Plastics and composites:** Various plastics and composites find application in EVs for interior components, trims, and other non-structural parts. These materials contribute to weight reduction and offer design flexibility.

It is important to note that the specific materials used in electric vehicles can vary among manufacturers and vehicle models. Advancements in technology and on-going research may also lead to the introduction of new materials in the future.

7.5 LITHIUM ION BATTERIES

Lithium-particle batteries, frequently alluded to as Li-particle or Li-ion batteries, are a kind of battery-powered battery ordinarily utilized in a large number of electronic gadgets. They have acquired critical notoriety and far-reaching reception because of their high energy thickness, long cycle life, and moderately low self-release rate. Li-particle batteries are utilized in different applications, including cell phones, personal computers (PCs), electric vehicles, and sustainable power frameworks (Tables 7.1 and 7.2).

Table 7.1 Positive Electrodes Material and Its Application with Benefits

Technology	*Application*	*Benefits*
Lithium nickel manganese cobalt oxide	Electric vehicles, power devices, network energy capacity	Great explicit energy and explicit power thickness
Lithium nickel cobalt aluminum oxide	Electric vehicles	High unambiguous energy, great life expectancy
Lithium manganese oxide	H-electric vehicle, wireless, PC	Great life expectancy
Lithium iron phosphate	Segway Personal Transporter, power tools, aviation products, automotive hybrid systems, PHEV conversions	Moderate thickness (2 A•h yields 70 amperes) High security contrasted with Cobalt/Manganese frameworks. Working temperature > 60 °C (140 °F)
Lithium cobalt oxide	Broad use, laptop	High unambiguous energy

Table 7.2 Negative Electrodes Material and Its Application with Benefits

Technology	*Application*	*Benefits*
Graphite	The predominant negative terminal material utilized in lithium particle batteries, restricted to a limit of 372 mAh/g	Minimal expense and great energy thickness. Graphite anodes can oblige one lithium iota for each six carbon molecules. The charging rate is administered by the state of the long, flimsy graphene sheets that comprise graphite. While charging, the lithium particles should make a trip to the external edges of the graphene sheet prior to stopping (intercalating) between the sheets. The meandering course takes such a long time that they experience clogs around those edges
Hard carbon	Home hardware	More prominent stockpiling limit
Lithium Titanate	Auto (Phoenix Motorcars), electrical network (PJM Interconnection Provincial Transmission Association control area), US Branch of Defense, transport (Proterra)	Worked on yield, charging time, toughness (wellbeing, working temperature −50–70 °C (−58–158 °F))
Silicon/carbon	Cell phones, giving 5000 mAh limit	Utilizes < 10% with silicon nanowires joined with graphite and folios. Energy thickness: ~74 mAh/g Another methodology utilized carbon-covered 15 nm thick gem silicon drops. The tried half-cell accomplished 1200 mAh/g in more than 800 cycles
Tin/cobalt alloy	Shopper hardware (Sony Nexelion battery)	Bigger limit than a cell with graphite (3.5 Ah 18650-type cells)

Here are some key characteristics and features of lithium ion batteries:

- **Energy density:** Li-particle batteries have a high energy thickness, and they can store a lot of energy compared to their size and weight. This makes them reasonable for versatile electronic gadgets where space and weight are significant variables (Figure 7.5).
- **Rechargeable:** One of the significant advantages of Li-ion batteries is their rechargeability. They can be charged and discharged multiple times, making them a more sustainable and cost-effective option compared to single-use disposable batteries.

- **Voltage:** Li-particle batteries commonly have an ostensible voltage of 3.6 to 3.7 volts per cell, which can differ depending upon the particular science utilized. Different cells are often associated in series to accomplish the ideal voltage for a specific application.
- **Cycle life:** Li-ion batteries have a relatively long cycle life, which is the number of charge and discharge cycles they can undergo before their capacity significantly decreases. The cycle life can vary depending on factors such as the battery chemistry, operating conditions, and charging/discharging protocols.
- **Self-discharge:** Li-particle batteries have a lower self-release rate contrasted with some other battery-powered battery sciences. This implies they can hold their charge for longer periods when not being used, which is worthwhile for gadgets that might be inactive for expanded periods between utilizes.
- **Charging time:** Li-ion batteries generally have faster charging times compared to other rechargeable batteries. However, the charging rate can be influenced by various factors, such as the battery's capacity, charger specifications, and temperature.

7.6 BATTERY MANAGEMENT SYSTEM (BMS)

Due to their specific charging and discharging requirements, Li-ion batteries often incorporate a battery management system. The BMS helps monitor and control various aspects of the battery's operation, including voltage levels, temperature, and safety features to prevent overcharging or over-discharging.

While Li-particle batteries offer various advantages, it is vital to take note that they additionally have a few constraints. They are delicate to high temperatures and can be defenseless to warming out of control, which might prompt overheating or even fire.

Specialists and producers keep on investigating headways in Li-particle battery innovation to upgrade their energy thickness, wellbeing, and life expectancy. These endeavors intend to satisfy the developing needs of arising advancements, for example, electric vehicles and network scale energy capacity frameworks.

7.7 BATTERY CALCULATION OF EV

Calculating the battery requirements for an electric vehicle (EV) involves considering several factors, including the desired driving range, vehicle efficiency, battery capacity, and energy consumption. Here's a general approach to calculating the battery requirements for an EV:

- **Determine the desired driving range:** Begin by deciding the ideal driving reach for your EV. This can shift depending on your particular requirements and the planned utilization of the vehicle. For example, if you want an EV with a range of 300 kilometers, this will be the target range for your calculations.

- **Estimate vehicle efficiency:** Estimate the vehicle's energy consumption or efficiency in terms of watt-hours per kilometer (W-h/km) or kilowatt-hours per 100 kilometers (kWh/100 km). This can be done by considering factors such as the vehicle's weight, aerodynamics, drivetrain efficiency, and driving conditions. It is important to note that vehicle efficiency can vary depending on driving style, terrain, and other factors, so it is best to use conservative estimates.
- **Calculate total energy required:** Multiply the desired driving range by the vehicle's energy consumption per kilometer. This will give you the total energy required for the desired range. For example, if your vehicle's energy consumption is estimated at 0.2 kWh/km and you want a range of 300 kilometers, the total energy required would be 60 kWh (0.2 kWh/km * 300 km).
- **Consider battery efficiency and usable capacity:** EV batteries have an efficiency factor that represents the energy lost during charging and discharging. Typically, this efficiency factor is around 90–95%. To account for this, divide the total energy required by the battery efficiency factor. For example, if you assume an efficiency factor of 90%, divide the total energy required (60 kWh) by 0.9 to get 66.7 kWh.
- **Determine usable battery capacity:** EV batteries are not typically discharged to their full capacity to preserve their lifespan. A common practice is to use a maximum of 80% of the battery's total capacity. Multiply the calculated battery capacity (66.7 kWh) by 0.8 to determine the usable battery capacity. In this example, the usable battery capacity would be 53.4 kWh (66.7 kWh * 0.8).
- **Consider safety margins:** It is advisable to include some safety margin to account for variations in driving conditions, battery degradation over time, and other factors. Adding an extra 10–20% to the usable battery capacity is a common approach. For example, if you add a 15% safety margin to the usable battery capacity, the final calculated battery capacity would be 61.4 kWh (53.4 kWh * 1.15).

It is important to note that these calculations provide a general approach, and actual battery requirements can vary depending on specific vehicle characteristics, driving conditions, and other factors. Additionally, it is recommended to consult with battery manufacturers or experts in EV engineering to ensure accurate and optimized battery sizing for your specific EV project.

7.8 ELECTRICAL MOTORS

Electrical motors are a critical component of electric vehicles (EVs), providing the necessary propulsion and driving force. They convert electrical energy from the vehicle's battery into mechanical energy to drive the wheels (Figure 7.6).

Figure 7.6 DC Series Motor.

Here are some key aspects of electric motors used in EVs:

7.8.1 Types of Electric Motors

There are several types of electric motors used in EVs, with the most common being:

a) DC Series Motor
b) Brushless DC Motor
c) Permanent Magnet Synchronous Motor (PMSM)
d) Three Phase AC Induction Motors
e) Switched Reluctance Motors (SRM)

a) **DC Series Motor:** The DC Series engine is a superb decision for foothold applications due to its high starting force limit. In the mid-1900s, it was the engine that was most often utilized for foothold applications. This engine's advantages incorporate basic speed control and the capacity to endure a fast ascent in load. It is the ideal footing engine as a result of this multitude of elements. High upkeep expected by brushes and commutators is a critical downside of DC series engines. Indian rail routes utilize these engines. DC brushed engines are the gathering wherein this engine has a place.

b) **Brushless DC Motor:** It looks like DC engines with super durable magnets in numerous ways. It misses the mark on the commutator and brush course of action, thus the name "brushless." Because of the electronic compensation of this engine, BLDC engines require no support. The foothold properties of BLDC engines incorporate high beginning force, high proficiency of 95–98%, and so on. BLDC engines are the most famous engines for use in electric vehicle applications. By standing out BLDC engines from standard brushed engines, you can find more about BLDC engines. The BLDC Motor has two types:

1. **Out runner type BLDC motor**

   Here, the stator is inside the engine while the rotor is outside. Since the wheel is straightforwardly connected to the outside rotor, it is otherwise called a center point engine. Such engines need not bother with an outside gear framework. In certain conditions, the actual engine comes furnished with planetary pinion wheels. This engine decreases the all-out size of the vehicle since it need not bother with a stuffed framework. Furthermore, it eliminates the requirement for a mounting region for the engine. The power yield in the in-sprinter configuration is obliged by a limitation on the engine size. Makers of electric cycles like Hullikal, Tronx, Spero, and light-speed bikes, among others, vigorously favor this motor (Figure 7.7).

Figure 7.7 Out Runner BLDC Motor.

2. **In runner type BLDC motor**
In this plan, unlike typical engines, the rotor of the engine is inside, and the stator is outside. The out-sprinter setup is marginally bulkier than the in-sprinter format on the grounds that these engines need an additional gearbox framework to send capacity to the wheels. BLDC engines are utilized by numerous three-wheeler producers, including Goenka Electric Engines, Speego Vehicles, Active Green, and Volta Car. Makers of low-and medium-execution bikes likewise use BLDC engines for impetus. These variables make it the engine of decision for some electric vehicle applications. The super durable magnet's significant expense is the greatest hindrance. Because of temperature conditions, super durable magnets have a more limited life expectancy when the engine is over-burdened beyond a specific point (Figure 7.8).

c) **Permanent Magnet Synchronous Motor**
This engine and the BLDC engine, which has long-lasting magnets on the rotor, are practically identical. These engines share foothold highlights with BLDC engines, like high power thickness and great effectiveness. The back EMF in PMSM is sinusoidal, though the back EMF in BLDC is trapezoidal. For more prominent power evaluations, super durable, magnet-coordinated engines are accessible. For elite execu-

Figure 7.8 In Runner BLDC Motor.

tion applications like vehicles and transports, PMSM is the best choice. Regardless of being more costly, PMSM is giving enlistment engines an intense battle due to its higher proficiency. Also, PMSM costs more than BLDC engines.

PMSM engines are used by most automakers in their hybrid and electric vehicles. Instances of vehicles that use PMSM engines include the Toyota Prius, Chevrolet Bolt EV, PortageCenter Electric, zero bicycles S/SR, Nissan Leaf, Hinda Accord, and BMW i3.

d) Three Phase AC Induction Motor

In fixed voltage and fixed recurrence activity, enlistment engines do not have a high beginning hat like DC series engines. Yet, by using different control strategies like FOC or v/f, this can be changed. These control strategies empower the engine to turn over with the best force, which is amazing for foothold applications. Since they require less upkeep, squirrel confine enlistment engines have a long life expectancy. Up to 92–95% proficiency can be accomplished while planning enlistment engines. An enlistment engine's disservice is that it requires a convoluted inverter circuit and is trying to control it (Figure 7.10).

In long-lasting magnet engines, the transition thickness B is impacted by the magnets. Subsequently, acceptance engines make changing the worth of B a lot less difficult than long-lasting magnet engines. It is subsequently on the grounds that, contingent upon the necessary force, the voltage and recurrence (V/f) of enlistment engines can be altered to change the worth of B. This adds to bringing down misfortunes, which raises productivity.

Figure 7.9 Permanent Magnet Synchronous Motor.

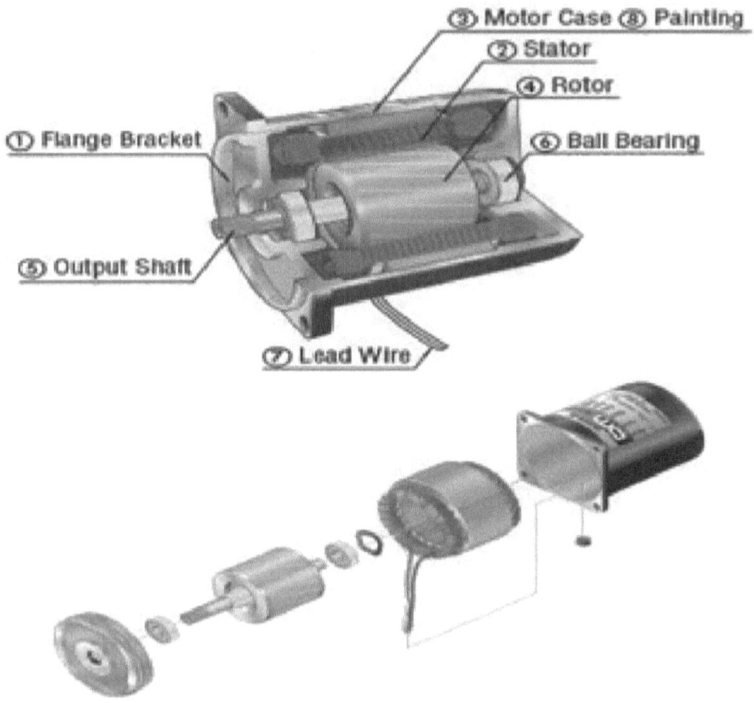

Figure 7.10 Three Phase Induction Motor.

The best delineation of the enlistment engines' better presentation thought about than contenders is the Tesla Model S. Tesla could have attempted to be free of the dependence on long-lasting magnets by utilizing enlistment engines. Indeed, even Mahindra Reva e2o's impetus framework is a three-stage enlistment engine. Enlistment engines will be utilized in cars and transports made by significant automakers like Goodbye Engines. The bike organization Televisions Engines will present an electric bike that is controlled by an enlistment engine. Because of their minimal expense, enlistment engines are inclined to be the choice for electric vehicles that emphasize execution. The capacity to endure cruel weather patterns is another advantage. The Indian Rail routes have started changing over from DC engines to AC enlistment engines because of these advantages.

e) **Switched Reluctance Motor**
Variable hesitance engines with twofold saliency fall inside the classification of exchanged hesitance engines. Exchanged Hesitance engines are reliable and have a clear plan. The SRM's rotor is a piece of covered steel

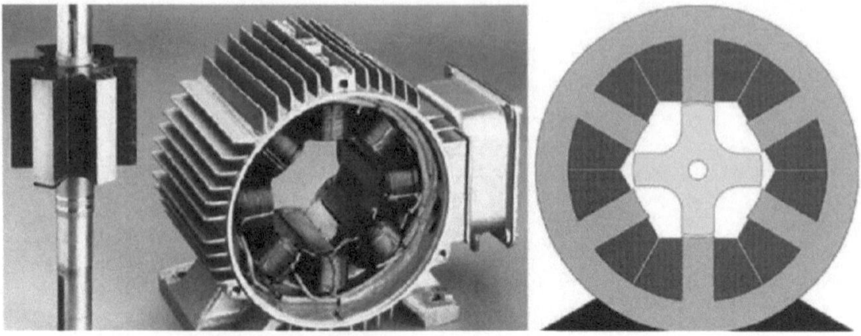

Figure 7.11 Switched Reluctance Motor.

that has neither super durable magnets nor windings connected to it. Subsequently, the rotor's latency is decreased, helping with fast speed increase. SRM is a practical decision for rapid application because of its power. High power thickness is another component that SRM conveys that is fundamental for electric vehicles. It is less difficult to cool the engine on the grounds that the stator creates the greater part of the intensity. The SRM's control intricacy and extending exchanging circuit are its two biggest defects. It has some clamor issues, too. SRM can ultimately supplant PMSM and enlistment engines once it enters the business area (Figure 7.11).

7.8.2 Selection of Electric Motor in Electrical Vehicle

- **Motor efficiency:** Electric motors used in EVs are designed to be highly efficient, converting a significant portion of electrical energy into mechanical output. Higher motor efficiency results in improved vehicle range and energy utilization. Efforts are made to optimize motor design, minimize losses, and improve control algorithms to enhance efficiency.
- **Power and torque:** Electric motors used in EVs are rated in terms of power output and torque. Power output determines the maximum speed a vehicle can achieve, while torque is responsible for acceleration and pulling power. EV manufacturers carefully select motors that provide sufficient power and torque for the desired vehicle performance.
- **Cooling and thermal management:** Electric motors generate heat during operation, and efficient cooling systems are essential to maintain optimal performance and prevent overheating. Cooling methods such as liquid cooling or air cooling are employed to dissipate heat from the motor.
- **Regenerative braking:** Electric motors in EVs can also function as generators during braking, capturing the kinetic energy and converting it into electrical energy. This regenerative braking process helps recharge the battery and improve overall energy efficiency.

- **Integration with power electronics:** Electric motors are integrated with power electronics components, such as inverters and motor controllers, to control motor operation. The power electronics system converts and manages the electrical energy between the battery and the motor, allowing for precise control of motor speed, torque, and direction.
- **Lightweight materials:** To enhance the efficiency and overall performance of electric motors, lightweight materials are used in their construction. These materials include high-strength alloys, composites, and advanced polymers, reducing the weight of the motor and improving the vehicle's energy efficiency.
- **Durability and reliability:** Electric motors in EVs undergo extensive testing to ensure their durability and reliability. They need to withstand high torque loads, operate efficiently over a wide range of speeds, and have a long lifespan to meet the demands of EV operation.
- Electric motors are continually evolving as technology advances, with ongoing research and development efforts focused on improving their efficiency, power density, and overall performance. These advancements play a crucial role in enhancing the driving range, acceleration, and sustainability of electric vehicles.

7.8.3 Specification of Electric Motor Used in Electric Vehicle

The specifications of electric motors used in electric vehicles can vary depending on the specific vehicle model, manufacturer, and intended application. However, I can provide you with a general overview of some common specifications:

- **Power rating:** The power rating of an electric motor is typically measured in kilowatts (kW) or horsepower (HP). It represents the amount of electrical power the motor can convert into mechanical power to drive the vehicle.
- **Torque:** Torque is the rotational force produced by the motor and is usually measured in Newton meters (Nm). It determines the acceleration and pulling power of the electric vehicle.
- **Voltage rating:** Electric motors in electric vehicles operate on high voltage systems, typically in the range of 200–800 volts. Higher voltage systems allow for more efficient power transmission and reduced energy losses.
- **Efficiency:** Productivity is a significant boundary that addresses the engine's capacity to change electrical energy into mechanical energy. It is generally communicated as a rate, and higher proficiency engines bring about superior reach and decreased energy utilization.
- **Cooling system:** Electric engines produce heat during activity, so a viable cooling framework is fundamental to keep up with ideal execution. Normal cooling strategies incorporate fluid cooling or air cooling.

- **Size and weight:** Electric motors for vehicles come in various sizes and weights, depending on the power requirements and packaging constraints of the vehicle. Motors used in electric cars are typically smaller and lighter than those used in larger vehicles like trucks or buses.
- **Regenerative braking:** Many electric motors used in electric vehicles support regenerative braking, which allows the motor to act as a generator during deceleration or braking, converting kinetic energy back into electrical energy to recharge the vehicle's battery.
- It is critical to take note that the particulars of electric engines can change essentially between various electric vehicle models and producers. Moreover, progressions in engine innovation are persistently being made, so fresher models might highlight unique or further developed details contrasted with more established ones.

7.9 BASIC TERMS ASSOCIATED WITH ELECTRIC MOTORS

- **Torque calculation for electric vehicle**: To calculate the torque required for an electric vehicle (EV), you need to consider several factors, including vehicle mass, desired acceleration, rolling resistance, aerodynamic drag, and grade (slope) resistance (Figure 7.12).
- **Determine the vehicle mass:** Begin by deciding the absolute mass of the vehicle, including the heaviness of the actual vehicle, travelers, freight, and any extra hardware. The mass is regularly estimated in kilograms (kg).
- **Calculate the force required for acceleration:** Decide the ideal speed increase execution of the vehicle. This can be communicated in meters each second squared (m/s2) or as far as the time it takes to arrive at a specific

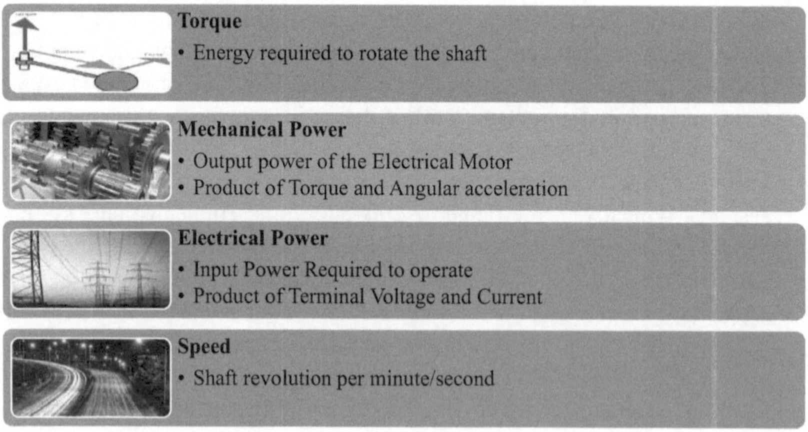

Figure 7.12 Basic Nomenclature.

speed (e.g., 0 to 100 km/h). Increase the vehicle mass by the ideal speed increase to work out the power expected for the speed increase.

- **Account for rolling resistance:** Rolling resistance refers to the force required to overcome the friction between the tires and the road surface. It depends on factors such as tire type, tire pressure, and road conditions. Rolling resistance can be estimated as a percentage of the vehicle weight. Increase the vehicle mass by the moving obstruction coefficient to ascertain the power expected to conquer moving opposition.
- **Consider aerodynamic drag:** Aerodynamic drag is the resistance experienced by the vehicle due to air pushing against it as it moves. It depends on the vehicle's shape, frontal area, and its coefficient of drag (C_d). Increase the front-facing region of the vehicle by the air thickness and the square of the vehicle's speed to work out the power because of streamlined drag.
- **Account for grade resistance:** Grade resistance refers to the force required to overcome the vehicle's movement uphill or downhill on a slope. Work out the grade opposition by increasing the vehicle mass by the gravitational steady (9.81 m/s2) and the sine of the point of the slant.
- **Sum the forces:** Add the value of all forces calculated above for finding the value of total force required to move the vehicle. This force is typically measured in Newton's (N).
- **Calculate torque:** Torque (T) is calculated by multiplying the total force by the effective radius (r) at which the force is applied. The effective radius is the radius at which the force from the wheel is transferred to the ground. Torque can be expressed in Newton-meters (Nm). Use the formula:

$$T = F \times r$$

Keep in mind that these calculations provide a general approach to torque estimation for an EV and may require adjustments based on specific vehicle characteristics and driving conditions. Consulting with EV engineering experts or using specialized simulation software can provide more accurate torque calculations for your specific EV design.

- **Mechanical power calculation in electric vehicle:** Calculating the mechanical power of a motor in an electric vehicle (EV) involves considering the motor's torque and rotational speed. Mechanical power is a measure of the motor's output power, representing the rate at which it can perform work.
- **Determine the torque:** Torque (T) is the rotational force produced by the motor and is typically measured in Newton-meters (Nm). The torque produced by the motor depends on various factors, including the motor design, current supplied, and operating conditions. The torque can be obtained from the motor's specifications or through motor testing.

- **Determine the rotational speed:** The rotational speed (ω) of the engine is the rate at which it pivots and is commonly estimated in cycles each moment (rpm) or radians each second (rad/s). The rotational speed can likewise be obtained from the engine's particulars or through engine testing.
- **Convert rotational speed:** Assuming that the rotational speed is given in rpm, convert it to radians each second by duplicating it by 2π/60. This transformation factor represents the way that there are 2π radians in a single upset and 60 seconds in a single moment.
- **Calculate mechanical power:** The mechanical power (P) of the motor can be calculated by multiplying the torque by the rotational speed. Use the formula:

  $$P = T \times \omega$$

  The resulting value represents the mechanical power output of the motor in watts (W) or kilowatts (kW). It indicates the rate at which the motor can deliver mechanical work or rotational energy to the drivetrain of the EV.

  It is important to note that the motor's mechanical power is influenced by factors such as motor efficiency, voltage, and current. Additionally, motor characteristics may vary depending on operating conditions and motor control strategies.
- **Battery efficiency:** Aluminum is used in the design and construction of battery components, such as battery housings and cooling systems. Aluminum has excellent thermal conductivity, allowing it to dissipate heat effectively and maintain optimal operating temperatures for the battery. Efficient thermal management is crucial for maintaining the performance, safety, and lifespan of the EV's battery system.
- **Regenerative braking systems:** Many electric vehicles incorporate regenerative braking systems, which convert kinetic energy during braking into electrical energy that can be stored in the battery. Aluminum is utilized in the construction of brake calipers and rotors, as it provides lightweight and efficient heat dissipation properties required for effective regenerative braking.
- **Chassis and Suspension Components:** Aluminum is commonly used in the manufacturing of chassis and suspension components of electric vehicles. These components include control arms, knuckles, and subframes. Aluminum's lightweight nature helps reduce unsprung weight, enhancing the vehicle's handling, responsiveness, and overall ride quality.
- **Aerodynamics:** Aluminum can be shaped and formed into complex structures, allowing automakers to design EVs with aerodynamic features. Improved aerodynamics minimize drag and improve energy efficiency, resulting in a longer driving range for electric vehicles.
- **Recyclability:** Aluminum is exceptionally recyclable, making it a harmless-to-the-ecosystem decision. Reusing aluminum requires altogether less

energy contrasted with essential creation, decreasing ozone harming substance emanations and protecting assets. It is important to note that while aluminum offers various advantages for electric vehicles, it also has some limitations. For instance, aluminum can be more expensive than steel, and the manufacturing processes for aluminum components may require specialized techniques and equipment. As a result, automakers often utilize a combination of materials, such as aluminum for certain components and steel for others, to strike a balance between cost, performance, and safety in electric vehicle design.

7.9.1 Copper

Copper is a fundamental material in electric vehicles (EVs) and assumes a basic part in different parts of their plan and usefulness. The application of copper can be identified in what follows:

- **Electric motors:** Copper is extensively used in the construction of electric motors, including both the stator and rotor windings. The high electrical conductivity of copper enables efficient transfer of electrical energy, leading to improved motor performance and higher power output. Copper windings also contribute to higher motor efficiency and reduced energy losses, thereby enhancing the overall range of the EV.
- **Power electronics:** Copper is utilized in the creation of force electronic parts like inverters, converters, and regulators in EVs. These parts are answerable for dealing with the progression of electrical energy inside the vehicle, changing it between various voltage levels, and controlling engine activity. Copper's low electrical resistance and high thermal conductivity enable efficient power transmission and help dissipate heat generated during operation, ensuring the reliability and performance of power electronics.
- **Wiring and interconnections:** Copper is utilized extensively in the wiring and interconnection systems of EVs. Copper wiring provides low electrical resistance, minimizing power losses during energy transmission within the vehicle's electrical system. It is used for high-current applications, including battery interconnections, charging infrastructure, and high-power charging cables.
- **Charging infrastructure:** Copper is vital in the infrastructure supporting EV charging. Charging cables and connectors often incorporate copper conductors due to their high electrical conductivity and durability. Copper's ability to handle high current levels allows for efficient and rapid charging of EVs.
- **Battery systems:** Copper is employed in various aspects of EV battery systems. It is used in the battery cells' current collectors, ensuring efficient current distribution and improving the battery's overall performance. Moreover, copper is used in the battery board frameworks, which screen and

control the battery's condition of charge, temperature, and voltage for ideal activity and wellbeing.

- **Heat management:** Copper's high thermal conductivity makes it suitable for heat dissipation and management within EVs. It is used in heat exchangers, cooling systems, and other thermal management components to dissipate heat generated by the electric drivetrain and power electronics, ensuring optimal operating temperatures and system efficiency.
- Copper's excellent electrical conductivity, thermal properties, and durability make it a preferred material for critical components in electric vehicles. The efficient use of copper in EVs contributes to their performance, energy efficiency, and overall sustainability.

7.9.2 Carbon Fiber Reinforced Polymer (CFRP)

Carbon Fiber Reinforced Polymer (CFRP) is a lightweight and high-strength composite material that finds applications in the manufacturing of electric vehicles (EVs). The use of CFRP is as follows:

- **Lightweight construction:** CFRP is eminent for its outstanding solidarity-to-weight proportion. By using CFRP parts in the vehicle's construction, for example, the skeleton, body boards, and inside parts, EV makers can fundamentally lessen the general load of the vehicle. The lightweight nature of CFRP helps improve the energy efficiency and range of EVs, as less energy is required to move a lighter vehicle.
- **Battery enclosures:** CFRP is utilized in the construction of battery enclosures and casings in EVs. CFRP's high strength and impact resistance provide robust protection to the battery pack, ensuring its safety and integrity. Additionally, CFRP's excellent thermal properties assist in managing the heat generated by the battery, enhancing its performance and lifespan.
- **Aerodynamics:** CFRP can be easily molded into complex shapes, allowing automakers to design aerodynamic components that reduce drag and improve the overall efficiency of EVs. By incorporating CFRP in aerodynamic features like splitters, diffusers, and spoilers, EVs can experience reduced air resistance, resulting in enhanced energy efficiency and extended range.
- **Suspension components:** CFRP is utilized in the creation of suspension parts, for example, control arms and springs. The high strength and firmness of CFRP add to responsiveness and ride quality in EVs. CFRP suspension parts additionally offer weight reserve funds and can assist with improving the general vehicle elements.
- **Interior design:** CFRP is utilized in the interior of EVs to create lightweight and aesthetically pleasing components. It can be used for panels, trims, and other interior parts, adding a touch of sophistication while reducing weight and enhancing overall comfort.

- It is worth noting that while CFRP offers several advantages, such as weight reduction and high strength, it can be relatively expensive compared to other materials like steel or aluminum. Additionally, the manufacturing processes for CFRP components may require specialized techniques and expertise. Therefore, CFRP is often used selectively in specific areas of EVs to maximize its benefits while managing costs effectively.

7.9.3 Thermal Management System

Thermal management is a critical aspect of electric vehicles (EVs) to ensure optimal performance, efficiency, and safety of various components. Several materials are used for thermal management in EVs [18–21]. Here are some commonly employed thermal management materials:

- **Thermal Interface Materials (TIMs):** TIMs are utilized to work with productive intensity move between two surfaces, ordinarily between an intensity-creating part and an intensity sink. Instances of TIMs incorporate warm lubes, warm cushions, and stage change materials (PCMs). These materials assist with filling holes, working on warmth conductivity and decreasing warmth opposition, guaranteeing successful intensity dispersal and forestalling overheating.
- **Heat transfer fluids:** Heat move liquids, like coolants or refrigerants, assume a critical part in the cooling frameworks of EVs. They flow through cooling channels or intensity exchangers to retain and divert heat from parts like the battery pack, power hardware, and engine. Normal-intensity move liquids incorporate ethylene glycol-based coolants, propylene glycol-based coolants, and refrigerants like R134a or R1234yf.
- **Phase Change Materials (PCMs):** PCMs are materials that can retain or deliver nuclear power during stage advances (strong to fluid or the other way around). They are utilized to store and control heat inside the EV's warmth administration framework. PCMs are especially valuable in applications where a high measure of intensity should be consumed or delivered during brief periods, for example, during quick charging or regenerative slowing down.
- **Heat sinks:** Heat sinks are usually utilized in EVs to disseminate heat from parts like power gadgets or engine regulators. They are normally made of materials with high warmth conductivity, like aluminum or copper. Heat sinks provide a huge surface region for warmth exchange and frequently integrate balances or different plans to upgrade heat dispersal through convection.
- **Insulation materials:** Insulation materials are utilized to prevent heat transfer between different compartments in an EV. They help contain heat within specific areas, such as the passenger cabin, preventing its loss to the surroundings. Common insulation materials include fiberglass, mineral wool, and ceramic fiber.

- **Thermal management coatings:** Specialized thermal management coatings can be applied to surfaces to enhance heat dissipation or insulation properties. These coatings can improve thermal conductivity or provide insulation barriers, depending on the specific requirements. The selection of thermal management materials depends on factors such as the specific component being cooled, desired heat transfer rates, operating temperatures, and safety considerations. EV manufacturers employ a combination of these materials and design strategies to maintain optimal temperatures, enhance efficiency, and prolong the lifespan of critical components.

7.9.4 Semiconductors

Semiconductors are essential in the working and activity of electric vehicles (EVs). They are fundamental for different frameworks and parts inside an EV. Here are a few vital utilizations of semiconductors in EVs:

- **Power gadgets:** Power gadgets are liable for controlling and dealing with the progression of electrical power inside an EV. Semiconductors, especially power semiconductors like protected door bipolar semiconductors (IGBTs) and metal-oxide-semiconductor field-impact semiconductors (MOSFETs), are utilized in inverters, converters, and engine drive frameworks. These semiconductors empower the proficient transformation of DC power from the battery to AC power for driving the electric engine.
- **Motor control:** Semiconductors are employed in motor control systems to regulate the speed, torque, and direction of the electric motor. Motor control units utilize microcontrollers and digital signal processors (DSPs) that incorporate semiconductors to process signals and execute control algorithms. These semiconductors provide precise and efficient motor control, resulting in improved performance and energy efficiency of the EV.
- **Battery Management Systems (BMS):** Semiconductors are necessary to BMS, which screens and controls the presentation, charging, and releasing of the EV's battery pack. BMS employs microcontrollers, analog-to-digital converters (ADCs), and other semiconductor components to measure battery parameters, manage cell balancing, and ensure optimal battery operation, safety, and longevity.
- **Charging infrastructure:** Semiconductors are used in EV charging stations and infrastructure to control charging processes and communication between the vehicle and the charger. Semiconductor devices facilitate power conversion, voltage regulation, and data communication protocols such as Controller Area Network (CAN) or ISO 15118 for charging interactions.
- **Safety systems:** Semiconductors assume a fundamental part in EV well-being frameworks, including non-freezing stopping mechanisms (ABS), foothold control frameworks (TCS), electronic soundness control (ESC),

and high-level driver-help frameworks (ADAS). These frameworks depend on semiconductors to deal with sensor information, settle on continuous choices, and enact wellbeing elements to upgrade vehicle strength, and tenant insurance [22–26].

- **Onboard diagnostics:** Semiconductors enable onboard diagnostics in EVs, allowing monitoring, diagnostics, and reporting of vehicle systems and components. Diagnostic systems utilize semiconductors for data acquisition, analysis, and communication with external diagnostic tools, aiding vehicle maintenance and troubleshooting.
- The headways in semiconductor innovation, for example, the advancement of wide-band hole semiconductors like silicon carbide (SiC) and gallium nitride (GaN), have prompted superior productivity, higher power thickness, and better warmth execution in EV frameworks, adding to improvement in general execution and reach.

7.9.5 Plastics and Composites

Plastics and composites play various roles in the manufacturing of electric vehicles (EVs). They offer unique properties and advantages that contribute to the performance, efficiency, and sustainability of EVs. Here are some key applications of plastics and composites in EVs:

- **Lightweighting:** Plastics and composites are lightweight materials that help reduce the overall weight of EVs. By replacing heavier metal components with lightweight plastics and composites, manufacturers can improve the vehicle's energy efficiency, range, and handling characteristics. These materials are commonly used in body panels, interior components, and structural elements to achieve weight reduction goals.
- **Aerodynamics:** Plastics and composites can be molded into complex shapes, allowing for streamlined designs that improve the aerodynamic performance of EVs. Smooth and contoured body panels and aerodynamic enhancements, such as front splitters, rear diffusers, and underbody panels, reduce drag and enhance energy efficiency by reducing air resistance.
- **Battery enclosures:** Plastics and composites are used in the development of battery nooks and housings. These materials give amazing protection and insurance to the battery pack while guaranteeing the wellbeing and respectability of the battery framework. Plastics and composites can be lightweight, influence-safe, and thermally protecting, adding to the general wellbeing and productivity of the EV's battery framework.
- **Interior components:** Plastics are extensively used in the interior of EVs for various components, including dashboards, door panels, seats, and trim pieces. These materials offer design flexibility, durability, and a wide range of surface finishes. Composites, such as carbon fiber-reinforced polymers

(CFRPs), may also be used for high-end interior trims, providing a lightweight and premium aesthetic.

- **Thermal management:** Plastics and composites can be designed with explicit warmth properties to aid the warmth administration of EVs. For instance, certain composite materials with high warmth conductivity can be utilized for heat sinks or warmth administration parts, assisting with dispersing heat produced by power gadgets, batteries, or engines.
- **Noise, Vibration, and Harshness (NVH) reduction:** Plastics and composites can contribute to reducing noise and vibrations within the vehicle cabin. These materials offer excellent damping properties, which help absorb and attenuate unwanted noise and vibrations, resulting in a quieter and more comfortable driving experience.
- **Recycling and sustainability:** Plastics and composites can be recycled, contributing to the overall sustainability of EVs. Recycled plastics can be used in various components, reducing the environmental impact associated with raw material extraction and manufacturing processes.
- It is important to note that the specific choice of plastics and composites in EVs can vary depending on factors such as cost, performance requirements, and manufacturing considerations. EV manufacturers often employ a combination of materials, including metals, plastics, and composites, to optimize the overall performance, safety, and sustainability of their vehicles

7.9.6 Payload Calculation of EV

Calculating the payload capacity of an electric vehicle (EV) involves considering the vehicle's gross vehicle weight rating (GVWR) and subtracting the vehicle's curb weight, passenger weight, and any additional equipment weight. Here's a general approach to calculating the payload capacity:

- **Determine the GVWR:** The GVWR is the most extreme weight that the vehicle is intended to convey, including the heaviness of the actual vehicle, travelers, freight, and any extra gear. It is regularly determined by the vehicle producer and can be tracked down in the vehicle's documentation or details.
- **Determine the curb weight:** The curb weight is the heaviness of the vehicle with practically no travelers, freight, or extra hardware. It incorporates the heaviness of the frame, body, powertrain, batteries, and other standard parts. The curb weight is also specified by the vehicle manufacturer and can be found in the vehicle's documentation.
- **Subtract the curb weight:** Subtract the curb weight from the GVWR to determine the available payload capacity. The excess weight addresses the greatest weight that the vehicle can convey regarding travelers, freight, and extra hardware.

- **Consider passenger weight:** Estimate the weight of the driver and passengers who will be traveling in the vehicle. This can differ depending on the quantity of travelers and their typical weight.
- **Consider additional equipment weight:** If there are any additional equipment or accessories installed in the vehicle, estimate their weight and add it to the payload capacity obtained in step 4. The subsequent worth addresses the accessible payload limit of the electric vehicle. It is critical to guarantee that the complete load of the travelers, freight, and extra gear does not surpass the payload ability to guarantee the protected and effective activity of the vehicle. Remember that these computations give a general way to deal with payload limit assessment for an EV.

7.10 CONCLUSIONS AND FUTURE SCOPE

In conclusion, this book chapter provided an insightful overview of electric vehicle design, highlighting its potential to revolutionize the automotive industry and contribute to a greener and more sustainable future. It served as a valuable resource for engineers, policymakers, and researchers seeking to contribute to the advancement of electric mobility and selection of motors while designing.

REFERENCES

[1]. Larminie, J.; Lowry, J. *Electric Vehicle Technology Explained*, 2012. John Wiley and Sons, USA.

[2]. U.S. Department of Energy (DOE), Advanced Vehicle Testing Activity (AVTA). https://www.energy.gov/eere/vehicles/advanced-vehicle-testing-activity-avta-data-and-results.

[3]. Olabi, G.; Abdelkareem, M. A. Energy storage systems towards 2050. *Energy* 2021, 219, 119634. https://doi.org/10.1016/j.energy.2020.119634.

[4]. Reddy, P. N.; Verma, V.; Kumar, A.; Awasthi, M. CFD simulation and thermal performance optimization of flow in a channel with multiple baffles. *Journal of Heat and Mass Transfer Research* 2023, 10(2), 257–268. https://doi.org/10.22075/JHMTR.2023.31108.1458.

[5]. Olabi, A. G.; Maghrabie, H. M.; Adhari, O. H. K.; Sayed, E. T.; Yousef, B. A. A.; Salameh, T.; Kamil, M.; Abdelkareem, M. A. Battery thermal management systems: Recent progress and challenges. *International Journal of Thermofluids* 2022, 15, 100171. https://doi.org/10.1016/j.ijft.2022.100171.

[6]. Patra, B.; Nema, P.; Khan, M. Z.; Khan, O. Optimization of solar energy using MPPT techniques and industry 4.0 modelling. *Sustainable Operations and Computers* 2023, 4, 22–28. https://doi.org/10.1016/j.susoc.2022.10.001.

[7]. Kumar, A.; Patil, P. FEA simulation based performance study of multi speed transmission gearbox. *International Journal of Manufacturing Materials and Mechanical Engineering* 2016, 6, 57–67. DOI: 10.4018/IJMMME.2016010103.
[8]. Singh, S.; Kumar, A.; Behura, S.K.; Verma, K. Challenges and Opportunities in Nanomanufacturing. In *Nanomanufacturing and Nanomaterials Design: Principles and Applications*. Singh, S., Behura, S.K., Kumar, A., Verma, K. Eds.; CRC Press: Boca Raton, FL, 2022; Chapter 2; pp. 17–30. DOI: 10.1201/9781003220602–2.
[9]. Srivastava, S.; Verma, D.; Thusoo, S.; Kumar, A.; Singh, V.P.; Kumar, R. Nanomanufacturing for Energy Conversion and Storage Devices. In *Nanomanufacturing and Nanomaterials Design: Principles and Applications*. Singh, S., Behura, S.K., Kumar, A., Verma, K. Eds.; CRC Press: Boca Raton, FL, 2022; Chapter 10; pp. 165–173. DOI: 10.1201/9781003220602–10.
[10]. Singh, V.P.; Dwivedi, A.; Karn, A.; Kumar, A.; Singh, S.; Srivastava S.; Srivastava, K. Nanomanufacturing and Design of High-Performance Piezoelectric Nanogenerator for Energy Harvesting. In *Nanomanufacturing and Nanomaterials Design: Principles and Applications*. Singh, S., Behura, S.K., Kumar, A., Verma, K. Eds.; CRC Press: Boca Raton, FL, 2022; Chapter 15; pp. 241–242. DOI: 10.1201/9781003220602–15.
[11]. Kundu, A.; Kumar, A.; Dutt, N.; Meena, C.S.; Singh, V.P. Introduction to Thermal Energy Resources and Its Smart Application. In *Thermal Energy Systems: Design, Computational Techniques and Applications*, Kumar, A., Singh, V.P., Meena, C.S., Dutt, N. Eds; CRC Press: Boca Raton, FL, 2023; Chapter 1; pp. 1–15.
[12]. Dewangan, A.K.; Moinuddin, S.Q.; Cheepu, M.; Sajjan, S.K.; Kumar, A. Thermal Energy Storage: Opportunities, Challenges and Future Scope. In *Thermal Energy Systems: Design, Computational Techniques and Applications*. Kumar, A., Singh, V.P., Meena, C.S., Dutt, N. Eds; CRC Press: Boca Raton, FL, 2023; Chapter 2; pp. 17–28.
[13]. Kundu, A.; Kumar, A.; Dutt, N.; Singh, V.P.; Meena, C.S. Modelling and Simulation of Thermal Energy System for Design Optimization. In *Thermal Energy Systems: Design, Computational Techniques and Applications*. Kumar, A., Singh, V.P., Meena, C.S., Dutt, N. Eds; CRC Press: Boca Raton, FL, 2023; Chapter 7; pp. 103–140.
[14]. Patra, B.; Nema, P. Advancements in Solar Thermal and Photovoltaic System. In *Thermal Energy Systems: Design, Computational Techniques, and Applications*. CRC Press (Taylor & Francis): Boca Raton, FL, 2023.
[15]. Patra, B. Introduction and Fundamentals of Solar Energy Collectors. In *Thermal Energy Systems: Design, Computational Techniques, and Applications*. CRC Press (Taylor & Francis): Boca Raton, FL, 2023.
[16]. Paul, A.K.; Prasad, A.; Kumar, A. Review on artificial neural network and its application in the field of engineering. *Journal of Mechanical Engineering: PRAKASH* 2022, 1 (1), 53–61. DOI: 10.56697/JMEP.2022.1107.
[17]. Sharma, V.K.; Kumar, V.; Joshi, R.S.; Kumar, A. Effect of REOs on tribological behavior of aluminum hybrid composites using ANN. In *Additive Manufacturing in Industry 4.0: Methods, Techniques, Modeling and Nano Aspects*. Sharma, V.K., Kumar, A., Gupta, M., Kumar, V., Sharma, D.K., Sharma, S.K. Eds.; CRC Press: Boca Raton, FL, 2022; Chapter 9; pp. 153–168. DOI: 10.1201/9781003360001–9
[18]. Rana, S.; Kumar, A.; Gori, Y.; Patil, P. Design and analysis of thermal contact conductance. *Journal of Critical Reviews* 2019, 6(5), 363–370.

[19]. Rana, S.; Kumar, A. FEA based design and thermal contact conductance analysis of steel and Al rough surfaces. *International Journal of Applied Engineering Research* 2019, 13 (16), 12715–12724.

[20]. Kumar, A.; Rana, S.; Gori, Y.; Sharma, N.K. Thermal Contact Conductance prediction Using FEM Based computational Techniques. In *Advanced Computational Methods in Mechanical and Materials Engineering*. Kumar, A., Gori, Y., Dutt, N., Singla, Y.K., Maurya, A. Eds.; CRC Press: Boca Raton, FL, 2022; Chapter 11; pp. 183–220. DOI: 10.1201/9781003202233-13.

[21]. Bansal, A.; Bhardwaj, H.K.; Sharma, V.K.; Kumar, A. Static and Dynamic Behavior Analysis of Al-6063 Alloy Using Modified Hopkinson Bar. In *Additive Manufacturing in Industry 4.0: Methods, Techniques, Modeling and Nano Aspects*. Sharma, V.K; Kumar, A., Gupta, M., Kumar, V., Sharma, D.K., Sharma, S.K. Eds.; CRC Press: Boca Raton, FL, 2022; Chapter 6; pp. 107–124. DOI: 10.1201/9781003360001–6.

[22]. Patil, P.; Kumar, A.; Gori, Y. FEA design and analysis of flow sensor. *Materialstoday Proceedings* 2021, 46 (20), 10563–10568. DOI: 10.1016/j.matpr.2021.01.315.

[23]. Patil, P.P.; Kumar, A. Design and FEA simulation of omega type coriolis mass flow sensor. *International Journal of Control Theory and Applications* 2017, 9 (40), 383–387.

[24]. Patil, P.; Sharma, S.C.; Paliwal, V.; Kumar, A. ANN modeling of Cu type omega vibration based mass flow sensor. *Procedia Technology* 2014, 14, 260–265. DOI: 10.1016/j.protcy.2014.08.034.

[25]. Patil, P.; Sharma, S.C.; Jaiswal, H.; Kumar, A. Modeling influence of tube material on vibration based EMMFS using ANFIS. *Procedia Material Science* 2014, 6, 1097–1103. DOI: 10.1016/j.mspro.2014.07.181

[26]. Meena, C.S.; Kumar, A.; Roy, S.; Cannavale, A.; Ghosh, A. Review on boiling heat transfer enhancement techniques. *Energies* 2022, 15, 5759. DOI: 10.3390/en15155759

Chapter 8

Lithium-Ion Battery for Electric Transportation

Types, Components, Pack Design, and Technology

Roushan Kumar

8.1 INTRODUCTION

The chapter covers a comprehensive overview of Lithium-ion batteries [1] for electric transportation, covering various aspects such as battery types, components, pack design, and technology [2, 3]. Lithium-ion batteries have gained popularity in electric vehicles due to their high energy density, low self-discharge rate, and long cycle life. The chapter covers types of lithium-ion batteries, components of a battery pack, design considerations, and the latest advances in battery technology. Types of Lithium-ion batteries include cylindrical, prismatic, and pouch cells [4, 5]. Each type has its own unique characteristics, advantages, and disadvantages. These lithium-ion batteries are extensively used in various applications, including electric transportation [6].

Components that make up a lithium-ion battery pack include individual cells, modules, cooling systems, and management systems. A battery pack for electric transportation consists of several key components that work together to store and deliver electrical energy [7]. The main components typically used in the development process of a lithium-ion battery pack are battery cells. The individual battery cells are the building blocks of the pack. These cells, such as cylindrical, prismatic, or pouch cells, contain the lithium-ion chemistry that stores and releases electrical energy [8]. The number and arrangement of cells within the pack determine its voltage and capacity. Battery modules are groups of interconnected battery cells. They provide structural integrity and facilitate the electrical connections between cells. Modules are designed to manage thermal effects and ensure balanced charging and discharging among the cells [9].

The battery management system (BMS) is a crucial component responsible for monitoring and controlling the battery pack's performance and safety [10]. It regulates the charging and discharging processes, monitors cell voltage and temperature, manages cell balancing, and provides important data for diagnostics and protection. Battery packs generate heat during operation, and a cooling system is necessary to maintain optimal operating temperatures and prevent overheating. This may include cooling fins, heat pipes, liquid cooling systems, or fans that

DOI: 10.1201/9781003464556-8

circulate air or coolant to dissipate excess heat [11]. The battery pack is enclosed within a protective housing to ensure mechanical strength and provide safety. The enclosure is typically made of materials that are resistant to impacts, vibrations, and environmental factors [12].

Electrical connectors and wiring are used to interconnect the battery cells, modules, and other components within the pack. These connections facilitate the flow of electrical current between the cells, BMS, and external systems, such as the vehicle's powertrain. Some battery packs require voltage regulation circuitry to convert the battery's output voltage to the appropriate level for the electric vehicle's electrical system. This circuitry ensures compatibility between the battery pack and the vehicle's components [13]. Battery packs often incorporate various safety features, such as thermal fuses, overcurrent protection, and voltage monitoring, to prevent potential hazards such as short circuits, overcharging, or excessive discharge. Each of these components plays a crucial role in the overall performance, safety, and longevity of the battery pack for electric transportation. Proper design, integration, and management of these components are essential for efficient and reliable operation [14].

The chapter covers important design considerations for lithium-ion battery packs, such as size, weight, thermal management, and safety. It highlights the importance of optimizing these factors to enhance the efficiency and reliability of the battery pack. The chapter explores the latest advancements in lithium-ion battery technology, including the development of solid-state electrolytes, silicon anodes, and recycling techniques. These advancements have the potential to significantly impact the future of electric transportation by improving battery performance, safety, and sustainability [15]. By providing a comprehensive guide to lithium-ion batteries for electric transportation, this chapter serves as a valuable resource for researchers, engineers, and students in the field. It covers the fundamental concepts, technological advancements, and design considerations necessary for understanding and developing effective battery systems for electric vehicles.

8.2 TYPES OF LITHIUM-ION BATTERIES

There are several types of lithium-ion batteries used in numerous applications, including consumer electronics, industrial automation, and electric transportation. Here are the three main types of lithium-ion batteries used in the modern era. It is worth noting that there are other specialized Lithium-ion battery types, such as Lithium Iron Phosphate (LiFePO4) batteries, which are known for their enhanced safety and longer cycle life but have lower energy density compared to other types [16]. Additionally, advancements in battery technology continue to introduce new types and variations to meet specific performance requirements, such as solid-state lithium-ion batteries, lithium-sulphur batteries, and lithium-air batteries.

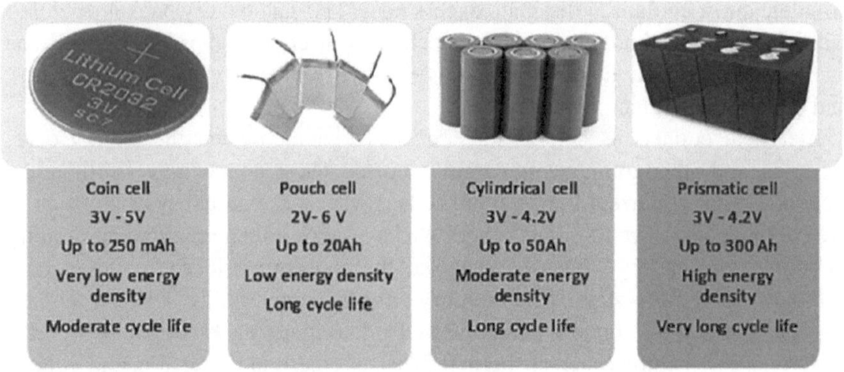

Figure 8.1 Different Types of Lithium-ion Batteries Used in Electric Vehicle Applications.

Figure 8.1 illustrates the different types of lithium-ion batteries used in electric vehicle applications, covering coin cells, pouch cells, cylindrical cells, and prismatic cells. Coin cells have low energy density and low ampere hour current rating, whereas prismatic cells have the highest energy density and high ampere-hour current rating in comparison with other types of lithium-ion battery technology.

8.2.1 A Coin Cell

A coin cell is also known as a button cell. They are a small cylindrical batteries commonly used in small electronic devices such as watches, calculators, and key fobs. Coin cells are compact and have a flat, disc-like shape resembling a coin or button, hence their name. They are often primary (non-rechargeable) batteries, although some rechargeable coin cells are also available. The key characteristics and components of a typical coin cell are electrodes, electrolytes, separators, housing, insulators, and safety features [17]. A coin cell has two electrodes—a cathode (positive terminal) and an anode (negative terminal). The cathode is typically made of a metal oxide, such as manganese dioxide, while the anode is composed of zinc or lithium. The electrolyte acts as a medium for the flow of ions between the cathode and the anode. It is usually a non-aqueous liquid or gel containing various salts. The separator is a thin, porous material that separates the cathode and the anode to prevent direct contact. It allows the flow of ions while preventing short circuits. The coin cell housing consists of a metal can, usually made of stainless steel, which serves as the outer casing of the battery. The housing provides structural support and acts as the negative terminal of the battery. An insulating gasket or ring made of a non-conductive material is used to prevent electrical contact between the positive and negative terminals of the coin cell. Some coin cells incorporate safety features such as a vent or pressure relief mechanism to prevent excessive pressure build-up or leakage in case of abuse or failure. Coin cells come in various sizes, with common types including CR2032, CR2025, and CR2016.

8.2.2 Pouch Cell

Pouch lithium-ion batteries, also known as flexible batteries, have a thin, flat, and flexible design. It is named after its pouch-like structure, which consists of layers of electrode materials and separators enclosed in a flexible, heat-sealed package. Pouch cells have gained popularity in various applications, including electric vehicles, portable electronics, and energy storage systems, due to their advantages in terms of form factor flexibility, energy density, and cost efficiency. The key characteristics and components of a pouch cell are electrodes, separators, electrolytes, pouch packaging, current collectors, tabs, and safety features. Pouch cells have two electrodes—a cathode (positive electrode) and an anode (negative electrode). The cathode is typically made of a lithium-containing metal oxide, such as lithium cobalt oxide (LiCoO2) or lithium iron phosphate (LiFePO4), while the anode is commonly made of graphite or a lithium titanate compound [18]. A thin, porous separator is placed between the cathode and the anode to prevent direct contact and short circuits while allowing the flow of lithium ions. The separator ensures the separation of the positive and negative electrodes while facilitating ion transport during charge and discharge cycles.

Pouch cells use a non-aqueous electrolyte, typically a lithium salt dissolved in an organic solvent. The electrolyte facilitates the transport of lithium ions between the cathode and the anode during battery operation. The electrode materials, separator, and electrolyte are contained within a flexible, laminated pouch made of polymer materials such as aluminium-laminated polymer or aluminium-plastic composite film. The pouch is designed to be lightweight, thin, and durable, offering advantages in terms of packaging efficiency and form factor flexibility [19]. Current collectors, usually made of aluminium or copper, are attached to the electrodes to facilitate the flow of electrical current. Tabs, also made of aluminium or copper, are typically connected to the current collectors and serve as the terminals for external electrical connections. Pouch cells often incorporate safety features such as pressure relief valves or thermal shutdown mechanisms to prevent excessive pressure or overheating in case of abuse or failure.

Pouch cells offer advantages such as higher energy density, lighter weight, and better packaging efficiency compared to other types of lithium-ion batteries. Their flexible design allows for custom shapes and sizes, enabling the efficient utilization of space in various applications. However, pouch cells may have slightly lower mechanical stability and require additional protection measures compared to more rigid battery designs. It is worth noting that while pouch cells are commonly used in electric vehicles, they are typically grouped together into modules and integrated into a larger battery pack for higher voltage and capacity requirements.

8.2.3 Cylindrical Cell

A cylindrical cell, also known as a cylindrical lithium-ion battery, is a type of rechargeable battery that features a cylindrical shape. It is one of the most common and widely used types of lithium-ion batteries due to its versatility, reliability, and widespread availability. Cylindrical cells are used in a variety of applications, including portable electronics, power tools, and electric vehicles. The key characteristics and components of a cylindrical cell are electrodes, separators, electrolytes, cylindrical cans, terminal collectors, and safety features [20]. A cylindrical cell consists of a central anode (negative electrode) and a surrounding cathode (positive electrode). The anode is typically made of graphite, while the cathode is composed of a lithium-containing metal oxide, such as lithium cobalt oxide (LiCoO2), lithium iron phosphate (LiFePO4), or lithium manganese oxide ($LiMn_2O_4$). A thin, porous separator is placed between the anode and cathode to prevent direct contact and short circuits while allowing the flow of lithium ions. The separator plays a critical role in maintaining the cell's electrochemical integrity and preventing internal short circuits. Cylindrical cells use a non-aqueous liquid electrolyte, typically a lithium salt dissolved in an organic solvent. The electrolyte facilitates the transport of lithium ions between the anode and cathode during charging and discharging cycles.

The electrode materials and electrolytes are enclosed within a cylindrical metal can, typically made of stainless steel. The can provides mechanical support, acts as the negative terminal, and protects the internal components from external impacts and damage. Cylindrical cells have metal tabs attached to the can, which serve as the positive and negative terminals for external electrical connections. These terminals enable the cell to be connected in series or parallel configurations to form battery packs [21]. Cylindrical cells often incorporate safety features such as venting mechanisms or thermal shutdown devices to prevent excessive pressure or overheating in case of abuse or failure. These safety mechanisms help protect the cell from damage and reduce the risk of hazardous events. Cylindrical cells come in standardized sizes denoted by numbers, such as 18650 or 21700, where the first two digits indicate the cell's diameter in millimetres and the last two digits represent its height. Different sizes offer varying capacities, voltage levels, and power outputs to suit different applications. They are widely used due to their reliability, availability, and compatibility with a range of electronic devices and systems. In electric vehicles, cylindrical cells are often grouped together into modules and combined into larger battery packs to meet the higher energy and power demands of the vehicle.

8.2.4 Prismatic Cell

A prismatic cell is a type of lithium-ion battery that features a rectangular or square shape. It is named after its prismatic or box-like structure, which distinguishes it from other battery cell shapes, such as cylindrical or pouch cells. Prismatic cells offer advantages in terms of space utilization, packing efficiency, and mechanical stability. They are commonly used in applications that require higher energy density and a more compact form factor. The key characteristics and components of a prismatic cell are electrodes, separators, electrolytes, prismatic enclosures, and safety features [22]. Prismatic cells have two electrodes—an anode (negative electrode) and a cathode (positive electrode). The anode is typically made of graphite, while the cathode is composed of a lithium-containing metal oxide, such as lithium cobalt oxide (LiCoO2), lithium iron phosphate (LiFePO4), or lithium nickel cobalt aluminium oxide (NCA). A thin, porous separator is placed between the anode and cathode to prevent direct contact and short circuits while allowing the flow of lithium ions. The separator is essential for maintaining the cell's electrochemical integrity and ensuring proper ion transport.

Prismatic cells use a non-aqueous liquid electrolyte, typically a lithium salt dissolved in an organic solvent. The electrolyte facilitates the movement of lithium ions between the anode and cathode during the battery's charge and discharge cycles. The electrode materials, separator, and electrolytes are enclosed within a rectangular or square metal or plastic casing. The casing provides structural support, protection, and insulation for the cell's internal components. Prismatic cells have metal tabs or leads attached to the casing, serving as the positive and negative terminals for external electrical connections. These terminals enable the cell to be connected in series or parallel configurations to form battery modules or packs [23]. Prismatic cells often incorporate safety features such as pressure relief mechanisms or thermal shutdown devices to prevent excessive pressure or overheating in case of abuse or failure. These safety mechanisms help protect the cell from damage and mitigate potential hazards.

Prismatic cells offer advantages such as higher energy density, improved packaging efficiency, and enhanced mechanical stability compared to cylindrical cells. Their flat shape allows for efficient space utilization, making them suitable for applications that require compact and high-capacity battery packs, such as electric vehicles and energy storage systems. It is important to note that prismatic cells are often grouped into modules and integrated into larger battery packs for higher voltage and capacity requirements in electric vehicles and other high-demand applications.

8.3 COMPONENTS OF A BATTERY PACK

A battery pack for electric transportation consists of several key components that work together to store and deliver electrical energy. The main components typically required in the development process of a battery pack are battery cells, battery modules, a battery management system (BMS), cooling systems, enclosures, connection diagrams, wiring, voltage regulation circuits, and safety features [24]. The individual battery cells are the building blocks of the pack. These cells, which can be cylindrical, prismatic, or pouch cells, contain the lithium-ion chemistry that stores and releases electrical energy. The number and arrangement of cells within the pack determine its voltage and capacity. Battery modules are groups of interconnected battery cells. They provide structural integrity and facilitate the electrical connections between cells. Modules are designed to manage thermal effects and ensure balanced charging and discharging among the cells.

8.3.1 Battery Management System

The BMS is a crucial component responsible for monitoring and controlling the battery pack's performance and safety. It regulates the charging and discharging processes, monitors cell voltage and temperature, manages cell balancing, and provides important data for diagnostics and protection. The BMS ensures the optimal operation and longevity of the battery pack. The key functions and features of a typical BMS are cell monitoring, state of charge (SOC) calculation, state of health (SOH) estimation, remaining useful life (RUL) calculation, providing safety and protection, cell balancing, charging control, communication, data logging, thermal management, and fault prognostics and diagnostic [25]. The BMS continuously monitors the voltage, current, and temperature of each individual battery cell within the pack. This information helps ensure that cells are operating within safe limits and allows for cell balancing, which equalizes the voltage levels of cells to maintain optimal performance and longevity. The BMS estimates the remaining capacity or state of charge of the battery pack based on voltage, current, and temperature measurements. Accurate SOC estimation is crucial for determining the available energy and predicting the battery's range and runtime.

The BMS assesses the state of health of the battery pack by analyzing various factors such as cell degradation, internal resistance, and capacity loss over time. SOH monitoring helps determine the battery's overall health, performance degradation, and remaining lifespan. The BMS incorporates safety features to prevent potentially hazardous conditions. It monitors for abnormalities such as overvoltage, under voltage, overcurrent, and over temperature, triggering protection mechanisms like cell balancing, current limiting, or thermal shutdown to safeguard the battery pack from damage or safety risks [26]. To ensure the cells within the pack have consistent voltage levels, the BMS performs cell balancing. It redistributes charge among cells during charging or discharging to equalize their states and prevent overcharging or over-discharging of individual cells. The BMS also manages the charging process to ensure safe and efficient charging. It controls the charging current, voltage limits, and charging profiles based on the battery manufacturer's specifications and charging algorithms. The BMS also monitors temperature during charging to prevent overheating. Figure 8.2 illustrates the block diagram of battery management system roles and responsivity in electric vehicle applications, covering data collection, normalization, classification, analysis, training, testing, validation, control, diagnostics, and prognostics.

The BMS provides communication interfaces, such as controller area network (CAN) bus or universal asynchronous receiver transmitter (UART), to exchange data with the vehicle's main control unit or external systems. It logs and stores

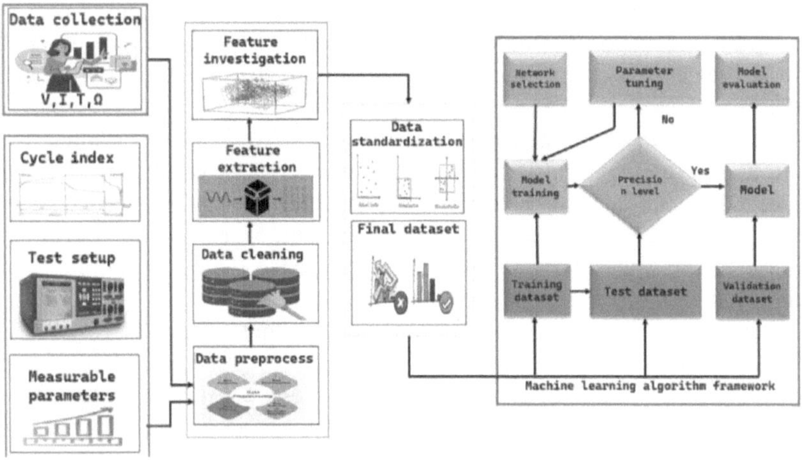

Figure 8.2 Block Diagram of Battery Management System Roles and Responsivity in Electric Vehicle Applications.

relevant battery performance data, fault codes, and diagnostic information for analysis, troubleshooting, and future optimizations. Thermal management in the BMS monitors and manages the temperature of the battery cells through temperature sensors integrated into the pack [27]. It activates cooling or heating systems, if present, to maintain the cells within the optimal temperature range and prevent thermal runaway or degradation. Fault diagnosis and reporting in the BMS detects and diagnoses faults or anomalies in the battery pack and generates error codes or alerts to notify the user or system operator. It provides real-time status information, including battery voltage, temperature, and alarms, allowing for timely intervention and preventive maintenance. The BMS is a critical component that ensures the safe and efficient operation of a lithium-ion battery pack. It helps optimize battery performance, prolong battery life, and protect against potential hazards, making it an essential part of electric transportation systems.

8.3.2 Cooling System

Battery packs generate heat during operation, and a cooling system is necessary to maintain optimal operating temperatures and prevent overheating. This may include cooling fins, heat pipes, liquid cooling systems, or fans that circulate air or coolant to dissipate excess heat. Efficient thermal management is essential for maintaining battery performance and extending its lifespan. The battery pack is enclosed within a protective housing to ensure mechanical strength and provide safety. The enclosure is typically made of materials resistant to impacts, vibrations, and environmental factors [28]. It also helps protect the battery cells and other components from external damage. Electrical connectors and wiring are used to interconnect the battery cells, modules, and other components within the pack. These connections facilitate the flow of electrical current between the cells, BMS, and external systems, such as the vehicle's powertrain. Proper wiring and connectors ensure reliable and efficient operation of the battery pack. Some battery packs require voltage regulation circuitry to convert the battery's output voltage to the appropriate level for the electric vehicle's electrical system [29]. This circuitry ensures compatibility between the battery pack and the vehicle's components. Figure 8.3 illustrates the schematic diagram of the cooling system used in an electric vehicle battery lithium-ion battery pack, covering air cooling, liquid cooling, and heat pipes approaches.

Battery packs often incorporate various safety features, such as thermal fuses, overcurrent protection, and voltage monitoring, to prevent potential hazards such as short circuits, overcharging, or excessive discharge. These safety features help ensure the safe operation of the battery pack and protect against accidents or damage. Battery packs may include monitoring and communication systems to provide real-time data on battery performance, health, and status. This information can be used for diagnostics, maintenance, and optimizing battery operation. Each of these components plays a crucial role in the overall performance, safety, and longevity of the battery pack for electric transportation. Proper design, integration, and management of these components are essential for efficient and reliable operation.

8.4 DESIGN CONSIDERATIONS

Designing a lithium-ion battery pack for electric transportation involves considering several important factors to ensure optimal performance, safety, and

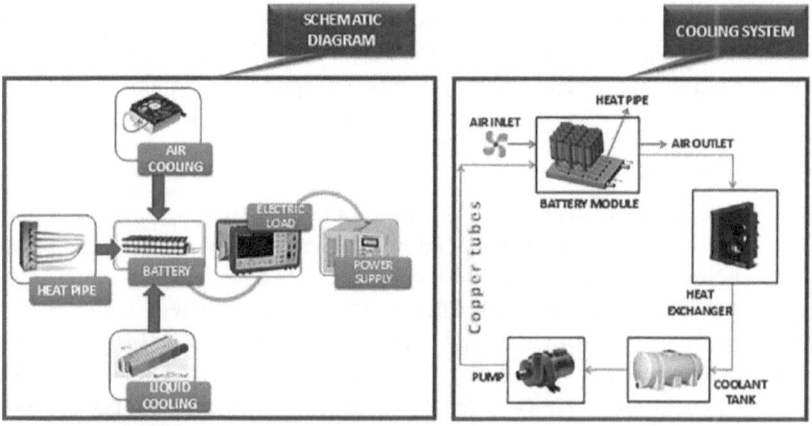

Figure 8.3 Schematic Diagram of Cooling System Used in Electric Vehicle Lithium-ion Battery Pack.

efficiency. The key design considerations for a lithium-ion battery pack are energy and power requirement, voltage and cell configuration, pack size and weight, thermal management, mechanical integrity, safety, cell balancing and monitoring, charging infrastructure compatibility, serviceability, maintenance, and environmental consideration [30]. Determining the energy capacity and power output required for specific electric vehicle applications, such as four-wheel or two-wheel, requires considering many factors, such as the desired driving range for electric vehicles or the power demands of the target device. This information helps determine the number and arrangement of battery cells needed in the pack. Selecting the appropriate voltage level for the battery pack based on the application's requirements, and this involves determining the series or parallel configuration of the battery cells to achieve the desired voltage and capacity [31].

Pack size and weight for different applications requires a balance between energy density and physical size. Considering space limitations, weight distribution, and packaging efficiency that optimize the pack size and weight can help maximize the vehicle's range and performance. The implementation of an effective thermal management system to control the temperature of the battery pack because of excessive heat can negatively affect battery performance, lifespan, and safety. Consider using cooling fins, heat pipes, liquid cooling, hybrid cooling, or other cooling methods to dissipate heat and maintain the cells within the recommended temperature range [32]. To design a robust and structurally sound battery pack enclosure to protect the cells from physical damage and ensure safety. Consider factors such as vibration, impact resistance, and environmental conditions, which include safety features like thermal fuses, overcurrent protection, and insulation to mitigate potential risks.

8.4.1 Cell Balancing and Monitoring

Cell balancing and monitoring implementation in a battery management system to monitor and balance the individual cells' voltage levels within the pack. The BMS helps prevent overcharging, over-discharging, and cell imbalance. It also provides important data for diagnostics, protection, and maintaining the pack's overall performance. The key functions performed for cell balancing and monitoring are individual cell balancing using active or passive methods. Cell balancing aims to equalize the voltage levels of individual battery cells within a pack. Due to variations in cell characteristics and aging, cells may have different voltage levels, which can lead to performance disparities and reduced overall pack capacity. The balancing methods performed by the BMS employ various balancing methods to redistribute charge among cells [33]. Common techniques include passive balancing and active balancing. Passive balancing uses resistors to bleed excess charge from higher voltage cells. It is a simple and cost-effective method but is less efficient for large voltage differences. Active balancing actively transfers charge between cells using additional circuitry and switches. It is more effective than passive balancing and can handle larger voltage differences. Active balancing can be performed during charging or discharging, depending on the BMS design.

Cell balancing control in the BMS monitors the voltage of each cell and determines the need for balancing. When a cell's voltage exceeds a predefined threshold, the BMS activates the balancing circuitry to initiate the balancing process. The BMS controls the balancing duration and frequency based on the pack's requirements and balancing algorithm. Cell monitoring involves continuous monitoring of the voltage, current, and temperature of each individual battery cell within the pack. Monitoring provides essential information about cell health, performance, and safety. Voltage monitoring in the BMS measures the voltage of each cell to ensure it operates within safe limits. Abnormal voltage levels can indicate cell overcharging, undercharging, or cell degradation. Current monitoring allows the BMS to track the charging and discharging currents of each cell. It helps prevent overcurrent conditions, monitor the state of charge, and estimate cell capacity. Temperature sensors are placed in or near the cells to monitor their temperature. Accurate temperature monitoring is crucial for maintaining safe operating conditions and preventing overheating or thermal runaway [34].

Data logging in the BMS logs and stores data related to cell voltages, currents, temperatures, and other parameters. Data logging enables performance analysis, fault diagnosis, and facilitates predictive maintenance and optimization of the battery pack. Diagnostic alerts in the BMS analyze the monitored data to detect anomalies or potential issues within the cells [35]. It generates alerts, error codes, or alarms to notify the user or system operator of any abnormalities. This allows for timely intervention and preventive actions. Cell balancing and monitoring are integral to maintaining the performance and safety of a lithium-ion battery pack. By equalizing cell voltages and monitoring their vital parameters, the BMS ensures that the pack operates efficiently, prolongs battery life, and minimizes the risk of cell failures or safety incidents.

8.4.2 Charging Infrastructure

Charging infrastructure refers to the network of charging stations and associated equipment that enables the recharging of electric vehicles. It plays a crucial role in supporting the widespread adoption and usability of electric transportation.

8.4.2.1 Charging Station Types

Three type of charging station exists for electric transportation system: Level 1, Level 2, and Level 3, also known as DC fast charging system [36]. Level 1 charging is the slowest charging option, typically using a standard household outlet (120V AC). Level 1 chargers have lower power output and are suitable for overnight charging at home or in locations where longer charging times are acceptable. Level 2 chargers operate at higher power (usually 240V AC) and provide faster charging compared to Level 1. They are commonly found in residential, workplace, and public charging stations. Level 2 chargers can fully charge an EV in a few hours. Level 3, also known as the DC fast charging system, has fast chargers that deliver high-power DC electricity directly to the EV's battery, enabling rapid charging. DCFC can charge an EV to 80% or more in 30 minutes or less, making it suitable for long-distance travel and high-traffic areas.

Homeowners can install Level 1 or Level 2 chargers in their garages or driveways for convenient overnight charging. Residential charging provides EV owners with the flexibility of having a fully charged vehicle each morning. Employers can install charging stations at workplaces to support employee EV charging during the day. Workplace charging encourages EV adoption and provides a convenient charging option for those who park at work for extended periods [37]. Public charging stations are available in various locations, such as shopping centres, parking garages, restaurants, and along highways. These stations offer Level 2 and DC fast charging options, allowing EV drivers to charge while running errands, travelling, or during longer stops [38].

8.4.2.2 Charging Network and Payment

Charging networks are service providers that operate and maintain a network of charging stations. They offer access to a large number of charging stations through membership or pay-as-you-go options, making it convenient for EV owners to locate and use charging stations across different locations. Charging stations may have different payment methods, including credit card readers, mobile apps, RFID cards, or payment through a charging network's membership. Standardization and interoperability of payment systems are being addressed to enhance user convenience.

8.4.2.3 Smart Charging and Grid Integration

Smart charging solutions enable the optimization of charging based on factors like electricity demand, grid load, and renewable energy availability. This helps manage electricity demand during peak hours and maximizes the use of renewable energy sources. Charging infrastructure can be integrated with the electrical grid to provide bidirectional power flow, allowing EVs to act as energy storage devices [39]. This Vehicle-to-Grid (V2G) integration can support grid stability, enable energy arbitrage, and provide backup power during grid outages. Fast-charging corridors are networks of DC fast charging stations strategically located along major highways and travel routes. These corridors enable long-distance travel for EVs, reducing range anxiety and supporting intercity and cross-country trips. The development and expansion of a robust and accessible charging infrastructure are vital for the widespread adoption of electric vehicles. It provides convenience, addresses range-anxiety concerns, and facilitates a smooth transition to sustainable transportation. Governments, private entities, and charging network operators are actively working to expand and enhance charging infrastructure to meet the growing demands of EV users [40, 41].

Design the battery pack with serviceability and maintenance in mind, considering different factors such as ease of access for maintenance, cell replacement, module repairs, etc [42]. Modular designs can facilitate easier maintenance and replacement of faulty components, minimizing downtime. Ensure compliance with environmental regulations and standards for the safe handling, disposal, and recycling of lithium-ion batteries. Consider incorporating sustainable and recyclable materials where possible and explore opportunities for end-of-life battery recycling [43, 44]. With careful design consideration, engineers can develop lithium-ion battery packs that meet the specific requirements of electric transportation applications while prioritizing safety, performance, and sustainability.

8.5. CONCLUSION AND FUTURE WORK

Electric vehicle battery technology is constantly evolving to enhance performance, range, charging speed, and overall efficiency. The latest advancements in EV battery technology cover battery with increased energy density, fast charging and discharging capability, solid-state batteries, longer cycle life batteries, sustainable material technology, advanced thermal management, and vehicle-to-smart grid integration.

a) Battery manufacturers are focusing on improving energy density, which refers to the amount of energy that can be stored in a given volume or weight of the battery. Higher energy density allows for longer driving ranges without the need for frequent recharging. Advancements in electrode materials, cell designs, and manufacturing processes contribute to increased energy density in EV batteries.
b) Fast charging is a key factor in making EVs more convenient and practical for long-distance travel. Advanced battery technologies, such as silicon anodes, solid-state electrolytes, and improved thermal management systems, are being developed to enable faster charging speeds. Ultra-fast charging stations capable of delivering high power to EVs are also being deployed to support rapid charging.
c) Solid-state batteries, which replace the liquid electrolyte in traditional lithium-ion batteries with a solid electrolyte, are gaining attention for their potential to offer higher energy density, improved safety, and faster charging. Solid-state batteries are also less prone to thermal runaway and can operate in a wider temperature range. Researchers and manufacturers are working to overcome technical challenges to commercialize solid-state battery technology for EVs.

d) Battery longevity is a crucial aspect of EVs, and advancements in electrode materials and cell chemistry, along with improved battery management systems, are extending the cycle life of EV batteries. This means that the batteries can endure more charge and discharge cycles before experiencing a significant decrease in capacity.
e) Battery manufacturers are focusing on using sustainable and responsibly sourced materials to reduce environmental impact. Efforts are underway to decrease the reliance on rare and expensive materials, such as cobalt, and explore alternative materials like nickel-rich cathodes and solid-state electrolytes. Recycling initiatives for EV batteries are also gaining traction to recover valuable materials and minimize waste.
f) Efficient thermal management is essential for maintaining optimal battery performance and extending its lifespan. Advancements in thermal management systems, such as liquid cooling and active cooling methods, are being implemented to regulate the temperature of EV batteries, ensuring safe and efficient operation even in extreme conditions.
g) Vehicle-to-Grid (V2G) technology allows bidirectional energy flow between EV batteries and the electrical grid. This technology enables EVs to not only consume electricity but also serve as energy storage devices, providing power back to the grid during peak demand periods. V2G integration holds the potential to enhance grid stability, enable renewable energy integration, and create new revenue streams for EV owners.

It is important to note that while these advancements are promising, they may be at different stages of development, and their commercial availability and widespread adoption may vary. However, the ongoing research and development in EV battery technology continue to drive improvements in performance, range, and charging speed, making electric vehicles increasingly practical and appealing to a broader range of consumers. Future work in lithium-ion battery technology for electric transportation focuses on advancing energy density, fast charging capabilities, safety, longevity, sustainability, and integration with renewable energy sources and the grid. These efforts aim to further improve the performance, usability, and environmental impact of electric vehicles, making them more competitive with conventional vehicles and accelerating the transition to sustainable transportation.

REFERENCES

[1] L. Gaines, "The future of automotive lithium-ion battery recycling: Charting a sustainable course," *Sustain. Mater. Technol.*, vol. 1, pp. 2–7, Dec. 2014, doi: 10.1016/j.susmat.2014.10.001.
[2] L. Driscoll, S. de la Torre, and J. A. Gomez-Ruiz, "Feature-based lithium-ion battery state of health estimation with artificial neural networks," *J. Energy Storage*, vol. 50, March 2022, doi: 10.1016/j.est.2022.104584.

[3] R. Kumar, R. K. Pachauri, P. Badoni, D. Bharadwaj, U. Mittal, and A. Bisht, "Investigation on parallel hybrid electric bicycle along with issuer management system for mountainous region," *J. Clean. Prod.*, vol. 362, p. 132430, 2022, doi: 10.1016/j.jclepro.2022.132430.

[4] A. G. Siani, M. Mousavi Badjani, H. Rismani, and M. Saeedimoghadam, "State of health estimation of lithium-ion batteries based on the CC-CV charging curve and neural network," *IETE J. Res.*, 2021, doi: 10.1080/03772063.2021.1908853.

[5] R. Kumar, A. Kumar, M. K. Gupta, J. Yadav, and A. Jain, "Solar tree-based water pumping for assured irrigation in sustainable Indian agriculture environment," *Sustain. Prod. Consum.*, vol. 33, pp. 15–27, 2022, doi: 10.1016/j.spc.2022.06.013.

[6] R. Kumar, N. Gupta, D. Bharadwaj, D. Dutt, and A. Joshi, "Design and development of electronic clutch control unit for manual transmission," *Mater. Today Proc.*, 2022, doi: 10.1016/j.matpr.2022.08.470.

[7] R. Kumar, K. Bansal, A. Kumar, J. Yadav, M. K. Gupta, and V. K. Singh, "Renewable energy adoption: Design, development, and assessment of solar tree for the mountainous region," *Int. J. Energy Res.*, pp. 1–17, 2021, doi: 10.1002/er.7197.

[8] A. Kumar, R. Kumar, A. Jain, A. Pandey, S. Thapliyal, and A. Sharma, *Automatic street lighting system*, vol. 624. Springer, 2018.

[9] K. Das, R. Kumar, and A. Krishna, "Supervised learning and data intensive methods for the prediction of capacity fade of lithium-ion batteries under diverse operating and environmental conditions," *Water Energy Int.*, vol. 66, no. 1, pp. 53–59, 2023.

[10] R. Kumar, M. Kumar Gupta, A. Kumar, P. Sharma, and R. Deorari, "Analysis of electronic clutch control unit for manual transmission vehicle oriented toward safety," *Mater. Today Proc.*, 2022, doi: 10.1016/j.matpr.2022.08.473.

[11] R. Kumar and A. Kumar, "Design and hardware development of power window control mechanism using microcontroller," *2013 International Conference on Signal Processing and Communication (ICSC)*, pp. 361–365, 2013, doi: 10.1109/ICSPCom.2013.6719813.

[12] L. Ungurean, M. V. Micea, and G. Cârstoiu, "Online state of health prediction method for lithium-ion batteries, based on gated recurrent unit neural networks," *Int. J. Energy Res.*, vol. 44, no. 8, pp. 6767–6777, 2020, doi: 10.1002/er.5413.

[13] R. Kumar, Divyanshu, and A. Kumar, "Nature based self-learning mechanism and simulation of automatic control smart hybrid antilock braking system," *Wirel. Pers. Commun.*, vol. 116, no. 4, pp. 3291–3308, 2021, doi: 10.1007/s11277-020-07853-7.

[14] A. Pal, R. Kumar, and V. R. Sanal Kumar, "Conceptual design of an automatic fluid level controller for aerospace applications," *2015 International Conference on Soft-Computing and Networks Security (ICSNS)*, 2015, doi: 10.1109/ICSNS.2015.7292433.

[15] L. Gaines, "Lithium-ion battery recycling processes: Research towards a sustainable course," *Sustain. Mater. Technol.*, vol. 17, p. e00068, 2018, doi: 10.1016/J.SUSMAT.2018.E00068.

[16] R. Zhan and L. Pan, "A cycling-insensitive recycling method for producing lithium transition metal oxide from Li-ion batteries using centrifugal gravity separation," *Sustain. Mater. Technol.*, vol. 32, 2022, doi: 10.1016/j.susmat.2022.e00399.

[17] J. Yadav, D. Maity, S. K. Kurre, A. Kumar, R. Kumar, and S. Chouhan, "Identification of damping prohibiting stick-slip form friction influenced dynamic system," *Proc. Inst. Mech. Eng. Part C J. Mech. Eng. Sci.*, pp. 1–8, 2022, doi: 10.1177/09544062221135517.

[18] J. Tian, R. Xiong, and W. Shen, "A review on state of health estimation for lithium ion batteries in photovoltaic systems," *eTransportation*, vol. 2, p. 100028, 2019, doi: 10.1016/j.etran.2019.100028.

[19] R. Kumar, N. J. Ahuja, M. Saxena, and A. Kumar, "Automotive power window communication with DTC algorithm and hardware-in-the loop testing," *Wirel. Pers. Commun.*, vol. 114, no. 4, pp. 3351–3366, 2020.

[20] J. Bi, J. C. Lee, and H. Liu, "Performance comparison of long short-term memory and a temporal convolutional network for state of health estimation of a lithium-ion battery using its charging characteristics," *Energies*, vol. 15, no. 7, 2022, doi: 10.3390/en15072448.

[21] M. K. Gupta, R. Kumar, V. Verma, and A. Sharma, "Robust control based stability analysis and trajectory tracking of triple link robot manipulator," *J. Eur. Syst. Autom.*, vol. 54, pp. 641–647, 2021.

[22] B. Gou, Y. Xu, and X. Feng, "State-of-health estimation and remaining-useful-life prediction for lithium-ion battery using a hybrid data-driven method," *IEEE Trans. Veh. Technol.*, vol. 69, no. 10, pp. 10854–10867, 2020, doi: 10.1109/TVT.2020.3014932.

[23] A. Kumar, K. Bansal, D. Kumar, A. Devrari, R. Kumar, and P. Mani, "FPGA application for wireless monitoring in power plant," *Nucl. Eng. Technol.*, vol. 53, no. 4, pp. 1167–1175, 2020, doi: 10.1016/j.net.2020.09.003.

[24] Z. Chen, H. Zhao, Y. Zhang, S. Shen, J. Shen, and Y. Liu, "State of health estimation for lithium-ion batteries based on temperature prediction and gated recurrent unit neural network," *J. Power Sources*, vol. 521, 2022, doi: 10.1016/j.jpowsour.2021.230892.

[25] A. S. Das, P. K. Dwivedi, A. K. Mondal, R. Kumar, R. M. Reddy, and A. Kumar, "Storage optimization of automated storage and retrieval systems using breadth-first search algorithm," 2017, doi: 10.1007/978-981-10-2999-8_18.

[26] H. Sharma, R. Kumar, A. Kumar, M. K. Gupta, and V. Kaundal, "Implementation of HMI using VISUAL BASIC. NET for induction hardening process," 2020, doi: 10.1007/978-981-13-8618-3_96.

[27] Y. Chen, Y. He, Z. Li, L. Chen, and C. Zhang, "Remaining useful life prediction and state of health diagnosis of lithium-ion battery based on second-order central difference particle filter," *IEEE Access*, vol. 8, pp. 37305–37313, 2020, doi: 10.1109/ACCESS.2020.2974401.

[28] Y. Che, X. Hu, X. Lin, J. Guo, and R. Teodorescu, "Health prognostics for lithium-ion batteries: mechanisms, methods, and prospects," *Energy Environ. Sci.*, vol. 16, no. 2, pp. 338–371, 2023, doi: 10.1039/d2ee03019e.

[29] Z. Yu, "SOH estimation method for lithium-ion battery based on discharge characteristics," *Int. J. Electrochem. Sci.*, vol. 17, p. ArticleID:220725, 2022, doi: 10.20964/2022.07.38.

[30] S. Khaleghi, D. Karimi, S. H. Beheshti, M. S. Hosen, H. Behi, M. Berecibar, and J. Van Mierlo, "Online health diagnosis of lithium-ion batteries based on nonlinear autoregressive neural network," *Appl. Energy*, vol. 282, p. 116159, 2021.

[31] A. Kumar, P. Sharma, M. K. Gupta, and R. Kumar, "Machine learning based resource utilization and pre-estimation for network on chip (NoC) communication," *Wirel. Pers. Commun.*, vol. 102, no. 3, pp. 2211–2231, 2018, doi: 10.1007/s11277-018-5376-3.

[32] R. Kumar, P. K. Dwivedi, D. Praveen Reddy, and A. S. Das, "Design and implementation of hydraulic motor based elevator system," 2015, doi: 10.1109/IICPE.2014.7115821.

[33] Q. Li, D. Li, K. Zhao, L. Wang, and K. Wang, "State of health estimation of lithium-ion battery based on improved ant lion optimization and support vector regression," *J. Energy Storage*, vol. 50, p. 104215, 2022, doi: 10.1016/j.est.2022.104215.

[34] H. Tian and P. Qin, "State of health prediction for lithium-ion batteries with a novel online sequential extreme learning machine method," *Int. J. Energy Res.*, vol. 45, no. 2, pp. 2383–2397, 2021, doi: 10.1002/er.5934.

[35] K. Rawat, D. Kumar, R. Kumar, D. Bharadwaj, V. Soti, G. Rathore, D. Biswas, and P. Singh, "Structural optimisation of go kart chassis with basic electronic driver assistance systems," *In 2023 2nd international conference on paradigm shifts in communications embedded systems*, Machine Learning and Signal Processing (PCEMS), pp. 1–6, 2023. IEEE.

[36] C. S. Meera, S. Sunny, R. Singh, P. S. Sairam, R. Kumar, and J. Emannuel, "Automated precise liquid transferring system," *2014 IEEE 6th India International Conference on Power Electronics (IICPE)*, 2015, doi: 10.1109/IICPE.2014.7115831.

[37] J. Yadav, S. K. Kurre, A. Kumar, and R. Kumar, "Nonlinear dynamics of controlled release mechanism under boundary friction," *Results Eng.*, vol. 11, p. 100265, 2021, doi: 10.1016/j.rineng.2021.100265.

[38] R. Kumar, N. J. Ahuja, and M. Saxena, "Improvement and approval of impediment recognition and activity for power window," in *Intelligent communication, control and devices*, Springer, 2018, pp. 855–864.

[39] R. Kumar, N. J. Ahuja, M. Saxena, and A. Kumar, "Modelling and simulation of object detection in automotive power window," *Indian J. Sci. Technol.*, vol. 9, no. 43, 2016, doi: 10.17485/ijst/2016/v9i43/104393.

[40] A. Jain, A. Kumar, A. P. Shukla, H. Alshazly, H. Elmannai, A. D. Algarni, R. Kumar, and J. Yadav, "Smart communication using 2D and 3D mesh network-on-chip," *Intell. Autom. Soft Comput.*, vol. 34, no. 3, 2022, pp. 2007–2021. https://doi.org/10.32604/iasc.2022.024770.

[41] K. Das, R. Kumar, and A. Krishna, "Analyzing electric vehicle battery health performance using supervised machine learning," *Renew. Sust. Energy Rev.*, vol. 189, p. 113967, 2024.

[42] K. Das and R. Kumar, "Electric vehicle battery capacity degradation and health estimation using machine-learning techniques: A review," *Clean Energy*, vol. 7, no. 6, pp. 1268–1281, 2023.

[43] K. S. Rawat and R. Kumar, "Insight into the novel low cost green air pollution monitoring and control systems: A technological solution from concept to market," *J. Clean. Prod.*, vol. 427, p. 139218, 2023. https://doi.org/10.1016/j.jclepro.2023.139218

[44] K. Das and R. Kumar, "Assessment of electric two-wheeler ecosystem using novel pareto optimality and TOPSIS methods for an ideal design solution," *World Electr. Veh. J.*, vol. 14, no. 8, p. 215, 2023.

Chapter 9

Intelligent Transport System in Smart Cities

Aspects and Challenges of Vehicular Networks and Cloud Computing

Dr. Shaik Khaja Mohiddin, Shaik Sharmila, Vadali Srinivas

9.1 INTRODUCTION

In order to create a smart city with efficient and sustainable transportation networks, Singapore is widely recognized as a pioneer in the use of Intelligent Transport Systems (ITS). The city-state has made major strides in traffic management, safety, and general mobility as a result of its innovative use of vehicular networks and cloud computing to increase its ITS capabilities. The Land Transport Authority (LTA) of Singapore has introduced the Intelligent Transport Systems Program, a complete ITS framework. Transportation operations are optimized, and the passenger experience is enhanced because of the program's use of cutting-edge technologies, communication systems, and data analytics.

Communication between vehicles, infrastructure, and traffic control systems occur in real-time because of Singapore's advanced vehicular network architecture. Using Dedicated Short-Range Communications (DSRC), the fifth generation of the Intelligent Transportation System (ITS) enables vehicles to communicate with one another and with roadside infrastructure [1]. The improved traffic management and safety that result from the use of this network's cooperative maneuvering, early warning systems, and dissemination of real-time traffic information. Singapore has also made the move to the cloud to aid with its ITS operations. The cloud's scalability, storage, and processing power are ideal for managing the massive amounts of data produced by the ITS ecosystem. Cloud-based collection and analysis of real-time data from a variety of traffic sensors, GPS devices, and security cameras yield actionable insights for traffic management and decision-making [2].

9.2 IMPORTANCE OF ITS IN SMART CITIES

Smart cities use data and technology to improve living conditions for locals while also fostering environmental responsibility and optimizing scarce municipal resources. Intelligent transport systems are crucial in accomplishing these goals

DOI: 10.1201/9781003464556-9

because they offer creative approaches to old problems in urban transportation. The incorporation of ITS into the infrastructure of smart cities has many potential advantages.

1. ***Improved Traffic Management:*** One benefit of ITS is better traffic management because authorities can use real-time data to adjust traffic lights and other controls on the fly. The result is improved traffic efficiency, less congestion, and shorter commute times.
2. ***Enhanced Safety:*** Another benefit of ITS is that it improves road safety through the use of technologies like intelligent traffic signal systems, collision warning systems, and vehicle-to-vehicle communication. Overall, road safety can be greatly enhanced through the early detection of dangerous circumstances and the prompt delivery of warnings.
3. ***Efficient Resource Utilization***: Fuel consumption, air pollution, and greenhouse gas emissions are all decreased thanks to ITS's ability to streamline traffic flow. It encourages people to take advantage of public transportation and bicycles instead of driving their own cars.
4. ***Improved Mobility***: Enhanced mobility is achieved because ITS users have access to up-to-the-minute data on traffic conditions, alternate routes, and transportation options. This enables people to make educated judgments, pick the best routes, and organize their travels with ease.
5. ***Integrated Transportation Systems***: ITS makes it possible to combine non-motorized and motorized modes of travel with those of private automobiles and public transportation. Facilitating multimodal travel and encouraging sustainable transportation options is the seamless communication and interchange across these modes.

9.3 OVERVIEW OF VEHICULAR NETWORKS AND CLOUD COMPUTING FOR ITS

Vehicular Networks: Vehicular ad hoc networks (VANETs) are the backbone of ITS, allowing vehicles and infrastructure to communicate with one another [3]. These networks facilitate communication between vehicles, infrastructure, and pedestrians by employing wireless technologies, including Dedicated Short Range Communication (DSRC) and Cellular Vehicle-to-Everything (C-V2X). In order to improve situational awareness and traffic management, VANETs allow for the instantaneous transmission of traffic data, safety alerts, and coordinated maneuvers.

Cloud Computing: When it comes to providing the necessary infrastructure and services for ITS, cloud computing plays a crucial role [4]. Data storage, processing, and analysis can all benefit from the cloud's scalability and adaptability. It allows for sophisticated analytics and decision-making algorithms to be run in real-time on data gathered in real-time from sources including sensors, GPS devices, and traffic cameras. Integrating cloud-based ITS platforms with energy management and public safety systems, among other examples of smart city applications, can yield synergistic benefits.

Numerous ITS uses are made possible by the merging vehicle networks with cloud computing.

1. ***Traffic Management Systems:*** Real-time traffic monitoring, congestion detection, and adaptive traffic signal control systems can be implemented to optimize traffic flow and reduce travel times.
2. ***Intelligent Parking Solutions***: Cloud-based parking systems can provide real-time information on available parking spaces, allowing drivers to locate and reserve parking spots in advance.
3. ***Connected and Autonomous Vehicles (CAVs):*** Vehicular networks enable communication between CAVs, infrastructure, and other road users, facilitating cooperative maneuvers, platooning, and safer autonomous driving [5].
4. ***Emergency Management:*** ITS can support emergency response systems by providing real-time traffic information, rerouting capabilities, and coordination between emergency services.
5. ***Mobility-as-a-Service (MaaS):*** Cloud-based platforms can integrate various transportation modes, payment systems, and user interfaces to provide seamless mobility solutions and promote the use of shared transportation [6].

9.4 VEHICULAR NETWORKS FOR ITS

Intelligent Transport Systems (ITS) heavily rely on vehicular networks to enable seamless communication and information exchange between vehicles, infrastructure, and other entities within the transportation ecosystem. Vehicular networks facilitate the flow of real-time data, support cooperative applications, and enhance the overall efficiency, safety, and reliability of transportation systems. Let's delve into the different types of vehicular networks, communication protocols, and the challenges they encounter.

9.4.1 Types of Vehicular Networks

9.4.1.1 Vehicular Ad Hoc Networks (VANETs)

VANETs are a subset of MANETs that have been optimized for use in vehicles. In order to share data about traffic patterns, road risks, and other pertinent information, they use V2V and V2I communication, respectively. VANETs provide collaborative applications, including traffic control, emergency services, and collision avoidance systems. They function using cellular-based technologies such as the IEEE 802.11p standard or dedicated short-range communication (DSRC) [7].

9.4.1.2 *Vehicle-to-Everything (V2X) Networks*

Communication between vehicles, infrastructure, pedestrians, and other networks are all included in the greater scope of V2X networks. Applications that involve pedestrians, traffic infrastructure, and the cloud are made possible by V2X networks in addition to those that use vehicles. Thanks to this all-encompassing network, we can implement cutting-edge safety features, streamline traffic, and raise our situational awareness [8].

9.5 COMMUNICATION PROTOCOLS FOR VEHICULAR NETWORKS

1. ***Dedicated Short-Range Communication (DSRC):*** DSRC is a communication protocol widely used in VANETs for vehicular communication. It operates in the 5.9 GHz frequency band and supports high-speed data exchange between vehicles and infrastructure. DSRC provides low-latency, reliable communication and is specifically designed for vehicular safety applications, including collision warning, intersection safety, and cooperative adaptive cruise control.
2. ***Cellular-Based Technologies (e.g., LTE-V2X and 5G):*** Cellular technologies are increasingly being deployed for V2X communication. LTE-V2X, based on Long-Term Evolution (LTE) cellular networks, enables direct communication between vehicles and infrastructure. It provides enhanced coverage, improved reliability, and low-latency communication. The advent of 5G networks further enhances V2X capabilities by offering ultra-low latency, massive device connectivity, and high data throughput, enabling more advanced ITS applications [9].

9.6 CHALLENGES OF VEHICULAR NETWORKS FOR ITS

1. ***Connectivity and Coverage:*** Ensuring reliable connectivity across a large geographic area remains a challenge for vehicular networks. Urban areas with high-density traffic may experience congestion and signal interference, impacting communication reliability. Deploying sufficient infrastructure and optimizing network coverage is crucial to mitigate connectivity challenges.
2. ***Interoperability:*** Vehicular networks often involve various stakeholders, including vehicle manufacturers, infrastructure providers, and service operators. Achieving interoperability and standardization among different systems and protocols is essential to ensure seamless communication and data exchange between different entities [10].
3. ***Security and Privacy:*** Vehicular networks are vulnerable to security threats, including unauthorized access, data manipulation, and cyber-attacks.

Ensuring the security and privacy of transmitted data, authentication mechanisms, and secure communication protocols are critical to maintaining the integrity and trustworthiness of vehicular networks.

4. ***Reliability and Quality of Service:*** Safety-critical applications rely on highly reliable vehicular networks. However, maintaining continuous and dependable communication can be difficult due to signal interference, network congestion, and dynamic topology. For ITS systems to function properly, it is essential to guarantee quality of service (QoS) and reduce communication delays so that data may be sent in real-time.
3. ***Scalability and Management:*** Scalability is becoming increasingly difficult in automotive networks as the number of linked automobiles and Internet of Things devices grows. The increasing number of connected vehicles need sophisticated methods of network administration, resource allocation, and load balancing to guarantee top-notch service for all users.

9.7 CLOUD COMPUTING FOR ITS

Cloud computing plays a significant role in supporting the infrastructure and services required for Intelligent Transport Systems (ITS). The cloud provides a scalable and flexible platform for data storage, processing, and analysis, which is crucial for the effective functioning of ITS applications.

Benefits of cloud computing for ITS include:

1. ***Scalability and Flexibility:*** Cloud-based ITS technologies are scalable and flexible, enabling effective resource allocation and cost minimization [11].
2. ***Data Storage and Processing:*** Cloud computing provides massive quantities of storage space for the massive volumes of data produced by ITS, allowing for effective data management and analysis [12].
3. ***Real-time Analytics:*** Insights into real-time data for better traffic management and decision-making [13] are made possible by cloud computing's support for advanced analytics and machine learning algorithms.

However, there are several challenges associated with cloud computing for ITS:

1. ***Data Privacy:*** Cloud-based ITS systems involve the collection and processing of sensitive personal and traffic data. Ensuring robust data privacy mechanisms and compliance with regulations is crucial to protect individuals' privacy [14].
2. ***Data Security:*** The security of data stored and transmitted in the cloud is a paramount concern. Robust encryption, access control, and threat detection mechanisms must be implemented to safeguard against cyber threats [15].
3. ***Connectivity and Reliability:*** Cloud-based ITS applications heavily rely on continuous and reliable internet connectivity. Any disruption in connectivity may impact the availability and effectiveness of the services [16].

Here are some common ITS cloud applications:

1. ***Traffic Prediction:*** Traffic patterns and congestion can be predicted by analyzing traffic data in the cloud, both historical and real-time. Using this data, traffic management tactics can be improved, and drivers' wait times can be better estimated [17].
2. ***Dynamic Routing:*** ITS platforms in the cloud can give you up-to-the-minute traffic updates and route recommendations that take into account the state of the roads. This aids vehicles in avoiding congestion, shortening their commutes, and enhancing overall traffic flow [18].
3. ***Incident Management:*** Cloud-based ITS systems can help emergency agencies, traffic authorities, and road users coordinate in real-time, which is crucial for managing incidents effectively. As a result, response times are improved, and accidents have less of an effect on traffic [19].

9.8 ASPECTS AND CHALLENGES OF ITS IN SMART CITIES

Intelligent Transportation Systems (ITS) are essential to the development of smart cities. Various challenges must be met before ITS can be effectively integrated and managed in smart cities. The need for collaboration between parties involved in ITS deployment and maintenance is highlighted by the fact that these challenges can be divided into technical, economic, and policy categories.

9.8.1 Technical Challenges

1. ***Communication Protocols***: Communication between cars, infrastructure, and other users is crucial to the success of ITS. To ensure smooth data sharing and cooperation, it is necessary to provide interoperability and standardization of communication protocols [20].
2. ***Data Management:*** Multiple types of sensors, cameras, and moving objects all contribute to the massive volumes of real-time data produced by ITS. Scalability, data integration, data quality, and real-time decision-making [21] are all obstacles to overcome when gathering, storing, processing, and analyzing this data effectively.
3. ***Infrastructure Compatibility***: Due to the wide variety of preexisting systems and technologies, ITS integration might be difficult. The successful implementation of ITS in smart cities depends on the interoperability, upgradability, and scalability of infrastructure components [22].

9.8.2 Economic and Policy Challenges

1. ***Cost:*** The installation of sensors, communication networks, and data storage facilities, all necessary for ITS, can add up to significant expenditures during deployment and maintenance. Key economic problems [23] include

developing sustainable funding sources and striking a balance between the costs and benefits of ITS development.

2. ***Regulatory Frameworks:*** To put ITS into action, a solid and encouraging regulatory frameworks must be created. Compliance, trust, and public acceptability of ITS systems depend on the resolution of legal and regulatory challenges such as data privacy, liability, and security norms [24].
3. ***Business Models***: For smart cities to survive in the long run, it is essential that sustainable business models be developed for ITS. For ITS systems to remain operational and attract private-sector investment, it is crucial to establish sustainable revenue sources, public-private partnerships, and collaborative finance methods [25].

9.8.2.1 Importance of Collaboration Among Stakeholders

The government, transportation agencies, technology providers, and residents of a smart city must all work together to implement ITS effectively. Cooperation makes it easier for:

1. ***Knowledge Sharing:*** Sharing information, insights, and best practices is facilitated by collaborative platforms. When people talk about what they have learned and how they have grown, it helps them solve problems, come up with new ideas, and encourage constant progress [26].
2. ***Coordinated Planning:*** When people work together, they are better able to make decisions and plans simultaneously. Strategic planning that includes a wide range of interested parties increases the likelihood that all relevant viewpoints will be taken into account, resulting in more successful and equitable ITS rollouts [27].
3. ***Resource Sharing:*** Cooperation permits the sharing of information and the use of combined knowledge, tools, and facilities. Smart city ITS deployments benefit from pooled resources and expertise, making them more efficient and less expensive [28].
4. ***User-Centric Solutions:*** It is essential to include customers and end-users in the process of creating and assessing ITS services. ITS solutions that are developed through public collaboration are more likely to meet actual needs, provide a better user experience, and boost quality of life [29].

Smart cities can overcome obstacles to the successful adoption of ITS by collaborating to address technical, economic, and policy difficulties, thereby producing transportation systems that are sustainable, efficient, and centered on the needs of the people who use them.

9.9 POTENTIAL SOLUTIONS FOR ITS IN SMART CITIES

There are a variety of approaches that can be used to solve the problems that arise with ITS in smart cities. These options include creating novel ITS applications, investing more in cloud computing infrastructure, and researching and developing new communication protocols for vehicle networks. When issues develop with intelligent transportation systems (ITS) in smart cities, many methods exist for addressing them. Novel ITS application development, increased cloud computing infrastructure spending, and the study and development of new communication protocols for vehicle networks are all viable choices.

1. ***Development of New Communication Protocols for Vehicular Networks:*** Because intelligent transportation systems (ITS) rely so largely on effective and dependable communication between vehicles, infrastructure, and other road users, the creation of novel communication protocols for vehicular networks is crucial. Enhanced connectivity, low latency, and greater dependability are available for ITS applications thanks to emerging protocols like Cellular Vehicle-to-Everything (C-V2X) and 5G-based Vehicle-to-Everything (V2X). These protocols allow for smooth communication, coordinated actions, and effective data transmission, all of which improve transportation systems in terms of safety and productivity.
2. ***Increased Investment in Cloud Computing Infrastructure:*** More money should be placed into the cloud computing infrastructure to support the processing needs and scalability of ITS. To process the ever-increasing amounts of real-time data being produced by sensors, cameras, and vehicles, cloud-based ITS platforms need reliable and scalable backend systems. The availability and performance needed for real-time data processing and analytics in ITS applications can be ensured by investing in data centers, edge computing resources, and high-speed networks [30].
3. ***Development of Innovative ITS Applications:*** When it comes to solving the problems that only smart cities face, innovation in ITS applications is crucial. ITS makes possible real-time traffic monitoring systems that report on things like congestion and accidents as they happen. These systems are able to foresee traffic patterns, locate congestion hotspots, and enhance traffic management tactics by employing cutting-edge data analytics and machine learning algorithms [31]. In addition, ITS-powered intelligent parking systems can direct motorists to open spots, easing traffic and enhancing convenience [32].

In addition, ITS-enabled smart traffic signal systems can optimize traffic flow and decrease travel times by adjusting signal timings based on real-time traffic circumstances. Users are able to make more educated judgments regarding their travel plans because of advanced traveler information systems' real-time updates on public transit timetables, traffic problems, and alternate routes [33]. These cutting-edge programs improve the effectiveness, security, and longevity of smart city transportation systems.

9.10 CASE STUDIES OF ITS IN SMART CITIES

Vehicle Networks and Cloud Computing for Intelligent Transportation Systems in Singapore

Significant benefits have resulted from the efficient use of vehicle networks and cloud computing for ITS in Singapore. Key figures and results include the following:

1. ***Improved Traffic Flow:*** Singapore's ITS activities have resulted in appreciable traffic flow improvements. The Land Transport Authority (LTA) reports that with the deployment of ITS measures, traffic speeds during peak hours have increased by 10% on average [34]. Congestion is lessened as a result of these upgrades, and commuters enjoy less hassle.
2. ***Increased Safety:*** Cloud computing and using vehicular networks for ITS have considerably increased road safety in Singapore. Vehicle networks' capacity for cooperative communication enables early warning systems, collision avoidance, and better situational awareness for drivers. Traffic fatalities and accidents have decreased as a result of this. The number of fatal accidents and overall traffic accidents decreased by 35% and 31%, respectively, in Singapore between 2010 and 2019 [35, 36].
3. ***Reliable Public Transportation:*** Singapore's public transportation system is now more reliable and efficient thanks to ITS applications like real-time bus arrival information and intelligent traffic signal control. Commuters may better plan their routes and cut down on waiting times thanks to accurate bus arrival timings. This has helped people use public transportation more frequently and rely less on private vehicles [37].

9.10.1 Lessons Learned and Best Practices

Future deployments can benefit from the lessons learned from Singapore's effective use of cloud computing and vehicle networks for ITS:

1. ***Integrated method:*** An integrated framework that combines multiple technologies, communication systems, and data analytics makes up Singapore's integrated approach to ITS. The transportation ecosystem's various elements are connected in a fluid and synergistic way thanks to this all-encompassing strategy.
2. ***Public-Private Collaboration:*** The success of ITS in Singapore has depended heavily on cooperation between government organizations, technology providers, and transportation companies. Public-private partnerships facilitate the sharing of resources, expertise, and data, leading to innovative solutions and efficient implementation.

3. ***Data-Driven Decision-Making:*** The utilization of real-time data and advanced analytics has been instrumental in Singapore's ITS initiatives. The collection, analysis, and utilization of data enable evidence-based decision-making, accurate predictions, and efficient traffic management strategies.
4. ***Continuous Monitoring and Evaluation:*** Regular monitoring and evaluation of ITS systems and applications are essential to identify areas for improvement and ensure ongoing effectiveness. Regular audits and assessments help optimize performance and address emerging challenges.

9.11 CONCLUSION

This chapter examined the several facets and difficulties of cloud computing and vehicular networks in the context of Intelligent Transport Systems (ITS) in smart cities. ITS has the ability to transform transportation and raise individuals' overall quality of life by merging cutting-edge technologies and communication systems. We emphasized the significance of ITS in smart cities throughout this chapter, acknowledging its function in increasing safety measures, encouraging sustainable transportation options, and optimizing traffic management. ITS promotes the seamless integration of diverse forms of transportation, increasing the overall efficiency and dependability of urban mobility. It does this by utilizing real-time data, advanced analytics, and intelligent systems. There are several prospects for improved data collecting, real-time traffic management, and increased safety and efficiency in transportation with the integration of vehicular networks and cloud computing inside ITS. Meanwhile, the obstacles that come with this integration include scalability, network administration, security, privacy issues, and dependence on dependable connectivity. For ITS in smart cities to reach its full potential, these issues must be resolved. Future-looking technologies like edge computing, 5G networks, and linked and autonomous cars are expected to further transform ITS. The landscape of ITS adoption is being shaped by ongoing standardization initiatives and regulatory considerations, assuring interoperability, security, and privacy in smart city transportation systems. Additionally, there are still unresolved research issues that offer potential for further investigation, such as improving network scalability, creating strong security systems, and enhancing data analytics methods.

The development of effective, secure, and sustainable transportation networks is greatly enhanced by intelligent transport systems in smart cities. Cities can efficiently solve the problems caused by urbanization and give residents convenient, dependable, and eco-friendly mobility options by adopting ITS. In the context of intelligent transportation systems in smart cities, this chapter sought to provide a thorough grasp of the aspects and issues linked to vehicular networks and cloud computing. It is our aim that this chapter will be a useful tool for academics, decision-makers, and industry experts, encouraging more study and development in the field of ITS. The incorporation of intelligent transport systems will continue to be a crucial pillar for developing effective, safe, and sustainable urban settings as authors move closer to establishing smarter, more connected cities. We can create a future where transportation is seamlessly integrated, cleverly controlled, and beneficial to the well-being of both persons and the communities they live in by embracing the possibilities of ITS and tackling the issues it provides.

REFERENCES

[1] Han, B., et al. (2019). Smart city traffic management system using the internet of vehicles and cloud computing. Journal of Sensors, 2019, 1–13.

[2] Lee, D. H., et al. (2019). Design and implementation of smart road infrastructure systems based on IoT and cloud computing. IEEE Access, 7, 11209–11223.

[3] Al-Fagih, A., Al-Fagih, M., & Al-Jaber, N. (2022). A comprehensive survey on vehicular communication systems. Wireless Communications and Mobile Computing, 10, 1–32.

[4] Chakraborty, C., et al. (2021). Cloud-based intelligent transportation systems: Recent advances and challenges. ACM Transactions on Intelligent Systems and Technology, 12(5), 1–31.

[5] Zanella, A., Bui, N., & Castellani, A. P. (2021). Internet of vehicles: From intelligent grid to autonomous cars and vehicular clouds. IEEE Transactions on Emerging Topics in Computing, 9(2), 1144–1159.

[6] Niu, B., et al. (2021). Intelligent transport systems for smart cities: From big data sensing to analytics. Future Generation Computer Systems, 121, 498–518.

[7] Altintas, O., Ozguner, U., & Ekici, E. (2016). Vehicular ad hoc networks. Cambridge University Press.

[8] Al-Fuqaha, A., Guizani, M., Mohammadi, M., Aledhari, M., & Ayyash, M. (2015). Internet of Things: A survey on enabling technologies, protocols, and applications. IEEE Communications Surveys & Tutorials, 17(4), 2347–2376.

[9] 3GPP. (2021). Cellular V2X (C-V2X) standards from 3GPP. Retrieved from www.3gpp.org/technologies/keywords-acronyms/98-c-v2x

[10] Tientrakool, P., & Hossain, E. (2016). Vehicular communications and networks: Architectures, protocols, operation and deployment. Cambridge University Press.

[11] Cui, J., et al. (2022). Cloud-enabled ITS for smart transportation systems: Challenges, solutions, and future directions. IEEE Transactions on Intelligent Transportation Systems, 23(1), 32–46.

[12] Gao, C., et al. (2022). A cloud-enabled traffic monitoring system for intelligent transportation systems. IEEE Transactions on Industrial Informatics, 18(5), 3366–3377.

[13] Chen, Y., et al. (2022). A comprehensive survey on cloud computing-based intelligent transportation systems: Applications, challenges, and trends. IEEE Access, 10, 47505–47524.

[14] Lai, L., et al. (2021). Privacy preservation in cloud-based intelligent transportation systems: A survey. IEEE Internet of Things Journal, 8(12), 10027–10047.

[15] Guo, L., et al. (2021). Security and privacy for cloud-based intelligent transportation systems: Challenges and solutions. Future Internet, 13(4), 94.

[16] Zhou, Q., et al. (2021). Challenges and opportunities of IoT-enabled vehicular cloud computing: A review. IEEE Internet of Things Journal, 8(6), 4157–4169.
[17] Li, C., et al. (2022). Cloud-based intelligent transportation system: Traffic prediction and congestion control. IEEE Transactions on Intelligent Transportation Systems, 23(4), 1531–1541.
[18] Sharmila, P. K., et al. (2022). Intelligent transportation system for urban traffic: An IoT-based cloud approach. IEEE Internet of Things Journal, 9(2), 1613–1625.
[19] Wang, R., et al. (2022). A cloud-based emergency management framework for traffic incidents in smart cities. IEEE Internet of Things Journal, 9(6), 5803–5816.
[20] Bai, F., et al. (2022). Vehicular communication protocols for intelligent transportation systems: A comprehensive survey. IEEE Communications Surveys & Tutorials, 24(3), 1505–1543.
[21] Huang, L., et al. (2022). Data management in smart transportation systems: A survey. IEEE Transactions on Intelligent Transportation Systems, 23(5), 1996–2011.
[22] Parwez, S. (2022). Infrastructure development challenges and opportunities for smart cities. Sustainable Cities and Society, 75, 103283.
[23] Ji, Y., et al. (2022). Cost-benefit analysis of urban intelligent transport systems (ITS) based on traffic data generated by mobile devices: A review. Journal of Cleaner Production, 335, 130712.
[24] Qu, Y., et al. (2022). Legal and regulatory challenges for intelligent transportation systems in smart cities: A review. IEEE Access, 10, 37935–37951.
[25] Kaddoura, I., et al. (2022). Business models for the internet of things: A systematic literature review. Journal of Cleaner Production, 332, 129859.
[26] Palmieri, F., et al. (2022). Collaborative platforms for smart cities: A comprehensive review. IEEE Transactions on Engineering Management, 69(1), 58–75.
[27] Barbosa, B., et al. (2022). Stakeholder engagement for smart city planning: A systematic literature review. Journal of Cleaner Production, 339, 130963.
[28] Zheng, D., et al. (2022). Collaborative mechanisms and enabling technologies for integrated mobility services in smart cities. Journal of Cleaner Production, 299, 126737.
[29] Zhang, Y., et al. (2022). A review of user-centered design approaches for intelligent transportation systems in smart cities. Journal of Cleaner Production, 335, 130607.
[30] Beloglazov, A., et al. (2022). Edge computing and cloud computing infrastructure for the internet of vehicles: A review. IEEE Transactions on Intelligent Transportation Systems, 23(4), 1402–1424.
[31] Zhang, X., et al. (2022). Real-time traffic prediction using spatiotemporal data: A deep learning approach. IEEE Transactions on Intelligent Transportation Systems, 23(1), 428–439.
[32] Su, Y., et al. (2022). Adaptive traffic signal control in connected vehicle environment: State-of-the-art and research challenges. IEEE Transactions on Intelligent Transportation Systems, 23(4), 1381–1401.
[33] Altayyar, A., et al. (2022). Intelligent public transport information system in smart cities: Requirements, architectures, and case studies. IEEE Transactions on Intelligent Transportation Systems, 23(1), 210–226.
[34] Land Transport Authority (2020). Land transport data—Traffic speeds. Retrieved from www.mytransport.sg/content/mytransport/home/dataMall/public-transport.html

[35] Land Transport Authority (2020). Land transport data—Traffic accidents. Retrieved from www.mytransport.sg/content/mytransport/home/dataMall/public-transport.html

[36] Mohiddin, S.K., & Babu, Y.S. (2020). Role of cloud forensics in cloud computing. In: Das, K., Bansal, J., Deep, K., Nagar, A., Pathipooranam, P., & Naidu, R. (eds) Soft Computing for Problem Solving. Advances in Intelligent Systems and Computing, vol. 1057. Springer, Singapore. https://doi.org/10.1007/978-981-15-0184-5_9.

[37] Kumar, P., et al. (2021). Internet of things and ITS: Emerging paradigm for smart transportation. Wireless Personal Communications, 117, 683–701.

Chapter 10

The Rise of Electric Vehicles–2023 Status and Future Directions

Karthikeyan V., Kirubakaran G., Varun Prakash R.

10.1 INTRODUCTION

Upgrading gasoline-powered automobiles with energy-efficient vehicles like electric vehicles (EVs) seems like a promising step towards imagining urban sustainability, given that GHG emissions from the automobile system are predominantly concentrated in metropolitan areas. These developments have demanded electric road transport vehicles.

10.1.1 Electric Vehicle

Instead of an internal combustion engine to increase torque, electromechanically-propelled electric vehicles (EVs) use electricity as the sole source of power. The electric vehicle's battery stores electricity to be used as the vehicle's primary source of propulsion. The electric motor's proposed system is advantageous since it converts power very efficiently. Several current global environmental issues are intolerable, and this calls for immediate action to cut down on carbon emissions. The automotive industry is a major contributor to climate change due to its heavy reliance on petroleum [1]. Electric buses may use both grid electricity and power from solar panels. A rechargeable battery is available for use. To help lower emissions, it's imperative that each auto manufacturer build no less than an individual hybrid EV style. By the year 2030, Sweden and China will no longer manufacture gasoline-energized vehicles. It is a further development towards the fight against toxic gas and promoting environmentally friendly electric automobiles [2]. The various types of electric vehicles (EVs) are compared to one another in terms of their batteries and other attributes in Table 10.1.

An ambulance that rides on a battery is an example of such an EV. Electricity from either an external source or a battery storage device can power the bus's motors. When compared to vehicles that function on petrol/diesel, their operative expenses are far lower. Vehicle fuel costs are reduced by a factor of two to three. The cost of electricity is typically more consistent than that of gasoline or liquid fuel. Many people rely on buses as their primary mode of transportation, which has

 DOI: 10.1201/9781003464556-10

Table 10.1 Various Types of Electrical Vehicles

Measures	*EV Type*			
	Combination EV	*Rechargeable-Type EV*	*Fuel Cell-Based EV*	*Renewable-Type EV*
Energy System	Battery, Supercapacitor, and Intelligent generators	Rechargeable Batteries and Super Capacitors	Fuel Cell	Solar or Wind-based Source
Momentum	Electric Motor Drives (EMD)	EMD	EMD	EMD
Issues	Combining multiple energy sources in a single system	Energy and Battery management and scheduling	Fuel processor implementation complexity and cost	Implementation cost

Table 10.2 Comparison of Different Kinds of Batteries Utilized in EVs

Battery Type	*Energy Density (Wh/kg)*	*Cost ($/kWh)*	*Performance (Cycle Life, Charge/Discharge Rate)*
Lithium-ion	100–250	100–200	Good cycle life, fast charge/discharge rate
Nickel-metal hydride	40–70	200–300	Moderate cycle life, slower charge/discharge rate
Lead-acid	30–40	100–150	Low cycle life, slower charge/discharge rate
Solid-state	300–400	200–400	Good cycle life, fast charge/discharge rate

a direct impact on carbon emissions. By switching to electric buses, we can lessen our reliance on finite resources like gas and diesel.

Electric vehicle efficiency and lifespan are also influenced by the battery technology employed [3]. Some of the most common battery technologies are assessed in Table 10.2.

Lithium-ion batteries are now the extremely prevalent energy source used by EVs, as seen in Table 10.2. They are low-cost and provide a great deal of energy. They aren't without their drawbacks, either, as their short shelf lives and potential safety hazards attest. Newer technologies like solid-state battery packs are promising because of their potential for higher security and higher density of energy, but they are still in the experimental stages.

Table 10.3 Comparison of Electric Vehicle Policies in Different Countries

Country	*Incentives for EV Buyers*	*Charging Infrastructure Development*	*Carbon Pricing Policy*
Norway	Tax exemptions, free public charging, free tolls	Strong development, high number of charging stations per capita	Yes, carbon tax
China	Tax exemptions, subsidies, license plate restrictions	Rapid development, high number of public charging stations	Yes, emissions trading system
United States	Tax credits, rebates, HOV lane access	Moderate development, uneven distribution of charging stations	Varies by state

The use of electric vehicles is a potentially useful approach to lowering emissions caused by transportation and fostering more environmentally responsible modes of travel. Table 10.3 contains a list of the numerous policies that have been implemented in a variety of countries [4]. When it comes to lowering the release of carbon from the automotive sector – responsible for 16% of the overall emissions worldwide – electric cars are the single most significant aspect of technology. Sales of battery-powered automobiles have skyrocketed in the past few years, thanks in large part to advancements in assortment, model variety, and overall profitability. In 2022, we anticipate that 13% of the total sales of new automobiles will be EVs. If the expansion of the past two decades continues, it will be possible to place automobile emissions of carbon dioxide on a trajectory that is compatible with the concept of net-zero emission (NZE) by 2050. However, electrically powered cars are still not commonplace in all parts of the globe. Sales have been low in countries that are developing or emerging because of the expensive cost of purchasing and the lack of charging facilities.

10.1.2 EV Sales Exploded Throughout Key Automobile Economies in 2021

In 2021, the purchase of EVs attained an all-time peak, although manufacturing tailbacks as well as the enduring COVID-19 outburst. By 2020, there will be almost 17 million electric vehicles on roads worldwide, compared with a total of 5.5 billion in 2020 (a garage sale of more than 9%). In 2021, EVs will make up 4% more of total vehicle sales than in 2020. In the NZE by 2040 development, there will be an additional 290 billion EVs on the road by 2030, including EVs construction, improving 65% of modern automobile trades. They need a yearly progress in sales share of < 5% to stay on pace with the NZE scenario. Figure 10.1 shows the International Energy Agency (IEA) authorization of EVs and their returns split in various countries [5].

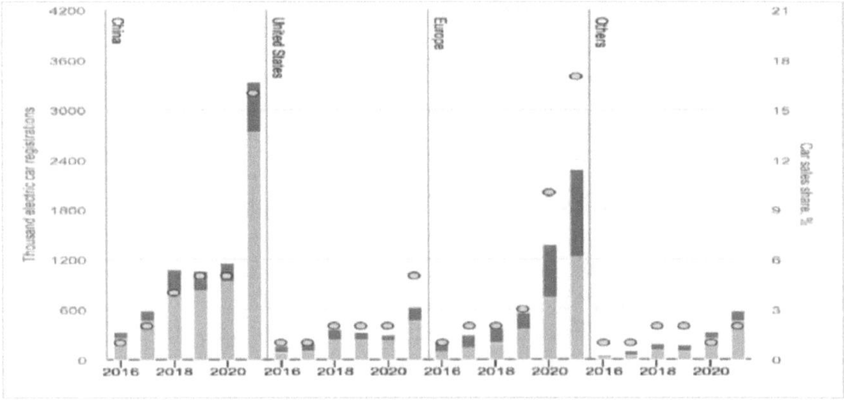

Figure 10.1 Licences of Electric Vehicles and Their Revenue Share in China, the US, Europe, and Asian Countries (International Energy Agency).

Sales of EVs have enriched in emerging nations, although they are still relatively uncommon. We predict that by the middle of 2022, international sales of electric cars will account for roughly 13% of all vehicle sales. In 2021, EVs will replace about 0.3 Million barrels per day (Mb/d) of fuel usage. To stay aligned with NZE, EVs will need to replace over 7 Mb/d of oil by 2030. To ensure that battery electric vehicles have enough range, energy density is essential. Many of the superlative-functioning energy packs can instantly attain a power level of more than 290 Wh/kg, a rise of approximately 95–155 Wh/kg a period earlier; it indicates that battery-powered automotives nowadays drive double the distance with a similar weight. Improvements in battery chemistry and cell design have allowed for this development. Two prominent forthcoming examples are the 4680 cells from Tesla and the Ultium cells from LG Energy Solution.

Still, there are other considerations besides energy density. Minimizing demand for scarce materials is a primary motivation behind EV advancement. The market share of nickel- and cobalt-free cathodes made from lithium iron phosphate (LFP) has doubled in the past year. This is because automakers have shifted to LFP to lessen their reliance on volatile commodity prices and because of novel compact strategies that allow a better box compactness with lowering compact dead mass. CATL, the largest EV manufacturer globally, recently declared a substantial enhancement to the production system for sodium-ion energy sources. This innovation has the potential to do away with the need for crucial metals altogether [6].

10.2 LITERATURE SURVEY

As worries about climate change and air pollution increase, so does the need for environmentally friendly modes of transportation. The advent of EVs offers hope for a reduction in transportation-related pollution. In this analysis, we will look at the pros, cons, and underlying technology of sustainable electric vehicles as they are right now.

Electric vehicles' breakout into the market has not eliminated the need for cars powered by traditional diesel. When compared to traditional, fossil fuel-powered automobiles, EVs fall short in areas such as life cycle evaluation, charging infrastructure, and usable distance. EV manufacturing generates 59% less CO_2 than ICEV manufacturing. While the ICEV's CO_2 emissions are just 120 g/km on a tank-to-wheels (TTW) basis, the LCA estimates that this number is closer to 170–180 g/km. The authors of Ref. [7] measure the typical carbon dioxide levels among the overall automobile relatively over an individual automobile, despite the fact that EVs have null radiations of carbon dioxide on a TTW base. Total CO_2 emissions over a vehicle's lifetime are very variable with respect to factors such as power source, country of manufacturing, and country of use [7].

When a greater aggregate of renewable supply is intermittent, the source can benefit from a flexible load like an electric vehicle [8]. Due to the small power of a particular operation, electric vehicle owners cannot take part in the automobile industry [9]. The complete perspective of an EV can only be realized with the implementation of flexible load and smart charging solutions. The battery level underpins the energy demand [10], whereas the timing constraint dictates the period by which a charging procedure needs to be finished. Similarly, Galus et al. [11] found that central and decentralized frameworks for electric vehicle pricing should overlap, with the former providing the pricing signal to owners [11]. The stochastic modelling methodology of an EV was investigated by Brady and O'Mahony (2016) [12]. They came to the conclusion that if parking time distribution conditions were made more stringent, the model's accuracy would improve overall. Some individuals who use electric vehicles desire charging them at a residence during evening hours of high power use, according to research published by Morrissey et al. (2016) [13]. When comparing the effects of charging electric vehicles, Foley et al. (2013) [14] discovered that peak charging has negative consequences. Steinhilber et al. (2013) [15] investigated the primary hurdles to an EV in two nations to determine the most effective methods for promoting new technology and innovation.

In accordance with the trip segmentation partition methodology, Yu et al. (2012 [16] presented an automobile pattern identification method for calculating EV range. Hayes et al. (2011) [17] constructed a vehicle model and studied its behaviour under a variety of weather and terrain situations. Higher dissemination levels as well as active tariffs enhance the likelihood of burdens at particular sites, according to research by Salah et al. (2015) [18]. The variety of types of these characteristics is then used to make comparisons between them. Framework-based active estimators for determining the speed of synchronous motors with stable magnets for blended EV models were investigated [19, 20], and their effects on the national power system and preservation utilization have been provided. Increase antiskid execution of the torque management architecture and electric stability

Table 10.4 Comparison of EV and Its Performance Measures

Author and Ref. No.	*Vehicle Type*	*Power Train Configuration*	*Range (km)*	*Top Speed (km/h)*	*Acceleration (0–100 km/h)*	*Efficiency (Wh/km)*
Kumar et al., 2017 [24]	Passenger	BEV	240	150	7.0	200
Gao et al., 2018 [25]	Light-duty	BEV	200	120	8.0	220
Tseng & Huang, 2019 [26]	Passenger	FCV	480	165	9.5	150
Xu et al., 2020 [27]	Bus	BEV	250	80	15.0	230
Shen et al., 2021 [28]	SUV	PHEV	50 (EV mode)	200	7.5	180 (EV mode)

control with the highest transferable torque approach was investigated [21]. In 2013, Lu et al. [22] compiled a summary of the most pressing problems associated with controlling Li-ion batteries in EVs. In [23], researchers reviewed the literature on EV simulation, energy management systems, and optimal management methodologies. Table 10.4 of Electric Vehicle Comparisons was produced using data analyzed from recent publications.

10.3 ADVANCED TECHNOLOGIES USED IN E-VEHICLES

The rapid advancement of technology has brought about significant changes in various industries, and the automotive sector is no exception. EVs enhanced as a balanced and environmentally responsive style of haulage, and advanced technologies have performed a significant function in their development and adoption. In this part, we will explore the various uses of advanced technologies in EVs, ranging from battery technology to autonomous driving, and how they have transformed the landscape of electric mobility.

10.3.1 Battery Technology (BT)

One of the key areas where advanced technologies have made significant strides in the EV industry is battery technology. The battery is the heart of an electric vehicle, providing the power needed for propulsion. Advanced battery technologies, i.e., lithium-ion batteries (LIB), have significantly improved the range and performance of EVs. LIBs are recognized for their extreme power storage, allowing EVs to travel longer distances on a single charge [29, 30]. Moreover, advancements in battery chemistry and manufacturing techniques have also increased durability, faster charging times, and improved safety features in EV batteries. For instance, solid-state batteries are emerging as a promising technology that could further enhance the performance and safety of EVs. These enhancements in power sources have provided for the wider adoption of EVs and have addressed one of the key concerns of consumers – range anxiety [31, 32].

Various manufacturers around the world are investing in research and development to improve battery technologies for EVs. The following part provides an overview of battery technologies used by different manufacturers, including key features, advancements, and challenges.

10.3.1.1 Tesla Inc.

Tesla Inc., a leading EV manufacturer, is known for its advanced battery technologies that have helped establish its dominance in the EV market. Tesla uses lithium-ion batteries, which are best for their thermal stability, allowing for longer ranges and better performance. Tesla's battery technology also includes proprietary features such as its BT managing that monitors as well as optimizes the performance of each battery cell.

Tesla has also been working on improving battery longevity and durability through research in areas like silicon anodes, solid-state batteries, and recycling technologies. The company's 4680 battery cells, announced in 2020, are expected to have higher energy density and be more cost-effective to produce, potentially revolutionizing the EV battery market.

10.3.1.2 LG Energy Solution

LG Energy Solution, an international director in battery production, supplies batteries to several EV manufacturers worldwide. The company's batteries use lithium-ion technology and are acknowledged for their maximum energy solidity and thermal stability. LG Energy Solution's batteries also incorporate advanced safety features such as thermal runaway protection and multiple layers of protection to prevent overcharging and short-circuits.

LG Energy Solution has been focusing on improving the energy density of its batteries and developing next-generation technologies like solid-state supplies, which provide even greater power storage, extensive lifespan, and quick charging capabilities. The company is also working on recycling technologies to promote sustainable battery production and reduce environmental impact.

10.3.1.3 Panasonic Corporation

Panasonic Corporation, a well-known manufacturer of consumer electronics, also produces batteries for EVs. The company's batteries use lithium-ion technology and are known for their reliability and performance. Panasonic's batteries are used by Tesla in its EVs, and the two companies have a long-standing partnership in battery production [33].

Panasonic has been working on advancements in battery chemistry, such as increasing the energy density and improving the thermal management of batteries to enhance their performance and safety. The company has also been researching solid-state batteries and recycling technologies to promote sustainable battery production and reduce waste.

10.3.1.4 BYD Company Limited

BYD Company Limited, a Chinese manufacturer, is known for its expertise in battery technologies for EVs. The company produces both lithium-ion and blade battery technologies. BYD's blade battery technology is unique, featuring a design that eliminates the need for a separate cooling system, reducing the risk of thermal runaway and improving safety [34].

BYD has been focusing on enhancing the energy density of its batteries and improving their thermal stability to optimize performance and safety. The company is also investing in research and development of solid-state batteries and recycling technologies to further advance its battery technologies [35].

10.4 AUTONOMOUS DRIVING

Another significant use of advanced technologies in EVs is in the field of autonomous driving. Advanced sensors, cameras, and computing systems are used in EVs to enable autonomous driving capabilities. These technologies allow EVs to perceive their surroundings, navigate, and make decisions without human intervention. Autonomous driving has the potential to significantly improve road safety by reducing human errors, optimizing traffic flow, and mitigating congestion. Additionally, it can raise the efficiency of EVs by optimizing routes, reducing idle time, and improving energy management. Companies like Tesla, Waymo, and Cruise are at the forefront of developing and testing autonomous driving technologies in EVs, and the advancements in this area are expected to have a transformative impact on the future of transportation [36, 37]. The following part provides an overview of autonomous driving technologies used by different manufacturers, including key features, advancements, and challenges.

10.4.1 Tesla Inc.

Tesla Inc., a leading electric vehicle (EV) manufacturer, is well-known for its advanced autonomous driving technology called Autopilot. Tesla's Autopilot is a semi-autonomous system that offers features such as dynamic drift control, path assistance, and smart parking.

Tesla's Autopilot is constantly revised with AI developments, which allow for continuous improvement and refinement of its autonomous driving capabilities. However, it is important to note that Tesla's connected vehicles are categorized as a second stage automotive coordination, which implies that a motorist is still necessary to navigate the automobile at any time.

10.4.2 Waymo LLC (Alphabet Inc.)

Waymo LLC, a subsidiary of Alphabet Inc., is a leader in autonomous driving technology and is focused on developing fully self-driving vehicles. Waymo's technology is based on an extensive suite of sensors, including lidar, radar, and cameras, along with powerful artificial intelligence (AI) algorithms for perception, mapping, and decision-making.

Waymo has been testing its autonomous driving technology on public roads for several years and has accumulated a significant amount of real-world driving data, which has helped improve the accuracy and safety of its system. Waymo's autonomous driving technology is classified as a Level 4 system, which means that the vehicle is capable of driving autonomously in most conditions without requiring driver intervention, although there may be some limitations.

10.4.3 General Motors (GM) and Cruise LLC

General Motors (GM), a major automotive manufacturer, has partnered with Cruise LLC, a self-driving technology company, to develop autonomous driving technology for its vehicles. Cruise, which is also backed by SoftBank and Honda, is focused on developing a Level 5 autonomous system that aims to achieve full self-driving capability without any human intervention.

GM and Cruise are using a combination of sensors, including lidar, radar, and cameras, along with machine learning algorithms to enable their autonomous driving technology. The companies have been testing their autonomous vehicles on public roads and have been making significant progress towards their goal of achieving full self-driving capability.

10.4.4 Mobileye (Intel Corporation)

Mobileye, a subsidiary of Intel Corporation, is a pioneer in advanced driver assistance systems and autonomous driving technology. Mobileye's technology is based on a camera-centric approach, using a single camera mounted on the windshield to capture data and analyze the vehicle's surroundings.

Mobileye's autonomous driving scheme incorporates the applications track suggestion, autonomous gear mechanism, and traffic jam assist, which enable the vehicle to operate autonomously in certain conditions. Mobileye's system also includes real-time mapping capabilities, which help with precise localization and path planning.

10.5 PERFORMANCE ENHANCEMENT

Apart from batteries and autonomous driving, advanced technologies are also being used in EVs to enhance their performance, comfort, and convenience. For instance, regenerative braking is a technology used in many EVs that allows the vehicle to recover energy while braking and decelerating, which is then stored in the battery for later use. This technology not only improves the general efficiency of the automobile but also extends the variety of the EV. Additionally, advanced materials, such as lightweight composites and carbon fibre, are being used in the manufacturing of EVs to reduce their weight, thereby increasing their range and performance. Moreover, advanced infotainment systems, connectivity features, and over-the-air updates are being integrated into EVs, providing a seamless and connected driving experience for users [38].

Tesla uses a combination of aluminum alloys and advanced polymers in the construction of its EVs to reduce weight and improve performance. For example, the Tesla Model S and Model X feature aluminum body structures that are lightweight yet highly durable. Additionally, Tesla uses carbon fibre-reinforced composites in certain components, such as the rear spoiler and interior trims, to reduce weight and enhance aerodynamics.

BMW's i3 and i8 models have a carbon fibre-reinforced plastic (CFRP) frame arrangement, which significantly reduces weight while maintaining structural rigidity. CFRP is a lightweight and high-strength material that offers excellent performance in terms of crash safety and energy efficiency. General Motors (GM) uses an aluminum-intensive body structure, which helps reduce weight and enhance fuel efficiency. Volkswagen is exploring the use of advanced composites, including CFRP, in its upcoming EV models to further reduce weight and improve performance.

10.5.1 Advancement in Sensors and Software

In addition to batteries and charging systems, advanced sensors and software are being developed and implemented in EVs to improve safety and performance. For example, advanced driver-assistance systems (ADAS) use sensors such as radar, lidar, and cameras to detect obstacles, pedestrians, and other vehicles, thereby reducing the risk of accidents [39]. Moreover, connectivity is another area where advanced technology is being implemented in EVs. Vehicle-to-vehicle and vehicle-to-infrastructure interaction schemes permit EVs to interconnect with added automobiles as well as traffic substructures, such as transportation indicators and highway symbols. This technology can help optimize traffic flow, decrease overcrowding, and enhance security by alerting drivers to potential hazards or traffic delays.

10.5.2 Charging Arrangement (CA)

Furthermore, CA is a crucial aspect of EV implementation, as well as advanced technologies are being used to improve the charging capabilities of EVs. Fast-charging stations equipped with advanced chargers using direct current (DC) fast-charging technology can charge an EV significantly faster compared to traditional alternating current (AC) chargers. Wireless charging technology is also gaining momentum, allowing EVs to charge without the need for physical connectors. This equipment practices EM fields to move energy from the charging infrastructure to the vehicle, making charging more convenient and user-friendly. The online electric vehicle (OLEV) established in Western countries is a commercially available active charging EV system [35, 36].

Figure 10.2 represents the sample online electric vehicle bus module implemented in South Korea.

10.5.2.1 India's Perspective on Charging Infrastructure

The convenience and approachability of CA are critical factors that impact the embracing of EVs in Asian countries. The Indian government has made numerous commencements to stimulate EV utilization as well as improve CA throughout

Figure 10.2 Sample OLEV Framework.

the nation. The Faster Adoption and Manufacturing of EVs in India (FAME India) arrangement was introduced in late 2010 to provide inducements for EVs and establish CA. Under FAME India, the government has sanctioned funds for setting up charging stations across major cities in India, focusing on public transportation hubs, highways, and urban areas. Several private and public companies have entered the charging infrastructure space in India. Companies like Tata Power, ABB, EV Motors, and Fortum have set up EV charging stations in various cities across India. Additionally, some fuel advertising corporations like Indian Oil, Bharat Petroleum, and Reliance are also venturing into EV charging infrastructures to adapt to the changing landscape of transportation.

Figure 10.3 represents the sample charging stations for the EVs implemented in India.

Charging stations in India typically offer both DC and AC chargers. DC chargers are generally faster and more suitable for long-distance travel, while AC chargers are slower but more appropriate for immediate charging at the house/workplace. DC chargers in India typically range from 15 kW to 50 kW, while AC chargers range from 3 kW to 7.2 kW.

In conclusion, the use of advanced technology in EVs is rapidly advancing, offering significant benefits such as improved battery performance, autonomous driving, regenerative braking, enhanced safety and performance, increased connectivity, and reduced weight. With continued research and development, these advancements will likely continue to advance the effectiveness, suitability, and constancy of EVs in the future.

Figure 10.3 A Rapid Charging Station in New Delhi.

10.6 COMPREHENSIVE ANALYSIS OF VARIOUS EV MANUFACTURERS

10.6.1 A Global Perspective

Electric Vehicles (EVs) are becoming progressively widespread worldwide owing to their minimal carbon footprint and extraordinary petroleum efficacy. The international EV trade is estimated to propagate substantially due to government initiatives and the growing demand for sustainable transportation. In this comprehensive analysis report, we will analyze various EV manufacturers, their market share, and key product offerings [40, 41]. Furthermore, Figure 10.4 shows the future forecast of EVs.

The following tables provide a detailed comparison of various EV manufacturers based on their market share, key product offerings, range, and pricing.

The following part aims to provide an overview of the top three global EV manufacturers, their key models and key specifications, along with a comparison table for easy reference.

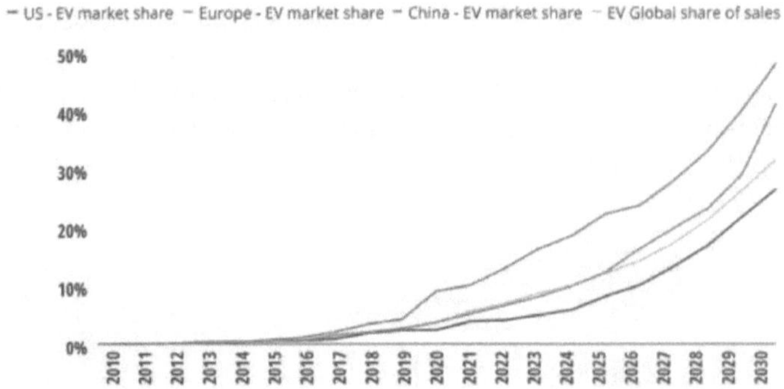

Figure 10.4 Future Forecast of EV Market Share.

Table 10.5 Global EV Market Share by Manufacturer

Rank	*Manufacturer*	*Market Share*
1	Tesla	25.1%
2	Volkswagen	10.5%
3	BYD	7.3%
4	Renault	5.5%
5	Nissan	4.9%
6	BMW	4.5%
7	General Motors	4.2%
8	Hyundai	3.7%
9	Ford	3.6%
10	Audi	3.3%

Table 10.6 Comparison of Key Product Offerings by EV Manufacturer

Manufacturer	*Model*	*Battery Size (kWh)*	*Range (miles)*	*Price (USD)*
Tesla	Model S	100	412	$79,990
	Model X	100	360	$89,990
	Model 3	75–82	263–353	$35,000
	Model Y	75–82	244–326	$39,990
Volkswagen	ID.3	45–77	215–341	$34,200
	ID.4	77	250	$39,995
BYD	e6	70	248	$44,700
	Tang	86.4	236	$45,800
Renault	Zoe	52–55	245–276	$32,810
Nissan	Leaf	40–62	149–226	$31,620
BMW	i3	42.2–120	153–200	$44,450
General Motors	Bolt EV	66	259	$31,995
Hyundai	Kona Electric	64	258	$37,390
	Ioniq Electric	38.3	170	$33,245
Ford	Mustang Mach-E	75–98	230–300	$42,895

Table 10.7 Key Specifications of Various Tesla Models

Model	*Range (miles)*	*Acceleration (0–60 mph)*	*Top Speed (mph)*	*Fast Charging (miles per minute)*
Model S	396–520+	1.99 seconds	200+	180
Model 3	358–358+	3.2 seconds	162	170
Model X	305–371+	2.5 seconds	163	130
Model Y	326–353+	4.8 seconds	155	158

10.6.1.1 Tesla Inc.

Tesla Inc., founded in 2003 and based in the US, is an inventor of electric vehicles (EVs), energy storage, and solar energy solutions. The following table provides Tesla's key EV models and key specifications.

10.6.1.2 Volkswagen AG

Volkswagen, originating in Germany, is an international automotive manufacturer known for its wide range of vehicles, including electric vehicles (EVs). The following table provides Volkswagen's key EV models and key specifications.

10.6.1.3 BYD Co., Ltd

BYD Co., Ltd., founded in 1995 and based in China, is an important industrialist of EVs, batteries, and supplementary renewable source-based substances. The following table provides BYD's key EV models and key specifications.

Table 10.8 Key Specifications of Various Volkswagen Models

Model	*Range (miles)*	*Acceleration (0–60 mph)*	*Top Speed (mph)*	*Fast Charging (miles per minute)*
ID.3	215–340	7.3 s	99–99+	160
ID.4	250–323	5.8 s	99–99+	125
e-Golf	144–186	9.6 s	93	60

Table 10.9 Key Specifications of Various BYD Models

Model	*Range (miles)*	*Acceleration (0–60 mph)*	*Top Speed (mph)*	*Fast Charging (miles per minute)*
Tang EV	265–326+	4.1 s	111	30
Han EV	373–405+	3.9 s	131	30
E2	155–187+	8.9 s	81	30
E6	250–301+	5.9 s	100	30
Qin EV	213–236+	4.5 s	111	30

10.6.2 India's Perspective

EVs are getting impetus in India, and numerous manufacturers have entered the market, offering a range of EVs that cater to different segments. In this part, we will examine leading EV manufacturers in India, including their product portfolio, features, specifications, and pricing, to provide a comprehensive overview of the Indian EV market [42].

10.6.2.1 Tata Motors (TM)

TM is one of the important automotive companies in India and has a significant manifestation in the EV market. The company offers several EVs, including the Nexon, Tigor, and Tiago EV. The Nexon is the most popular EV in India, with a distance of 312 km on a particular charge and a fast charging time of 60 minutes for 0–80% charge. The Tigor EV and Tiago EV are more affordable options, with a range of 306 km and 213 km, respectively.

10.6.2.2 Mahindra & Mahindra (M & M)

M & M is another principal car company in India that has been at the forefront of EV development. The company offers several EVs, including the eKUV100, e-Verito, and e2oPlus. The eKUV100 is an affordable compact SUV with a coverage of 150 km, while the e-Verito is a sedan with a range of 110 km. The e2oPlus is a compact hatchback with a range of 120 km.

Table 10.10 Comparison of EVs from Various Manufacturers in India

Manufacturer	*Model*	*Range (km)*	*Top Speed (km/h)*	*Battery Capacity (kWh)*	*Price (INR Lakh)*
Tata Motors	Nexon EV	312	120	30.2	14.99
	Tigor EV	306	80	26	10.58
	Tiago EV	213	80	16.2	6.54
Mahindra & Mahindra	eKUV100	150	80	15.9	8.25
	e-Verito	110	86	18.7	11.90
	e2oPlus	120	100	16.6	6.07
MG Motors	MG ZS EV	419	140	44.5	21.00
Hyundai	Kona Electric	452	167	39.2	23.79
Ather Energy	Ather 450X	116	80	2.9	1.59
	Ather 450 Plus	116	80	2.7	1.49

10.6.2.3 MG Motors

MG Motors is a relatively new entrant in the Indian automotive market but has gained a significant presence in the EV segment with the launch of the MG ZS EV. The MG ZS EV is a compact SUV with a distance of 419 km on a particular charge and features like i-SMART EV 2.0, which provides real-time battery and range status updates and remote charging capabilities.

10.6.2.4 Hyundai

Hyundai is a leading automotive manufacturer worldwide and has a significant presence in the Indian EV market. The company offers the Kona Electric, India's first long-range EV, with a coverage of 452 km on a distinct charge. The Kona Electric comes with several features such as regenerative braking, smart cruise control, and wireless charging.

10.6.2.5 Ather Energy (AE)

AE is a startup that offers battery-based scooters, including the Ather 450X and Ather 450 Plus. The Ather 450X is a superior e-bike with a range of 116 km and features like 4G connectivity, Bluetooth, and navigation.

10.7 ELECTRIC VEHICLE SCENARIO IN DEVELOPING COUNTRY – ANALYSIS IN INDIA

This section discusses some of the difficulties encountered by electric vehicles in India. More attention must be paid to encourage EV adoption as a whole rather than just as a niche market. The discrepancy between planned and actual actions also needs to be taken into account.

10.7.1 Plan for Buying EVs in India

The Indian government is actively reassuring the utilization of EVs by implementing various initiatives and incentives around the country. Some of these plans are described in more detail in what follows. The government has declared the National Electric Mobility Mission Plan (NEMMP) to develop power protection, lessen the unconstructive impact of diesel-powered automobiles on the atmosphere, and foster the growth of indigenous engineering [43]. FAME II is a proposal introduced by the government to facilitate the widespread use of battery-based and hybrid automotives [44]. With the advent of EV technology and the necessity to decrease power requirements in the automobile sector, NITI Aayog's 2017 transformative movement study has established a pathway for deploying wholesome EVs.

10.7.2 Commercial Obstacles for EVs Across India

Technical constraints, policy obstacles, and a lack of infrastructure are all aspects of the Indian market that may be improved to make electric vehicles more accessible.

Market: Repairing, maintaining, and troubleshooting an electric vehicle should be performed by a professional technician who should be available to guarantee that the EV is properly cared for. The battery packs of an electric car are not only costly, but they also want to be interchanged multiple times throughout the course of the automobile's lifespan. The discernment of the consumer has a substantial impression on both the acquisition of different customers and the maintenance of current ones. The basic components that go into the production of electric vehicle batteries are the exceptional soil reserves [45].

Figure 10.5 illustrates the drivers and hurdles that hinder the widespread implementation of electric vehicles.

Technical: To construct an electric vehicle, mechanics remove the petrol tank and engine and replace them with electric components such as a motor, batteries, chargers, and controllers. The restricted range that EVs give on a particular charge or chamber of petrol is a fundamental obstacle to their broad adoption. Unless meticulous preparations have been made, the driver may be unable to undertake a lengthy trip. Because of this, there is now an obstacle equal in size to the required driving distance. To what extent a battery takes to charge depends largely on its capacity. Longer charging times are required to bring greater capacity car batteries up from empty to full. Even worse, the period consumed to charge a battery depends on how fast the charging point works.

Policy: India's shift to EVs should be aided by the government's announcement to subsidize charging facilities nationwide. Good news for the rollout of EV charging stations across India: the country's power ministry has explicitly stated that such facilities do not require permission to operate.

Figure 10.5 Key Drivers and Barriers to EV Adoption.

Infrastructure: A growth in the quantity of EVs necessitates the construction of more charging stations because of the resulting spike in demand for electricity. The low rate of electric car sales in India can be attributed in large part to the country's inadequate charging infrastructure. Manufacturers don't always provide accurate information about battery replacement costs, but replacing a dead battery outside of its warranty term can be very costly [46].

10.7.3 Government Supports to Improve EV Market

The government has introduced numerous measures to foster the expansion of the country's EV market. Some of the major steps the government has implemented to encourage the EV market are listed in Table 10.11.

Table 10.11 Government Supports to Promote EVs in India

Government Policy/Program	*Inauguration Year*	*Goals*
Electric Vehicle Policy, 2018 [47]	2018	Promotes EV adoption and achieve 30% EV penetration by 2030.
FAME India Scheme [48]	2015 (extended in 2018)	Provides incentives for EV adoption and manufacturing.
Electric Vehicle Charging Infrastructure	2019	Set up 2,636 EV charging points through more than 60 metropolises in the country.
Electric Two-Wheeler Market	-	Witnessed significant growth, dominated by Hero Electric, Okinawa Autotech, and Ather Energy.
Niti Aayog [49]	2018	India's EV market has the potential to grow to 30% of new vehicle sales by 2030 through policy interventions.
Lower GST rate	2020	The Goods and Services Tax for electric vehicles has been reduced from more than 10% to less than 5%, making EVs more reasonable for customers.
Exemption from registration fees	2021	Electric vehicles are exempted from payment of registration fees under the Central Motor Vehicles Rules, 1989, to enhance the EV market.
Tax incentives	2021	The government has provided income tax reductions of a maximum 1.5 lakh on the concern remunerated on credits taken for the buying of electric vehicles to improve EV purchases.

Niti Aayog published a report in 2018, titled "India's Electric Vehicle Roadmap," outlining government's strategy to motivate the implementation of EVs in the country through a variety of policy initiatives [49]. These initiatives included financial incentives for EV manufacturers and consumers, investments in charging infrastructure, and research and development of new technologies. "Electric Vehicle Outlook 2020" (2020), published by BloombergNEF, is a report that examines the EV industry in India and makes projections for the growth of EV sales in the country over subsequent years [50]. According to this estimate, in the year 2030, India will take its place as the fourth largest consumer of electric vehicles.

10.8 CONSUMER PERSPECTIVE OF ELECTRIC VEHICLES

Global warming has impacted many aspects of the climate crisis. CO2 is the predominant factor of global warming. Though CO2 emissions fell during the COVID-19 pandemic, they are expected to grow by 4.8% [51] as the economy recovers by 2021. Therefore, most nations have resorted to decreasing the impact of global warming. One main decision is to decrease CO2 emissions from vehicle transportation. Moreover, the decrease in fuel availability has also led to a search for alternatives. Therefore, global attention is focussing on electric vehicles [52]. With increasing awareness among consumers regarding global warming and other environmental risks, the interest in electric vehicles (EVs) is also on the rise. However, their acceptance in daily life is dependent on the performance and cost of EVs. As the global EV price dropped, the sales of EVs peaked by 43% [51, 52].

Any new product will thrive depending on the consumer's demand. So, it is important that manufacturers take a consumer-centric approach to get the general public's long-term acceptance of EVs. Lack of economic sense and unreliability makes consumers hesitant about electric vehicles [53]. Consumer perspective is important to identify key obstacles to adoption. By understanding these concerns, manufacturers can work to address them through targeted marketing campaigns, educational materials, or even by redesigning the vehicles themselves. By taking into consideration the public perspective, manufacturers ultimately help to drive widespread adoption and uptake of EVs [54].

In general, social acceptance of a product will be influenced by different factors like community, institutions, and individual perspective. Based on the previously mentioned factors, obstacles to electric vehicle adoption involve driving range and other ecological factors [55, 54].

10.8.1 Driving Range Anxiety

Driving range plays a pivotal role in hesistance among the general public to choose EVs. Sixty percent of EVs do not satisfy consumers' driving range requirements [56]. Moreover, current driving range needs to be improved a lot more for long distance travel. This factor is mainly dependent on battery capacity. So, with this prime obstacle, different EV models are evolving with variations in battery types. However, in many countries, like the United States, Finland, and India, the daily driving range is below 50 km. This implies that even the currently available EVs are capable of meeting users' drive range needs. Moreover, accurate knowledge about driving range increases users' acceptance.

10.8.2 Charging Infrastructure

10.8.2.1 Charging Location

As discussed previously, driving range plays an important role in the social acceptance of electric vehicles. That challenge of low driving range is eased by the availability of appropriate changing stations. In Qatar [51, 56], electric vehicles were more acceptable despite their low driving range. This is because of the availability of a large number of fuelling stations. Furthermore, the acceptance increases in a reduction in distances between charging stations [13]. The availability of charging stations on highways will boost heavy vehicles. Many users still prefer home-based charging features because they are economical and convenient. The presence of charging features in parking lots will further increase user acceptance.

10.8.2.2 Charging Duration

The second important issue related to charging is the time period required for charging. Duration for charging increases with an increase in battery range. To the benefit of the industry, many users are willing to accept a charging duration of up to three hours [57]. Still, many want instant charging facilities that provide the same experience as fuel engines. Taking into consideration the previously mentioned contradicting factors, it can be concluded that slow charging time is acceptable by users during night time, and during the day, users expect fast charging.

10.8.2.3 Charging Type

Charger type has a significant impact on user acceptance. Level 1 charging is the simplest and cheapest option, allowing drivers to plug their electric vehicle into a standard household outlet. While this may be convenient, the long charging time limits the range of the vehicle. Level 2 chargers are faster and more expensive, with an output of up to 240 volts, reducing the charging time needed. However, the installation of these chargers on two levels is not widely available, limiting usability for many. The most modern option for charging, Level 3 or DC fast chargers, can provide an 80% charge in just 30 minutes.

10.8.2.4 Environmental Factors

During a regular drive, electric vehicle emissions are very low. This contrasts with existing combustion engines in which drive emission is high. However, there are a few concerns related to the environmental impact of the manufacturing process of electric vehicles. This is mainly caused by sourcing and making Lithium-ion batteries. The majority of electric vehicles depend on Li-ion batteries. In these batteries, the cathode is normally prepared using a metal oxide, like cobalt, nickel, or manganese, which stores the positively charged ions. The anode is typically made up of graphite, which stores the negatively charged ions. The electrolyte is a non-aqueous solution that facilitates the movement of ions between the cathode and anode during discharge and charging. The main reason for its choice is great energy density and the charge-discharge cycle [3].

However, excavating lithium necessitates an enormous quantity of aquatics, raising concerns over depleting the environmental resources. During the manufacturing of Li-ion batteries, CO2 is released at a high level [58]. Furthermore, the improper disposal of the battery composition elements can cause environmental pollution [59, 60]. The previous are the major parameters impacting social acceptance of electric vehicles. Apart from these, factors like consumers' willingness to pay, government incentives, and so on will also impact consumers accordingly. Finally, other common factors like aesthetic appearance and impact on users driving style play a crucial role. Electric vehicles, however, scored better in these miscellaneous aspects. A great number of consumers had a positive perspective in relation to smooth driving, speeding within town limits, and economic factors [3, 61].

10.9 CONCLUSION

The rise of EVs is a promptly emergent trend in the automotive industry. With the increasing concern for the environment and the need for cleaner shipping, EVs have appeared as a popular resolution. In 2023, the adoption rate of electric vehicles has continued to increase with more consumers opting for these vehicles as their primary mode of transportation. This tendency is anticipated to endure as governments and automakers work together to reduce carbon emissions and encourage the usage of EVs. However, the widespread implementation of electric vehicles is quite hindered by several challenges. One of the major issues is the non-existence of charging architecture, which limits the coverage distance and convenience of EVs. Another challenge is the high price tag of EVs compared to conventional gas- and oil-based automobiles, although this gap is expected to close as battery technology improves and production costs decrease.

Despite these challenges, the future of electric vehicles looks bright. The development of more efficient and powerful batteries, coupled with the expansion of charging infrastructure, will make electric vehicles more accessible and convenient for consumers. Governments and automakers are also working together to incentivize the acceptance of EVs and to create a more sustainable transportation system. In conclusion, the rise of EVs in 2023 and beyond represents a significant shift in the automotive industry towards cleaner and more sustainable transportation. While challenges remain, the future looks promising for electric vehicles as technology continues to improve and more consumers adopt this new mode of transportation.

REFERENCES

[1]. Chikhi, F., El Hadri, A., & Cadiou, J. C. (2005, June). ABS control design based on wheel-slip peak localization. In *Proceedings of the Fifth International Workshop on Robot Motion and Control, 2005 RoMoCo'05* (pp. 73–77). IEEE.

[2]. Beier, J., Neef, B., Thiede, S., & Herrmann, C. (2016). Integrating on-site renewable electricity generation into a manufacturing system with intermittent battery storage from electric vehicles. *Procedia CIRP*, *48*, 483–488.

[3]. Sambamurthy, S., Raghuvanshi, S., & Sangwan, K. S. (2021). Environmental impact of recycling spent lithium-ion batteries. *Procedia CIRP*, *98*, 631–636.

[4]. Alkaabi, K., & Abuelgasim, A. (2017). Applications of unmanned aerial vehicle (UAV) technology for research and education in UAE. *International Journal of Social Sciences Arts and Humanities*, *5*(1), 4–11.

[5]. IEA. Electric car registrations and sales share in China, United States, Europe and other regions, 2016–2021. Paris: IEA. www.iea.org/data-and-statistics/charts/electric-car-registrations-and-sales-share-in-china-united-states-europe-and-other-regions-2016-2021

[6]. IEA (2022). Electric vehicles. IEA. www.iea.org/reports/electric-vehicles

[7]. https: www.goldmansachs.com/insights

[8]. Kempton, W., & Tomić, J. (2005). Vehicle-to-grid power fundamentals: Calculating capacity and net revenue. *Journal of Power Sources*, *144*(1), 268–279.

[9]. Bessa, R. J., & Matos, M. A. (2012). Economic and technical management of an aggregation agent for electric vehicles: A literature survey. *European Transactions on Electrical Power*, *22*(3), 334–350.

[10]. Kang, J. E., & Recker, W. W. (2009). An activity-based assessment of the potential impacts of plug-in hybrid electric vehicles on energy and emissions using 1-day travel data. *Transportation Research Part D: Transport and Environment*, *14*(8), 541–556.

[11]. Galus, M. D., Vayá, M. G., Krause, T., & Andersson, G. (2019). The role of electric vehicles in smart grids. *Advances in Energy Systems: The Large-scale Renewable Energy Integration Challenge*, *18*, 245–264.

[12]. Brady, J., & O'Mahony, M. (2016). Modelling charging profiles of electric vehicles based on real-world electric vehicle charging data. *Sustainable Cities and Society*, *26*, 203–216.

[13]. Morrissey, P., Weldon, P., & O'Mahony, M. (2016). Future standard and fast charging infrastructure planning: An analysis of electric vehicle charging behaviour. *Energy Policy*, *89*, 257–270.

[14]. Foley, A., Tyther, B., Calnan, P., & Gallachóir, B. Ó. (2013). Impacts of electric vehicle charging under electricity market operations. *Applied Energy*, *101*, 93–102.

[15]. Steinhilber, S., Wells, P., & Thankappan, S. (2013). Socio-technical inertia: Understanding the barriers to electric vehicles. *Energy Policy*, *60*, 531–539.

[16]. Yu, H., Tseng, F., & McGee, R. (2012, March). Driving pattern identification for EV range estimation. In *2012 IEEE International Electric Vehicle Conference* (pp. 1–7). IEEE.

[17]. Hayes, J. G., De Oliveira, R. P. R., Vaughan, S., & Egan, M. G. (2011, September). Simplified electric vehicle power train models and range estimation. In *2011 IEEE Vehicle Power and Propulsion Conference* (pp. 1–5). IEEE.

[18]. Salah, F., Ilg, J. P., Flath, C. M., Basse, H., & Van Dinther, C. (2015). Impact of electric vehicles on distribution substations: A Swiss case study. *Applied Energy*, *137*, 88–96.

[19]. Hartmann, N., & Özdemir, E. D. (2011). Impact of different utilization scenarios of electric vehicles on the German grid in 2030. *Journal of Power Sources*, *196*(4), 2311–2318.

[20]. Yang, F., Sun, Y., & Shen, T. (2007, October). Nonlinear torque estimation for vehicular electrical machines and its application in engine speed control. In *2007 IEEE International Conference on Control Applications* (pp. 1382–1387). IEEE.

[21]. Yin, D., Oh, S., & Hori, Y. (2009). A novel traction control for EV based on maximum transmissible torque estimation. *IEEE Transactions on Industrial Electronics*, *56*(6), 2086–2094.

[22]. Lu, L., Han, X., Li, J., Hua, J., & Ouyang, M. (2013). A review on the key issues for lithium-ion battery management in electric vehicles. *Journal of Power Sources, 226*, 272–288.

[23]. Yong, J. Y., Ramachandaramurthy, V. K., Tan, K. M., & Mithulananthan, N. (2015). A review on the state-of-the-art technologies of electric vehicle, its impacts and prospects. *Renewable and Sustainable Energy Reviews, 49*, 365–385.

[24]. Kumar, V., Kumar, S., & Sood, Y. R. (2017). Comparative analysis of electric vehicle powertrain configurations for passenger cars. *Energy Conversion and Management, 138*, 52–62.

[25]. Xiao, M., Tang, L., Zhang, X., Lun, I. Y. F., & Yuan, Y. (2018). A review on recent development of cooling technologies for concentrated photovoltaics (CPV) systems. *Energies, 11*(12), 3416.

[26]. Tseng, C. C., & Huang, Y. L. (2019). Comparative study on electric vehicle powertrain systems: Fuel cell versus battery. *International Journal of Hydrogen Energy, 44*(34), 18630–18642.

[27]. Li, Y., Wang, J., & Liu, Y. (2020). Comparative study of electric vehicle powertrain configurations for buses in different road conditions. *International Journal of Automotive Technology, 21*(2), 387–398.

[28]. Shen, W., Cao, Y., & Zhang, X. (2021). Comparative analysis of powertrain system design for electric vehicles with different energy storage systems. *IEEE Access, 9*, 120975–120984.

[29]. Hong, J., Wang, Z., Qu, C., Ma, F., Xu, X., Yang, J., Zhang, J., Zhou, Y., Shan, T., & Hou, Y. (2023). Fault prognosis and isolation of lithium-ion batteries in electric vehicles considering real-scenario thermal runaway risks. *IEEE Journal of Emerging and Selected Topics in Power Electronics, 11*(1), 88–99. doi: 10.1109/JESTPE.2021.3097827.

[30]. Kim, W., Lee, P. Y., Kim, J., & Kim, K. S. (2021). A robust state of charge estimation approach based on nonlinear battery cell model for lithium-ion batteries in electric vehicles. *IEEE Transactions on Vehicular Technology, 70*(6), 5638–5647.

[31]. Baek, D., & Chang, N. (2018). Runtime power management of battery electric vehicles for extended range with consideration of driving time. *IEEE Transactions on Very Large Scale Integration (VLSI) Systems, 27*(3), 549–559.

[32]. Cui, D., Wang, Z., Zhang, Z., Liu, P., Wang, S., & Dorrell, D. G. (2021). Driving event recognition of battery electric taxi based on big data analysis. *IEEE Transactions on Intelligent Transportation Systems, 23*(7), 9200–9209.

[33]. Yu, Z., Gan, C., Ni, K., Chen, Y., & Qu, R. (2020). Dual-electric-port bidirectional flux-modulated switched reluctance machine drive with multiple charging functions for electric vehicle applications. *IEEE Transactions on Power Electronics, 36*(5), 5818–5831.

[34]. Tran, M. K., Bhatti, A., Vrolyk, R., Wong, D., Panchal, S., Fowler, M., & Fraser, R. (2021). A review of range extenders in battery electric vehicles: Current progress and future perspectives. *World Electric Vehicle Journal, 12*(2), 54.

[35]. Lukic, S., & Pantic, Z. (2013). Cutting the cord. *IEEE Electrification Magazine, 1*(1), 57–64.

[36]. Jeong, S., Jang, Y. J., & Kum, D. (2015). Economic analysis of the dynamic charging electric vehicle. *IEEE Transactions on Power Electronics, 30*(11), 6368–6377.

[37]. Okasili, I., Elkhateb, A., & Littler, T. (2022). A review of wireless power transfer systems for electric vehicle battery charging with a focus on inductive coupling. *Electronics, 11*(9), 1355.

[38]. Sanguesa, J. A., Torres-Sanz, V., Garrido, P., Martinez, F. J., & Marquez-Barja, J. M. (2021). A review on electric vehicles: Technologies and challenges. *Smart Cities*, *4*(1), 372–404. https://doi.org/10.3390/smartcities4010022

[39]. Lewicki, W., Stankiewicz, B., & Olejarz-Wahba, A. A. (2020). The role of intelligent transport systems in the development of the idea of smart city. In *Smart and Green Solutions for Transport Systems: 16th Scientific and Technical Conference "Transport Systems. Theory and Practice 2019" Selected Papers 16* (pp. 26–36). Springer International Publishing.

[40]. Electric Vehicle Comprehensive Analysis and Investment Opportunities by 2027: The Future of Automotive Industry. Globe Newswire, February 26, 2020. www.globenewswire.com/en/news-release/2020/02/26/1990724

[41]. Tiwari, A., & Farag, H. (2022). Analysis and modeling of value creation opportunities and governing factors for electric vehicle proliferation. *Energies*, *16*(1), 438.

[42]. Parihar, A., & Urele, M. K. (2021). Status Quo Analysis of Various Segments of Electric Mobility and Low Carbon Passenger Road Transport in India. National Institution for Transforming India (NITI Aayog), India. https://policycommons.net/artifacts/2423747/status-quo-analysis-of-various-segments-of-electric-mobility-and-low-carbon-passenger-road-transport-in-india/3445339/.

[43]. Ministry of Power, Government of India. (2018). National Electric Mobility Mission Plan (NEMMP) 2020. Retrieved from https://powermin.nic.in/en/content/national-electric-mobility-mission-plan-nemmp-2020

[44]. Department of Heavy Industry, Government of India. (2018). FAME India Scheme. Retrieved from https://fame2.heavyindustry.gov.in/

[45]. Coffman, M., Bernstein, P., & Wee, S. (2017). Electric vehicles revisited: A review of factors that affect adoption. *Transport Reviews*, *37*(1), 79–93.

[46]. Goel, S., Sharma, R., & Rathore, A. K. (2021). A review on barrier and challenges of electric vehicle in India and vehicle to grid optimisation. Transportation Engineering, *4*, 100057.

[47]. Ministry of Power, Government of India. (2019). EV Charging Infrastructure. Retrieved from https://powermin.nic.in/en/content/ev-charging-infrastructure

[48]. Society of Manufacturers of Electric Vehicles. (2021). Sales of Electric Vehicles in India. Retrieved from www.smev.in/sales-dashboard/

[49]. Aayog, N. (2018). India's electric vehicle roadmap. https://e-amrit.niti.gov.in/national-level-policy

[50]. McKerracher, C., Izadi-Najafabadi, A., O'Donovan, A., Albanese, N., Soulopolous, N., Doherty, D., Boers, M., Fisher, R., Cantor, C., Frith, J., & Mi, S. (2020). Electric vehicle outlook 2020. Bloomberg New Energy Finance. Technical Report 2020. https://about.bnef.com/electric-vehicle-outlook-2020/

[51]. Ramachandaramurthy, V. K., Ajmal, A. M., Kasinathan, P., Tan, K. M., Yong, J. Y., & Vinoth, R. (2023). Social acceptance and preference of EV users: A review. *IEEE Access*, *11*, 11956–11972. doi: 10.1109/ACCESS.2023.3241636

[52]. Hwang, R. (2021). The next generation of hybrid cars: Plug-in hybrids can help reduce global warming and slash oil dependency. New York: U. S. National Resources Defense Council, 2007.

[53]. Irle, R. (2021). Global plug-in vehicle sales reached over 3, 2 million in 2020. EV-volumes.com (accessed on 23 June 2021).

[54]. Tesla—Statistics & Facts. Retrieved May 2021, from www.statista.com/topics/2086/tesla/

[55]. Proost, S., & van Dender, K. (2010). What sustainable road transport future?: Trends and policy options, OECD/ITF Joint Transport Research Centre Discussion Papers, No. 2010/14, OECD Publishing, Paris, https://doi.org/10.1787/5km4q8jnn7q4-en.

[56]. IEA. (2019). Global EV Outlook 2019: Scaling-up the transition to electric mobility. Paris: OECD Publishing. https://doi.org/10.1787/35fb60bd-en

[57]. Jang, S., & Choi, J. Y. (2021). Which consumer attributes will act crucial roles for the fast market adoption of electric vehicles?: Estimation on the asymmetrical & heterogeneous consumer preferences on the EVs. *Energy Policy*, *156*, 112469.

[58]. Lai, X., Gu, H., Chen, Q., Tang, X., Zhou, Y., Gao, F., Han, X., Guo, Y., Bhagat, R., & Zheng, Y. (2022). Investigating greenhouse gas emissions and environmental impacts from the production of lithium-ion batteries in China. *Journal of Cleaner Production*, *372*, 133756.

[59]. Jiang, S., Hua, H., Zhang, L., Liu, X., Wu, H., & Yuan, Z. (2022). Environmental impacts of hydrometallurgical recycling and reusing for manufacturing of lithium-ion traction batteries in China. *Science of the Total Environment*, *811*, 152224.

[60]. Priessner, A., Sposato, R., & Hampl, N. (2018). Predictors of electric vehicle adoption: An analysis of potential electric vehicle drivers in Austria. *Energy Policy*, *122*, 701–714.

[61]. Habich-Sobiegalla, S., Kostka, G., & Anzinger, N. (2018). Electric vehicle purchase intentions of Chinese, Russian and Brazilian citizens: An international comparative study. *Journal of Cleaner Production*, *205*, 188–200.

Chapter 11

Techno-Economic and Future Aspects of the HEV-EV-FCV

Decarbonization, Digitalization, and Sustainability

Varun Pratap Singh, Ashwani Kumar

11.1 INTRODUCTION TO HEV-EV-FCV TECHNOLOGIES

Hybrid Electric Vehicles (HEVs), Electric Vehicles (EVs), and Fuel Cell Vehicles (FCVs) are at the forefront of a growing movement towards more eco-friendly and sustainable modes of transportation throughout the world [1]. Reduced emissions of greenhouse gases, improved air quality, and lower fuel costs are just a few of the advantages of these automobiles. In this chapter, we will take a look at the inner workings of three different technologies [2].

11.1.1 Hybrid Electric Vehicles (HEVs)

Vehicles that use both an electric motor and a conventional internal combustion engine are known as hybrid electric vehicles (HEVs). The battery can be charged by an internal combustion engine or via regenerative braking. Hybrid electric vehicles (HEVs) are cleaner and more environmentally friendly than conventional gasoline-powered automobiles. To augment the internal combustion engine and provide the wheels more power, HEVs use an electric motor. When the vehicle slows down, the engine turns into a generator, using the brakes to supply energy to the batteries. When the car is stopped or idling, the engine shuts off and the electric motor takes control [3].

There are three primary classifications of HEV categorization, including series HEV, parallel HEV, and series-parallel HEV. In series (extended-range) hybrid vehicles, a combustion engine is employed to operate an electric generator. This generator is responsible for recharging a battery and providing power to an electric motor, which ultimately propels the vehicle forward. The Cadillac ELR, Chevrolet Volt, and Fisker Karma are hybrid vehicles that have garnered attention in the automotive industry [4]. In contrast, parallel hybrids incorporate a mechanical gearbox, an internal combustion engine, and an electric motor [5]. The electric motor has the capability to operate as a generator when necessary, and the allocation of power between the engine and the motor is optimized to provide optimal

 DOI: 10.1201/9781003464556-11

performance. In instances where more power is required, the electric motor inside a parallel mild hybrid system primarily functions as an augmentative force. Ultimately, series-parallel hybrids have the capability to operate using only the internal combustion engine, solely the electric motor, or a combination of both simultaneously. The power distribution of the engine has been meticulously designed to optimize its operation within its optimal range [6].

Other varieties of hybrid electric vehicles (HEVs) include full hybrid electric vehicles (FHEVs), mild hybrid electric vehicles (MHEVs), and plug-in hybrid electric vehicles (PHEVs). FHEVs are equipped with both an internal combustion engine and an electric motor, which may function independently or in conjunction with each other. MHEVs include a modest electric motor to boost the efficiency of the petrol engine. Due to their bigger battery capacity, PHEVs can go further on electric power alone than MHEVs [7].

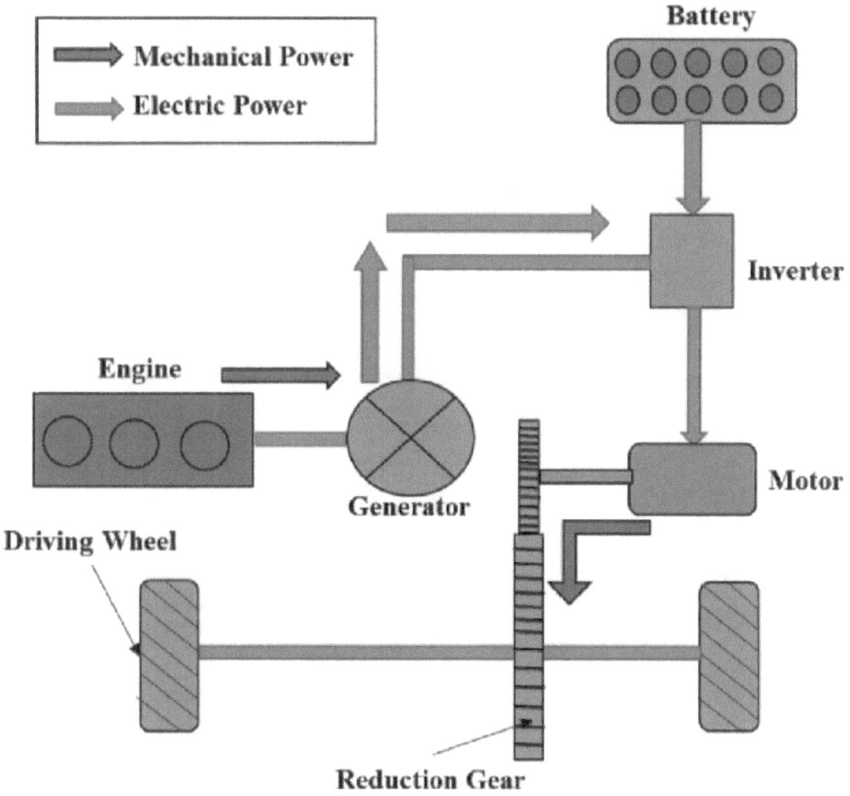

Figure 11.1 (a, b, c): Schematic Diagram of Different Hybrid Electric Vehicles: (a) Series (Extended-Range) Hybrid EV; (b) Parallel Hybrid EV; (c) Series-Parallel Hybrid EV.

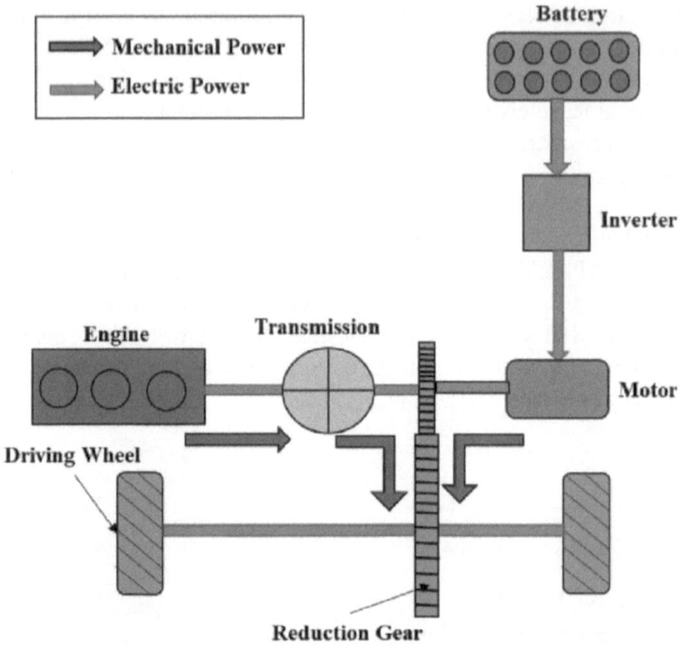

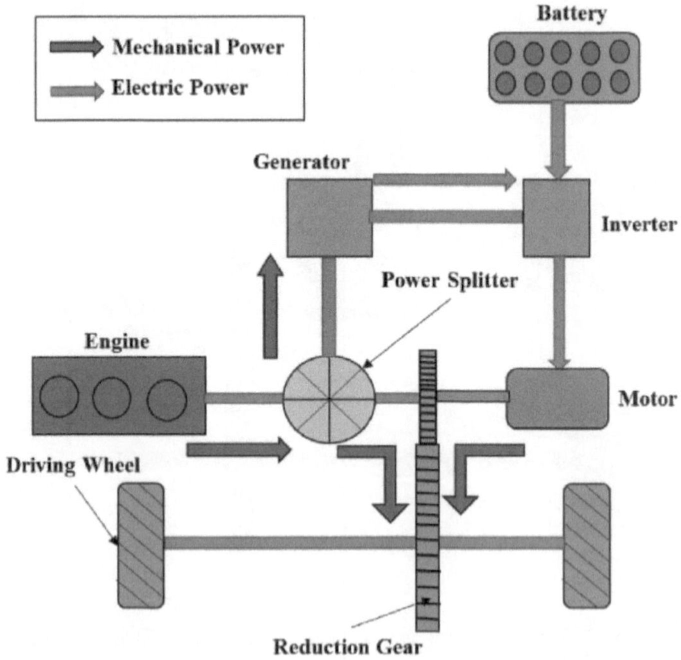

Figure 11.1 (Continued)

11.1.2 Electric Vehicles (EVs)

Electric vehicles (EVs) are propelled by electric motors that obtain their energy from stored electricity in batteries. The batteries have the capability to undergo recharging through the utilization of external power sources, such as charging outlets within residential settings or dedicated charging stations. To commence the movement of the vehicle, the utilization of electric motors is employed to propel the wheels. EVs are widely acknowledged for their ability to minimize emissions, achieve high levels of efficiency, and operate silently [8]. The electric motor of an EV produces a magnetic field through the utilization of energy derived from the vehicle's battery. The magnetic field is accountable for the production of torque, which subsequently propels the wheels of the vehicle. The battery charging process takes place when the vehicle is connected to an external power source. The process of recharging a battery is essential once it has exhausted its power supply. The duration of this depletion can vary depending on factors such as the specific type of battery and the rate at which it is being charged [9].

Battery Electric Vehicles (BEVs) can be classified based on the arrangement and positioning of their motors, which include single-motor, dual-motor, tri-motor, and quad-motor all-wheel drive configurations. BEVs with a single motor possess a sole electric motor that is tasked with the propulsion and provision of power to either the front or rear wheels [10]. Dual-motor BEVs, often denoted as all-wheel drive (AWD) BEVs, are furnished with a pair of electric motors, whereby one motor propels the front wheels and the other motor propels the rear wheels. The current design offers improved traction, performance, and handling by independently regulating the distribution of torque to each axle. Tri-motor BEVs are outfitted with a trio of electric motors, with a pair situated at the rear axle and one positioned at the front axle. This particular arrangement allows these vehicles to demonstrate exceptional power, torque, and AWD attributes, making them highly suitable for scenarios involving high-performance and off-road usage. The automobile is outfitted with a quadruple-motor system that facilitates the distribution of power to each of the four wheels, thereby granting the vehicle the capability of all-wheel drive. BEVs further enhance this concept by employing four electric motors, with each motor specifically allocated to propel an individual wheel [11]. This particular configuration provides unparalleled levels of traction, torque vectoring, and performance, allowing for precise control and exceptional manoeuvrability in a wide range of driving situations. Quad-motor systems are commonly observed in the context of high-performance electric supercars or sophisticated off-road vehicles. The categorization of BEVs according to their motor design facilitates the choice of vehicles with distinct performance attributes, hence offering a range of alternatives to accommodate diverse driving requirements and individual preferences [12, 13].

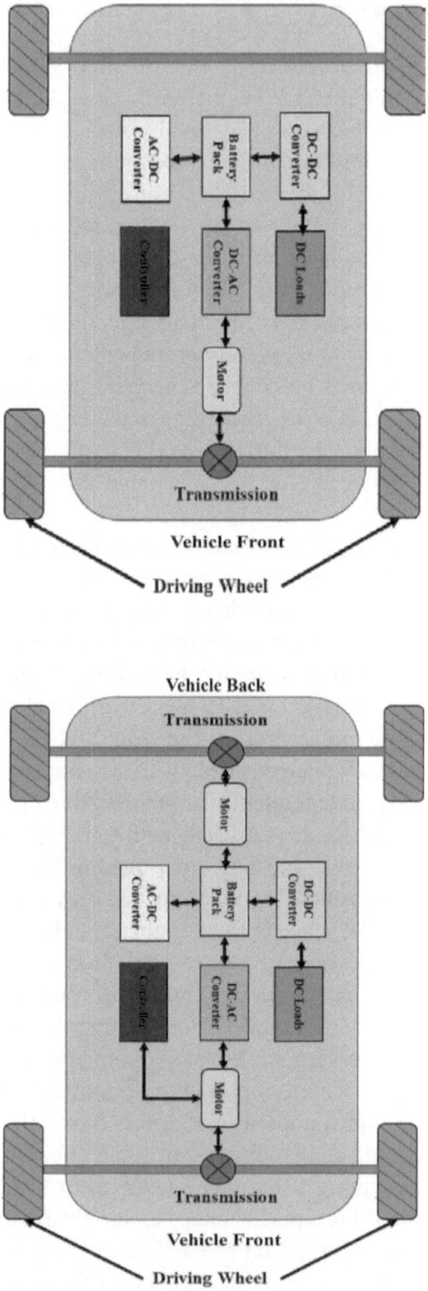

Figure 11.2 (a, b, c, d): Schematic Diagram of Different Electric Drive-Train Topologies: (a) Single-Motor Front-Wheel Drive; (b) Dual-Motor All-Wheel Drive; (c) Triple Motor-Wheel Drive; (d) Quad-Motor All-Wheel Drive.

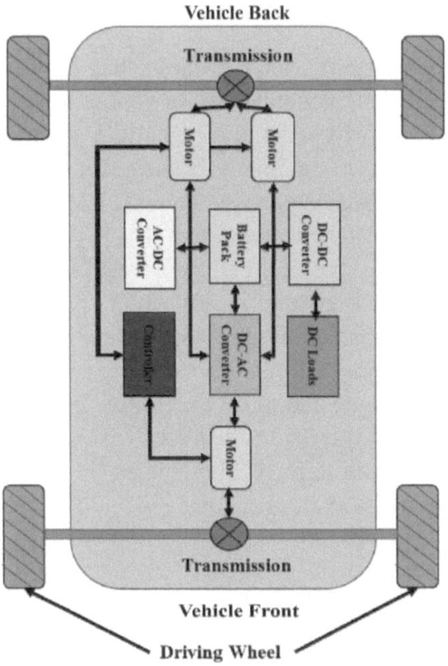

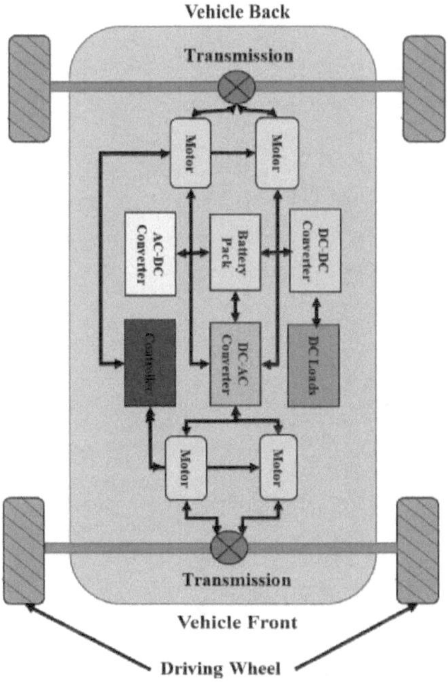

Figure 11.2 (Continued)

11.1.3 Fuel Cell Vehicles (FCVs)

Fuel cell vehicles (FCVs) employ electric motors that derive their power from fuel cell stacks. The fuel cell stack generates water vapour as its sole emission during the process of converting hydrogen and oxygen into electrical energy. The commendable efficiency and environmentally friendly operation of fuel cell vehicles (FCVs) are well acknowledged. Fuel cell vehicles (FCVs) utilize energy generated by a fuel cell stack to power the electric motor [14, 15]. The process of combining hydrogen and oxygen within the fuel cell stack results in the generation of both electricity and water vapour. Like traditional gasoline-powered vehicles, hydrogen is stored in high-pressure tanks within the vehicle and can be refuelled in a couple of minutes. Fuel cell vehicles (FCVs) are available in two distinct types: those that employ reformulated hydrogen and those that utilize pure hydrogen. Pure hydrogen fuel cell vehicles (FCVs) utilize reformed hydrogen as their primary source of energy storage and utilization. Fuel cell vehicles (FCVs) have the capability to convert various fuel sources, such as natural gas, into hydrogen through the use of a reformer [16].

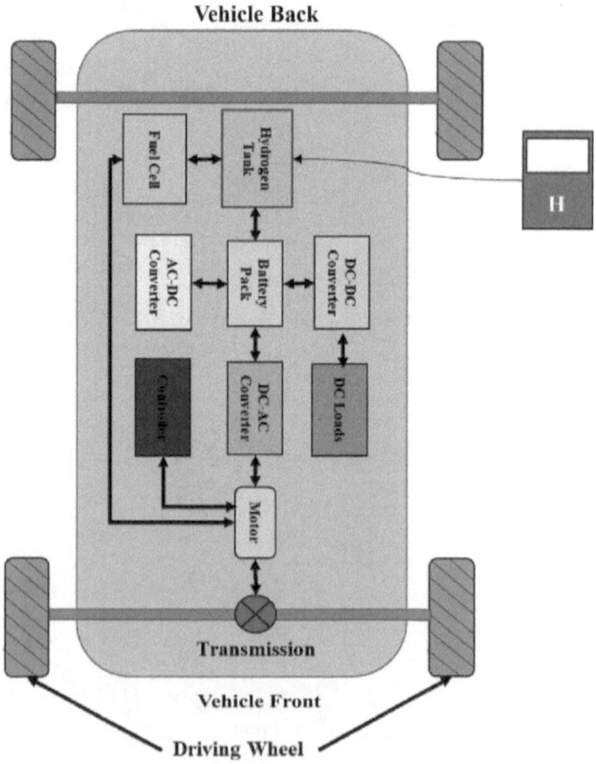

Figure 11.3 Schematic Diagram of a Fuel Cell Vehicle.

11.2 EVOLUTION OF THE ELECTRIC VEHICLE MARKET

The sales figures for electric vehicles are experiencing a steady upward trend. The electric vehicle market has had positive effects as a result of governmental initiatives aimed at promoting environmentally friendly automobiles. Numerous countries have a strong inclination to substitute conventional gasoline-powered automobiles with electric vehicles [17].

The declining cost of batteries is an additional factor that contributes to the increasing demand for electric vehicles. According to a report by Bloomberg NEF, it was projected that the cost of battery packs would decrease to $137 per kilowatt-hour (kWh) by the year 2020. This is a significant reduction from the price of $1,100 per kWh recorded in 2010. According to a study, it is projected that by the year 2030, the cost of a battery pack will have significantly decreased to $58 per kWh. This reduction in cost is expected to make electric cars economically comparable to conventional gas-powered automobiles [18, 19]. The electric vehicle market is seeing growth and expansion. The increasing demand for electric vehicles (EVs) can be attributed to the concurrent promotion of eco-friendly transportation policies by governments and the decreasing prices of batteries. Consequently, it is anticipated that the proportion of sales attributed to electric vehicles will continue to increase, bringing about a transformative impact on the transportation industry [20].

The electric vehicle sector in India has experienced substantial growth due to governmental support and initiatives [21]. According to the Society of Manufacturers of Electric Vehicles, there was a 20% rise in the sales of electric vehicles in India, reaching a total of 156,000 units in the fiscal year 2020–2021. According to the data, there was a significant surge of 140% in the sales of electric two-wheelers in 2018 [22].

Several European governments are actively pursuing a future scenario whereby the exclusive sale of electric vehicles is realized. According to the European Alternative Fuels Observatory, there is an anticipated surge of 137% in electric vehicle sales in 2020 compared to the figures recorded in 2017 [23]. The market share of electric vehicles (EVs) had a notable growth, rising from 3.7% in 2017 to 10.5% in 2020 [24].

The electric vehicle market in the United States has the most rapid growth among North American countries. According to the International Energy Agency, there was a 4% growth in the sales of electric vehicles in the United States for the year 2020. The market share of electric vehicles in the United States had growth from 1.8% in 2019 to 2.4% in 2020 [25].

China possesses the largest electric vehicle market in the Asia-Pacific region. According to the China Association of Automobile Manufacturers, there is an anticipated 10% growth in the sales of electric vehicles in China for the year 2020. According to a study conducted in 2020, the market share of electric vehicles in China witnessed a growth from 4.5% in 2019 to 5.4% in 2020 [26].

According to the International Energy Agency, it is anticipated that the sales volume of electric vehicles will see a notable surge in 2020, reaching a record-breaking figure of 3 million units. This projection reflects a substantial increase of 41% compared to the sales volume observed in the preceding year, 2019. The market penetration of electric vehicles (EVs) had a notable rise, escalating from 2.5% in 2019 to 4.6% in the subsequent year of 2020. According to the International Energy Agency (IEA), the projected global number of electric vehicles (EVs) is expected to increase significantly from 11 million in 2020 to a remarkable 145 million by the year 2030 [27].

The electric vehicle industry on a worldwide scale is seeing significant growth, as several countries set ambitious goals to replace conventional gasoline-powered automobiles with electric alternatives. The transportation sector is currently experiencing substantial changes as a result of the growing adoption of electric vehicles. This can be attributed to the declining costs of batteries and the advancements made in supporting infrastructure [28].

11.3 ENVIRONMENTAL BENEFITS OF EV-HEV-FCV

Compared to conventional gasoline-powered automobiles, EVs, HEVs, and FCVs are considered more environmentally friendly as they emit significantly fewer greenhouse gases than conventional gasoline-powered automobiles. According to research [29], electric vehicles exhibit reduced levels of greenhouse gas emissions compared to their gasoline-powered counterparts. In contrast to conventional gasoline-powered vehicles, HEVs, and FCVs have a reduced carbon footprint due to their enhanced fuel efficiency. EVs, HEVs, and FCVs generate significantly lower levels of particulate matter, nitrogen oxides, and volatile organic compounds compared to conventional vehicles. Improving air quality is of particular significance in urban regions [30–34].

Vehicles powered by gasoline have higher energy consumption per mile compared to EVs, HEVs, and FCVs. In addition to saving resources, this reduces harmful pollutants and fossil fuel use in transportation. The quieter electric motors of EVs, HEVs, and FCVs contribute to a reduction in ambient noise [35]. This improves urban dwellers' quality of life by lowering ambient noise. When compared to conventional gasoline-powered vehicles, EVs, HEVs, and FCVs significantly reduce emissions and improve air quality. These environmental benefits will grow as the technology advances and is used more broadly, ultimately leading to a more sustainable transportation infrastructure [36].

11.4 ECONOMIC ASPECTS OF EV-HEV-FCV

Compared to conventional gasoline-powered vehicles, electric vehicles, hybrids, and fuel-cell vehicles provide substantial financial benefits. Prices for EVs, HEVs, and FCVs are higher than for gasoline-powered vehicles because of the vehicles' more complex technology and batteries. With more efficient machinery and greater output, costs will gradually decrease. Vehicle operating expenses for EVs, HEVs, and FCVs are lower than gasoline-powered vehicles because of their greater energy efficiency and reduced repair and replacement costs. Since electric vehicles have fewer moving parts, they require less upkeep than their gasoline-powered counterparts. Fuel costs are reduced since electricity is less expensive than petrol [37].

Subsidies for EVs, HEVs, and FCVs are available in several countries. Financial incentives include things like tax refunds and credits. There has been a rise in the number of zero-emissions vehicles on the road thanks to mandates in some nations. The availability of EV charging stations has a monetary impact [38–40]. As more electric vehicles enter the market, there will be a greater need for public charging stations. Governments and private companies are spending money on charging stations, but the high cost of getting them up and running might discourage consumers. EVs, HEVs, and FCVs all use large batteries that need to be recycled. It may be difficult and costly to get recycled batteries. Costs for recycling batteries should go down as demand goes up because of economies of scale [41].

Manufacturing, charging infrastructure, and battery recycling are all industries that will benefit from the rise of EVs, HEVs, and FCVs. However, typical automotive employment may be lost if people move away from gasoline-powered automobiles. Reduced reliance on imported oil is one of the benefits of EVs, HEVs, and FCVs. For countries that need to import oil, this is critical [42].

Electric vehicles, hybrid electric vehicles, and fuel cell vehicles all have higher resale values than gas-powered vehicles due to lower operating costs and longer lifespans. As consumers learn about the environmental benefits of EVs, HEVs, and FCVs, demand for used versions of these vehicles will increase [43]. The total cost of ownership over the course of an EV's lifespan is an important factor to consider. In the long run, electric, hybrid, and FCVs are the superior purchase due to their higher resale values and reduced operating costs. As technology advances and prices decrease, these vehicles are expected to compete with gasoline-powered vehicles.

11.5 DECARBONIZATION AND DIGITALIZATION OF THE TRANSPORTATION INDUSTRY

Transportation must be the focal point of initiatives to reduce greenhouse gas emissions. Decarbonization reduces carbon emissions by transitioning to low-carbon energy and decreasing energy consumption [44].

11.5.1 The Importance of Decarbonization in the Transportation Industry

The transportation industry is responsible for nearly a quarter of the world's CO2 emissions, with road transportation being the largest contributor. Decarbonizing the transport sector is, therefore, essential for limiting global warming [45]. Significant environmental, social, and economic benefits such as enhanced air quality, reduced reliance on fossil fuels, and job creation in the renewable energy and transportation sectors may result from decarbonizing the business [46].

11.5.2 Challenges in Decarbonization

The decarbonization of the transportation sector is hindered by technological limitations, infrastructure development, and consumer acceptance. Developing cost-effective, reliable, and energy-dense battery solutions for electric vehicles is one of the main challenges [47]. Moreover, establishing charge infrastructure, especially in rural areas and developing nations, is a formidable challenge. Lastly, attaining consumer acceptability and comprehension of low-carbon transit options, as well as resolving range anxiety issues, continues to be challenging [48–51].

11.5.3 Strategies to Achieve Decarbonization

Several decarbonization methods are available. Increasing public transportation, transitioning to electric vehicles, employing alternative fuels such as hydrogen and biofuels, enhancing energy efficiency, and investing in a sustainable transportation infrastructure are a few examples. Moreover, regulations such as carbon pricing, fuel economy standards, and subsidies can expedite the transition to low-carbon transportation [52].

11.5.4 The Role of Digitalization in the Transportation Industry

Digitalization refers to the use of digital technology to improve the efficiency, effectiveness, and sustainability of transportation networks. Digitalization is essential for enhancing transportation efficacy, reducing energy consumption, and hastening the transition to low-carbon modes of transportation.

11.5.5 The Need for Digitalization for Sustainable Transportation

Because of digitalization, the transportation industry can utilize low-carbon energy sources such as renewables, making it more environmentally friendly. Digitalization may also enhance the effectiveness of conveyance systems, thereby conserving energy and resources. Smart transport systems enhance traffic flow, reduce congestion, and promote the sustainability of transport networks through digitalization [53].

11.5.6 The Future of Digitalization in HEV-EV-FCV Technologies

For the development of HEV-EV-FCV technology, digitization is necessary. There is the potential for digitalization to increase the effectiveness, range, and energy efficacy of these vehicles. The development of smart charging infrastructure and the incorporation of renewable energy into automobiles may also be aided by digitalization. Lastly, digitization can facilitate remote vehicle monitoring, autonomous driving, and integrated services, which all enhance the user experience of these vehicles [54].

11.6 TECHNOLOGICAL ADVANCES AND FUTURE PROSPECTS OF HEV-EV-FCV

To promote sustainable transportation, hybrid electric vehicles (HEVs), electric vehicles (EVs), and fuel cell vehicles (FCVs) have all been developed. These technologies decrease pollution, energy usage, and expenses. Hybrid electric vehicles combine an internal combustion engine with an electric motor and battery to enhance fuel economy. Hydrogen fuel cells power the electric motors of fuel cell-powered automobiles [55].

11.6.1 Technological Advances in HEV-EV-FCV

Because of developments in battery technology, electric motors, and charging infrastructure, HEV-EV-FCV technologies have advanced. Enhanced vehicle range, performance, and charging times have lowered the bar for widespread adoption [56].

11.6.2 Future Prospects and Challenges of HEV-EV-FCV

As more governments and businesses commit to low-carbon mobility, HEV-EV-FCV technologies have a promising future. However, we must improve battery technology, vehicle prices, and the charging infrastructure. Renewable energy and eco-friendly fuels, such as hydrogen, are necessary for the long-term viability of these technologies [57].

11.6.3 Impact of HEV-EV-FCV on Sustainable Transportation

These technologies have the potential to mitigate climate change and enhance air quality due to the impact of HEVs, EVs, and FCVs on sustainable mobility. Thanks to these innovations, green energy and transport jobs may be created, along with greater energy security and less reliance on fossil fuels. The use of HEVs, EVs, and FCVs has the potential to drastically improve the transportation industry [58].

11.7 ECONOMIC AND SOCIAL IMPLICATIONS

Greater economic and social benefits may result from the widespread adoption of HEV-EV-FCV technologies, which enable more environmentally favourable modes of transportation.

11.7.1 Economic Implications

Widespread use of HEV, EV, and FCV technologies may result in employment and economic growth in the renewable energy and transportation industries. In addition, purchasers can save a substantial amount of money over the life of the vehicle due to reduced operating costs and the possibility of greater energy efficiency. There may be a need for public policy measures to encourage consumers and corporations to invest in these technologies, but the high cost of entry for these vehicles continues to be a significant barrier to adoption [59].

11.7.2 Social Implications

In a major way, HEVs, EVs, and FCVs affect the future of eco-friendly transportation. Because of their ability to lessen emissions and increase energy efficiency, these developments are vital for fighting climate change and improving air quality. These methods can lessen the world's need for fossil fuels, boost energy reliability, and open up a plethora of new career paths in the renewable energy and transportation sectors. The convergence of HEV, EV, and FCV technologies might usher in a greener, more environmentally friendly transportation age [60].

11.8 FACTORS AFFECTING THE ADOPTION OF HEV-EV-FCV TECHNOLOGIES

Price, range, charging infrastructure, customer awareness, and education all play a role in the adoption of HEV-EV-FCV technology. Although these technologies have come a long way, they still have a way to go before they are fully developed and ready for widespread usage [61].

11.8.1 The Role of Public Policy in Promoting Sustainable Transportation

Adopting HEV-EV-FCV technology and making the switch to sustainable transportation may be aided by public policy. These factors include but are not limited to monetary incentives for customers, investments in charging infrastructure, monies for new technological research and development, as well as rules and standards to assure the safety and reliability of these cars. Public education and awareness efforts may increase consumer demand and ease the transition to sustainable transportation [62].

11.9 SUSTAINABILITY IN THE AUTOMOTIVE INDUSTRY

The automotive industry must emphasize sustainability due to the transportation sector's large share of global greenhouse gas emissions. This is crucial to lessen the impact of climate change and provide a better future for humanity.

11.9.1 Importance of Sustainability

The automotive sector must emphasize sustainability to effectively handle the environmental, economic, and social difficulties associated with vehicle production and operation. For the automotive sector, sustainability is working to lessen the negative effects of vehicles and their production on the environment. It also involves ensuring economic viability and competitiveness, as well as promoting social responsibility and ethical practices [63].

11.9.2 Challenges in Achieving Sustainability

The complexity of supply networks and regulatory bodies, the high resource and energy demands of manufacturing processes, and the scarcity of sustainable materials and technologies all work against the auto industry's efforts to become more sustainable. In addition to achieving environmental targets, the automobile sector must solve problems associated with vehicle design and performance [64].

11.9.3 Strategies to Achieve Sustainability

Measures that solve issues and encourage sustainable practices are needed to achieve sustainability in the car sector. Increased use of renewable energy in manufacturing, decreased waste and increased efficiency in production, the development and use of sustainable materials, improved vehicle design and performance to lower emissions and increase efficiency, and the promotion of sustainable practices throughout the supply chain are all examples of such strategies. Furthermore, businesses, governments, and other stakeholders need to work together to overcome complicated hurdles and accomplish long-term goals [65].

11.10 CONCLUSION AND RECOMMENDATIONS

The implementation of Hybrid Electric Vehicle (HEV), Electric Vehicle (EV), and Fuel Cell Vehicle (FCV) technologies has the potential to contribute significantly to the reduction of carbon emissions, the integration of digital advancements, and the promotion of sustainable practices within the transportation sector. In recent years, there have been substantial advancements in the field of technology. However, despite these advancements, certain obstacles continue to impede its general adoption and successful use.

It is imperative for policymakers, industry stakeholders, and other pertinent actors to collaborate to formulate and execute facilitative policies, incentives, and regulations that foster the advancement and adoption of these technologies, ultimately leading to the attainment of sustainable mobility. The allocation of resources towards the advancement of cutting-edge technology and materials research and development, alongside the establishment of essential infrastructure, such as charging stations and hydrogen refuelling stations, has the potential to facilitate the shift towards environmentally friendly modes of transportation. In order to promote equitable distribution of sustainable mobility advantages, with a particular focus on low-income populations and rural regions, it is imperative that these technologies are made readily available and economically feasible. This objective may be achieved by implementing strategies such as discounted purchase and lease programmes, finance initiatives, and the promotion of shared transit alternatives.

In summary, the implementation of HEV-EV-FCV technologies has the potential to enhance the sustainability of the transportation sector. However, the successful adoption and optimization of these technologies necessitate a collaborative endeavour and a comprehensive, forward-thinking approach.

11.11 FUTURE OUTLOOK FOR HEV-EV-FCV

The decarbonization, digitization, and sustainability of transportation may be achieved via the use of hybrid electric vehicle (HEV), electric vehicle (EV), and fuel cell vehicle (FCV) technologies. Despite recent advances, there are still barriers to the general adoption and successful use of these technologies.

For these technologies to advance and be adopted in the context of sustainable mobility, policymakers, industry stakeholders, and other relevant actors must work together to create and implement enabling policies, incentives, and laws. Sustainable mobility may be easier to attain with the help of cutting-edge technology, high-tech materials, and solid infrastructure like charging stations and hydrogen refuelling stations. It is crucial that these technologies be reasonably priced so that underprivileged people and residents of rural regions may have access to them and benefit from them. Renting, borrowing, and ridesharing programmes that get government funding subsidies can help accomplish this goal. In conclusion, the development of HEVs, EVs, and FCVs has the potential to improve the environmental friendliness of the transportation sector. To overcome challenges and maximize these technologies' potential will require concerted teamwork and an all-encompassing, future-focused strategy.

REFERENCES

[1] A. G. Olabi, T. Wilberforce, and M. A. Abdelkareem, "Fuel cell application in the automotive industry and future perspective," *Energy*, vol. 214, p. 118955, 2021, https://doi.org/10.1016/j.energy.2020.118955.

[2] S. J. Skerlos and J. J. Winebrake, "Targeting plug-in hybrid electric vehicle policies to increase social benefits," *Energy Policy*, vol. 38, no. 2, pp. 705–708, 2010, https://doi.org/10.1016/j.enpol.2009.11.014.

[3] K. Jyotheeswara Reddy and S. Natarajan, "Energy sources and multi-input DC-DC converters used in hybrid electric vehicle applications – A review," *Int. J. Hydrogen Energy*, vol. 43, no. 36, pp. 17387–17408, 2018, https://doi.org/10.1016/j.ijhydene.2018.07.076.

[4] B. Xiao, J. Ruan, W. Yang, P. D. Walker, and N. Zhang, "A review of pivotal energy management strategies for extended range electric vehicles," *Renew. Sustain. Energy Rev.*, vol. 149, p. 111194, 2021, https://doi.org/10.1016/j.rser.2021.111194.

[5] V. P. Singh, S. Jain, and A. Kumar, "Establishment of correlations for the thermo-hydraulic parameters due to perforation in a multi-V rib roughened single pass solar air heater," *Exp. Heat Transf.*, vol. 35, no. 5, pp. 1–20, 2022, https://doi.org/10.1080/08916152.2022.2064940.

[6] V. Krithika and C. Subramani, "A comprehensive review on choice of hybrid vehicles and power converters, control strategies for hybrid electric vehicles," *Int. J. Energy Res.*, vol. 42, no. 5, pp. 1789–1812, 2018, https://doi.org/10.1002/er.3952.

[7] C. Mi and M. A. Masrur, *Hybrid electric vehicles: Principles and applications with practical perspectives*. UK: John Wiley & Sons, 2017.

[8] V. P. Singh, S. Jain, A. Karn, A. Kumar, G. Dwivedi, C. S. Meena, N. Dutt, and A. Ghosh, "Recent developments and advancements in solar air heaters: A detailed review," *Sustain.*, vol. 14, no. 19, pp. 1–57, 2022, https://doi.org/10.3390/su141912149.

[9] K. L. Maheswari, S. Kavitha, and M. Kathiresh, "Introduction to Electric Vehicles and Hybrid Electric Vehicles BT", in *E-Mobility: A New Era in Automotive Technology*, M. Kathiresh, G. R. Kanagachidambaresan, and S. S. Williamson, Eds. Cham: Springer International Publishing, 2022, pp. 1–29.

[10] V. P. Singh, S. Jain, A. Karn, G. Dwivedi, A. Kumar, S. Mishra, N. K. Sharma, M. Bajaj, H. M. Zawbaa, and S. Kamel, "Heat transfer and friction factor correlations development for double pass solar air heater artificially roughened with perforated multi-V ribs," *Case Stud. Therm. Eng.*, vol. 39, p. 102461, 2022, https://doi.org/10.1016/j.csite.2022.102461.

[11] V. P. Singh, S. Jain, and J. M. L. Gupta, "Analysis of the effect of variation in open area ratio in perforated multi-V rib roughened single pass solar air heAter-PArt A," *Energy Sources, Part A Recover. Util. Environ. Eff.*, vol. 44, pp. 1–21, 2022, https://doi.org/10.1080/15567036.2022.2029976.

[12] T. Sivasakthivel, V. Verma, R. Tarodiya, C. S. Meena, V. P. Singh, and R. Kumar, "Chapter 11: Analysis of Optimum Operating Parameters for Ground Source Heat Pump System for Different Cases of Building Heating and Cooling Mode Operations," in *Thermal Energy Systems: Design, Computational Techniques, and Applications*, 1st ed., A. Kumar, N. Dutt, V. Singh, and C. Meena, Eds. Boca Raton, FL USA: CRC Press, 2023, pp. 183–207.

[13] S. F. Tie and C. W. Tan, "A review of energy sources and energy management system in electric vehicles," *Renew. Sustain. Energy Rev.*, vol. 20, pp. 82–102, 2013, https://doi.org/10.1016/j.rser.2012.11.077.

[14] K. V. Kordesch and G. R. Simader, "Environmental impact of fuel cell technology," *Chem. Rev.*, vol. 95, no. 1, pp. 191–207, 1995, https://doi.org/10.1021/cr00033a007.

[15] S. Mekhilef, R. Saidur, and A. Safari, "Comparative study of different fuel cell technologies," *Renew. Sustain. Energy Rev.*, vol. 16, no. 1, pp. 981–989, 2012, https://doi.org/10.1016/j.rser.2011.09.020.

[16] M. Nayal, A. K. Sharma, S. Jain, and V. P. Singh, "Chapter 11 Green Hydrogen Production Methods, Designs and Applications," in Highly Efficient Thermal Renewable Energy Systems Design, Optimization and Applications, 1st ed., R. W. V. Verma; S Thangavel, N Dutt, A Kumar, Ed. CRC Press, 2024, pp. 178–193.

[17] A. K. Sharma, M. Nayal, S. Jain, and V. P. Singh, "Chapter 07 Optimization Techniques of Solar Thermal and Hybrid Energy Systems," in Highly Efficient Thermal Renewable Energy Systems Design, Optimization and Applications, 1st ed., R. W. V. Verma; S Thangavel, N Dutt, A Kumar, Ed. CRC Press, 2024, pp. 1–14.

[18] M. Saini, A. Sharma, V. P. Singh, G. Dwivedi, and S. Jain, "Solar thermal receivers – A review," *Adv. Mater. Manuf. Energy Eng.*, vol. II, pp. 310–325, 2022, https://doi.org/10.1007/978-981-16-8341-1.

[19] E. Shafiei, H. Thorkelsson, E. I. Ásgeirsson, B. Davidsdottir, M. Raberto, and H. Stefansson, "An agent-based modeling approach to predict the evolution of market share of electric vehicles: A case study from Iceland," *Technol. Forecast. Soc. Change*, vol. 79, no. 9, pp. 1638–1653, 2012, https://doi.org/10.1016/j.techfore.2012.05.011.

[20] S. Carley, S. Siddiki, and S. Nicholson-Crotty, "Evolution of plug-in electric vehicle demand: Assessing consumer perceptions and intent to purchase over time," *Transp. Res. Part D Transp. Environ.*, vol. 70, pp. 94–111, 2019, https://doi.org/10.1016/j.trd.2019.04.002.

[21] M. Nilsson and B. Nykvist, "Governing the electric vehicle transition – Near term interventions to support a green energy economy," *Appl. Energy*, vol. 179, pp. 1360–1371, 2016, https://doi.org/10.1016/j.apenergy.2016.03.056.

[22] A. Ahmad, Z. A. Khan, M. Saad Alam, and S. Khateeb, "A review of the electric vehicle charging techniques, standards, progression and evolution of EV technologies in Germany," *Smart Sci.*, vol. 6, no. 1, pp. 36–53, 2018, https://doi.org/10.1080/23080477.2017.1420132.

[23] M. Nayal, A. K. Sharma, S. Jain, and V. P. Singh, "Chapter 06 Design and Modelling of Solar, Geothermal and Hybrid Energy Systems.pdf," in Highly Efficient Thermal Renewable Energy Systems Design, Optimization and Applications, 1st ed., R. W. V. Verma; S Thangavel, N Dutt, A Kumar, Ed. CRC Press, 2024, pp. 1–15.

[24] P. R. Fayziyev, I. A. Ikromov, A. A. Abduraximov, and Q. M. Dehqonov, "Timeline: History of the electric car, trends and the future developments," *Eurasian Res. Bull.*, vol. 6, no. SE-Articles, pp. 89–94, 2022, [Online]. Available: www.geniusjournals.org/index.php/erb/article/view/888.

[25] J. Li, "Compatibility and investment in the us electric vehicle market," *Unpubl. manuscript*, MIT, 2019.

[26] Q. Tan, M. Wang, Y. Deng, H. Yang, R. Rao, and X. Zhang, "The cultivation of electric vehicles market in China: Dilemma and solution," *Sustainability*, vol. 6, no. 8. pp. 5493–5511, 2014, https://doi.org/10.3390/su6085493.

[27] E. Figenbaum, N. Fearnley, P. Pfaffenbichler, R. Hjorthol, M. Kolbenstvedt, R. Jellinek, B. Emmerling, G. M. Bonnema, F. Ramjerdi, L. Vågane, and L. M. Iversen, "Increasing the competitiveness of e-vehicles in Europe," *Eur. Transp. Res. Rev.*, vol. 7, no. 3, p. 28, 2015, https://doi.org/10.1007/s12544-015-0177-1.

[28] V. P. Singh and S. Jain, "Economic analysis of a large scale solar updraft tower power plant," Sustain. Energy Technol. Assessments, vol. 58, p. 103325, 2023, doi: https://doi.org/10.1016/j.seta.2023.103325.

[29] V. P. Singh, A. Dwivedi, A. Karn, A. Kumar, S. Singh, S. Srivastava, and K. Srivastava, "Nanomanufacturing and Design of High-Performance Piezoelectric Nanogenerator for Energy Harvesting," in *Nanomanufacturing and Nanomaterials Design: Principles and Applications*. S. Singh, S. K. Behura, A. Kumar, and K. Verma, Eds. CRC Press, 2022, pp. 241–272. https://doi.org/10.1201/9781003220602

[30] G. Pistoia, *Electric and Hybrid Vehicles: Power Sources, Models, Sustainability, Infrastructure and the Market*. Netherlands: Elsevier, 2010.

[31] C. C. Chan, "The state of the art of electric, hybrid, and fuel cell vehicles," *Proc. IEEE*, vol. 95, no. 4, pp. 704–718, 2007, https://doi.org/10.1109/JPROC.2007.892489.

[32] R. Kumar, A. Kumar, Laxmikant, N. Dutt, and V. P. Singh, "Cutting parameters optimization in turning operation using taguchi method," *Int. J. Mater. Manuf. Sustain. Technol.*, vol. 1, no. 2, pp. 22–30, 2022, https://doi.org/10.56896/IJMMST.2022.1.2.009.

[33] A. R. Singh, S. K. Singh, and V. P. Singh, "Process parameters optimization of carbon nano tube based catalytic transesterification of algal oil," *Mater. Today Proc.*, 2023, https://doi.org/10.1016/j.matpr.2023.01.418.

[34] A. Kundu, A. Kumar, V. P. Singh, C. S. Meena, and N. Dutt, "Chapter 1: Introduction to Thermal Energy Resources and Their Smart Applications," in *Thermal Energy Systems: Design, Computational Techniques, and Applications*. A. Kumar, V. P. Singh, C. S. Meena, and N. Dutt, Eds. Boca Raton, FL USA: CRC Press, 2023, pp. 1–15. https://doi.org/10.1201/9781003395768 (Annexure 12).

[35] R. K. Shubham Srivastava, Deepti Verma, Shreya Thusoo, Ashwani Kumar, and Varun Pratap Singh, "Nanomanufacturing for Energy Conversion and Storage Devices," in *Nanomanufacturing and Nanomaterials Design: Principles and Applications*. S. Singh, S. K. Behura, A. Kumar, and K. Verma, Eds. Boca Raton, FL USA: CRC Press, 2022, pp. 165–174. https://doi.org/10.1201/9781003220602

[36] M. Kebriaei, A. H. Niasar, and B. Asaei, "Hybrid electric vehicles: An overview," in *2015 International Conference on Connected Vehicles and Expo (ICCVE)*, 2015, pp. 299–305, https://doi.org/10.1109/ICCVE.2015.84.

[37] L. Gao and Z. C. Winfield, "Life cycle assessment of environmental and economic impacts of advanced vehicles," *Energies*, vol. 5, no. 3. pp. 605–620, 2012, https://doi.org/10.3390/en5030605.

[38] V. P. Singh and G. Dwivedi, "Technical analysis of a large – scale solar updraft tower power," *Energies*, vol. 16, no. 1, pp. 1–29, 2023, https://doi.org/10.3390/en16010494.

[39] W. Li, P. Wei, and X. Zhou, "A cost-benefit analysis of power generation from commercial reinforced concrete solar chimney power plant," *Energy Conversion and Management*, vol. 79, pp. 104–113, 2014, https://doi.org/10.1016/j.enconman.2013.11.046.

[40] A. Kumar, V. P. Singh, C. S. Meena, and N. Dutt, *Thermal Energy Systems: Design, Computational Techniques, and Applications*, 1st ed. A. Kumar, V. P. Singh, C. S. Meena, and N. Dutt, Eds. Boca Raton FL, USA: CRC Press, 2023. https://doi.org/10.1201/9781003395768

[41] R. R. Heffner, K. S. Kurani, and T. S. Turrentine, "Symbolism and the adoption of fuel-cell vehicles," *22nd Int. Batter. Hybrid Fuel Cell Electr. Veh. Symp. Expo. EVS 2006*, vol. 1, pp. 1112–1123, 2006.

[42] S. Jain, N. Kumar, V. P. Singh, S. Mishra, and N. K. Sharma, "Transesterification of algae oil and little amount of waste cooking oil blend at low temperature in the presence of NaOH," *Energies*, vol. 16, no. 1, pp. 1–12, 2023, https://doi.org/10.3390/en16031293.

[43] A. Datta, A. Kumar, A. Kumar, A. Kumar, and V. P. Singh, "Advanced Materials in Biological Implants and Surgical Tools," in *Advanced Materials for Biomedical Applications*, 1st ed., N. D. Ashwani Kumar, Yatika Gori, Avinash Kumar, Chandan Swaroop Meena, Eds. Boca Raton, FL USA: CRC Press, 2022, pp. 283–298.

[44] F. Qian, W. Gao, Y. Yang, and D. Yu, "Economic optimization and potential analysis of fuel cell vehicle-to-grid (FCV2G) system with large-scale buildings," *Energy Convers. Manag.*, vol. 205, p. 112463, 2020, https://doi.org/10.1016/j.enconman.2019.112463.

[45] S. C. A. de Almeida and R. Kruczan, "Effects of drivetrain hybridization on fuel economy, performance and costs of a fuel cell hybrid electric vehicle," *Int. J. Hydrogen Energy*, vol. 46, no. 79, pp. 39404–39414, 2021, https://doi.org/10.1016/j.ijhydene.2021.09.144.

[46] G. Pant, C. S. Meena, A. Saxena, A. Kumar, V. P. Singh, and N. Dutt, "Study the Temperature Variation in Alternate Coils of Insulated Condenser Cum Storage Tank: Experimental Study," in Lecture Notes in Mechanical Engineering. Springer, Singapore, 2023, pp. 627–638, doi: https://doi.org/10.1007/978-981-99-2382-3_52.

[47] V. P. Singh, C. S. Meena, A. Kumar, and N. Dutt, "Double pass solar air heater: A review," *Int. J. Energy Resour. Appl.*, vol. 1, no. 2, pp. 22–43, 2022, https://doi.org/10.56896/IJERA.2022.1.2.009.

[48] A. Saxena, A. N. Prajapati, G. Pant, C. S. Meena, A. Kumar, and V. P. Singh, "Water Consumption Optimization of Hybrid Heat Pump Water Heating System," in Lecture Notes in Mechanical Engineering. Springer, Singapore, 2023, pp. 721–732, doi: https://doi.org/10.1007/978-981-99-1894-2_61.

[49] G. S. Rathore, S. Rathor, N. Arif, A. Karn, and V. P. Singh, "Recent progress in solar dryers using phase changing material: A review," *Int. J. Energy Resour. Appl.*, vol. 2, no. 1, pp. 57–77, 2023, https://doi.org/10.56896/IJERA.2023.2.1.005.

[50] N. Dutt, A. Binjola, A. J. Hedau, A. Kumar, V. P. Singh, and C. S. Meena, "Comparison of CFD results of smooth air duct with experimental and available equations in literature," *Int. J. Energy Resour. Appl.*, vol. 1, no. 1, pp. 40–47, 2022, https://doi.org/10.56896/IJERA.2022.1.1.006.

[51] A. K. Yadav, V. P. Singh, A. Kumar, Y. Gori, and A. Kumar, "Study the properties of high-density polyethylene with nano particles silica composite," *Int. J. Mater. Manuf. Sustain. Technol.*, vol. 1, no. 1, pp. 29–36, 2022, https://doi.org/10.56896/IJMMST.2022.1.1.005.

[52] C. Jin, J. D. Ampah, S. Afrane, Z. Yin, X. Liu, T. Sun, Z. Geng, M. Ikram, and H. Liu, "Low-carbon alcohol fuels for decarbonizing the road transportation industry: A bibliometric analysis 2000–2021," *Environ. Sci. Pollut. Res.*, vol. 29, no. 4, pp. 5577–5604, 2022, https://doi.org/10.1007/s11356-021-15539-1.

[53] C. Llopis-Albert, F. Rubio, and F. Valero, "Impact of digital transformation on the automotive industry," *Technol. Forecast. Soc. Change*, vol. 162, p. 120343, 2021, https://doi.org/10.1016/j.techfore.2020.120343.

[54] R. Chinoracký and T. Čorejová, "Impact of digital technologies on labor market and the transport sector," *Transp. Res. Procedia*, vol. 40, pp. 994–1001, 2019, https://doi.org/10.1016/j.trpro.2019.07.139.

[55] M. S. Hossain, L. Kumar, M. El Haj Assad, and R. Alayi, "Advancements and future prospects of electric vehicle technologies: A comprehensive review," *Complexity*, vol. 2022, p. 3304796, 2022, https://doi.org/10.1155/2022/3304796.

[56] I. López, E. Ibarra, A. Matallana, J. Andreu, and I. Kortabarria, "Next generation electric drives for HEV/EV propulsion systems: Technology, trends and challenges," *Renew. Sustain. Energy Rev.*, vol. 114, p. 109336, 2019, https://doi.org/10.1016/j.rser.2019.109336.

[57] A. Sachdeva and D. D. Mohite, "Recent development in green fuel vehicles and their future advancements," *Int. Res. J. Eng. Technol.*, vol. 7, pp. 1175–1192, 2020.

[58] B. K. Bose, *Power Electronics and Motor Drives: Advances and Trends*. Netherlands, Elsevier Science, 2020.

[59] R. Diaz and A. Felipe, "Life cycle analysis of electric and hydrogen buses to improve decision making," MS thesis. Universitat Politècnica de Catalunya, 2022.

[60] C. Mendez, M. Contestabile, and Y. Bicer, "Hydrogen fuel cell vehicles as a sustainable transportation solution in Qatar and the Gulf cooperation council: A review," *Int. J. Hydrogen Energy*, 2023, https://doi.org/10.1016/j.ijhydene.2023.04.194.

[61] K. Romejko and M. Nakano, "Portfolio analysis of alternative fuel vehicles considering technological advancement, energy security and policy," *J. Clean. Prod.*, vol. 142, pp. 39–49, 2017, https://doi.org/10.1016/j.jclepro.2016.09.029.

[62] H. Pohl and M. Yarime, "Integrating innovation system and management concepts: The development of electric and hybrid electric vehicles in Japan," *Technol. Forecast. Soc. Change*, vol. 79, no. 8, pp. 1431–1446, 2012, https://doi.org/10.1016/j.techfore.2012.04.012.

[63] Y. Zhang, Y. Yu, and B. Zou, "Analyzing public awareness and acceptance of alternative fuel vehicles in China: The case of EV," *Energy Policy*, vol. 39, no. 11, pp. 7015–7024, 2011, https://doi.org/10.1016/j.enpol.2011.07.055.

[64] E. Shafiei, B. Davidsdottir, J. Leaver, H. Stefansson, and E. I. Asgeirsson, "Comparative analysis of hydrogen, biofuels and electricity transitional pathways to sustainable transport in a renewable-based energy system," *Energy*, vol. 83, pp. 614–627, 2015.

[65] J. Zheng, S. Mehndiratta, J. Y. Guo, and Z. Liu, "Strategic policies and demonstration program of electric vehicle in China," *Transp. Policy*, vol. 19, no. 1, pp. 17–25, 2012, https://doi.org/10.1016/j.tranpol.2011.07.006.

Chapter 12

Modern Electric, Hybrid Electric Vehicles, and Hydrogen Fuel Cell Vehicles

K. Balamurugan, T. Jerry Alexander

12.1 INTRODUCTION TO MODERN ELECTRIC VEHICLES

Transportation is an important essential aspect of modern society, playing a vital role in connecting people and communities, facilitating trade and employment, and thereby supporting social and economic development. The utilization and popularity of the transportation mode differ based on the location, culture, and infrastructure of the region. However, in the universe, the most common and comfortable means of transportation include the car, bus, train, motorcycle or bicycle, plane, boat, etc., among which cars are the most popular means. Even though various fuels such as diesel, ethanol, biodiesel, CNG, hydrogen, and others are available, gasoline (petrol) is the most often used in automobiles. The availability and popularity of various fuels vary depending on the geographical location. As the accessible reserves of gasoline, a non-renewable resource, shrink, various adverse consequences emerge, including higher gasoline costs, a decreased gasoline supply, and increased environmental pollution as alternative, less efficient fuels are used. To address the problem of gasoline depletion, many techniques have been proposed. One such strategy is boosting the use of alternative fuels, such as bio fuels or electric vehicles, which can assist in reducing the gasoline demand. Internal combustion engine (ICE) automobiles are major contributors to global warming because they emit carbon dioxide and other hazardous gases into the atmosphere, trapping heat. Global warming creates the following effects: (1) glacier melting, (2) sea level rise, (3) weather pattern fluctuation, (4) changes to ecosystems, and (5) effects on human health. Hence, it is very much necessary to cease polluting the environment. To avoid such a dangerous situation in the future, the entire automobile industry is switching to electric vehicle manufacturing. Electric vehicles are harmless to the environment compared to conventional ICE vehicles. Electric vehicles do not emit harmful gases and protect the world from the dangerous situation. The future automobile industry will manufacture only electric vehicles or hybrid electric vehicles. The important advantages of electric vehicles are environmental concerns, technological advances, reliability, efficiency, and consumer demands (Limon et al. 2023). Modern electric vehicles

DOI: 10.1201/9781003464556-12

(EVs) are vehicles that work with only electric power. These vehicles have highly efficient rechargeable battery units to provide an uninterrupted power supply to the electric motors. The electric motors are used to propel the vehicle. Such types of vehicles are also called Battery Electric Vehicles (BEVs) (Sanguesa et al. 2021). BEVs have no internal combustion engine and are charged using a wall socket. Examples of BEVs are the Tesla Model S, Nissan Leaf (InsideEVs accessed April 20, 2023), BMW i3, etc. Typically, most BEVs can cover an average distance of 160 to 250 km and a maximum of 500 km with one fully charged battery.

12.1.1 Types of Electric Vehicles

There are several types of electric vehicles (Types of Electric Vehicles, accessed April 2023), discussed in what follows:

(1) Battery Electric Vehicles (BEVs): These are also known as all-electric vehicles (EVs), and they run entirely on electricity stored in rechargeable battery packs. BEVs have no ICE and, hence, do not emit pollutants. On a single charge, the traversing range varies from approximately 80 miles to 400 miles.
(2) Plug-in Hybrid Electric Vehicles (PHEVs): These vehicles have both an electric motor and an internal combustion engine. When the battery runs out of charge, PHEVs can run on either diesel or gasoline in addition to charging from an external power source. They have a typical electric-only range of 20 to 50 miles before switching to a gasoline engine.
(3) Hybrid Electric Vehicles (HEVs): These vehicles have an internal combustion engine and an electric motor, but they cannot be plugged in to charge. Regenerative braking and other methods are used by HEVs for electricity generation and recharging the battery. A portion of the vehicle's power is provided by the electric motor, which lowers emissions and fuel consumption.
(4) Hydrogen Fuel Cell Electric Vehicles (HFCVs): These vehicles use hydrogen as fuel, which is converted into electricity in a fuel cell. The electricity powers an electric motor that propels the vehicle. FCEVs emit only water vapour and have a range of about 300–400 miles on a single tank of hydrogen.

Compared to traditional automobiles, BEVs have extra systems and components that make them identical to what is shown in Figure 12.1.

The important components of BEVs are the battery unit, electric motors, power electronics system, charging port, regenerative braking unit, thermal management system, infotainment system, etc.

(a) Battery Unit: The battery unit provides the necessary power to the electric motors and other accessories in BEVs. Lithium-ion batteries are employed

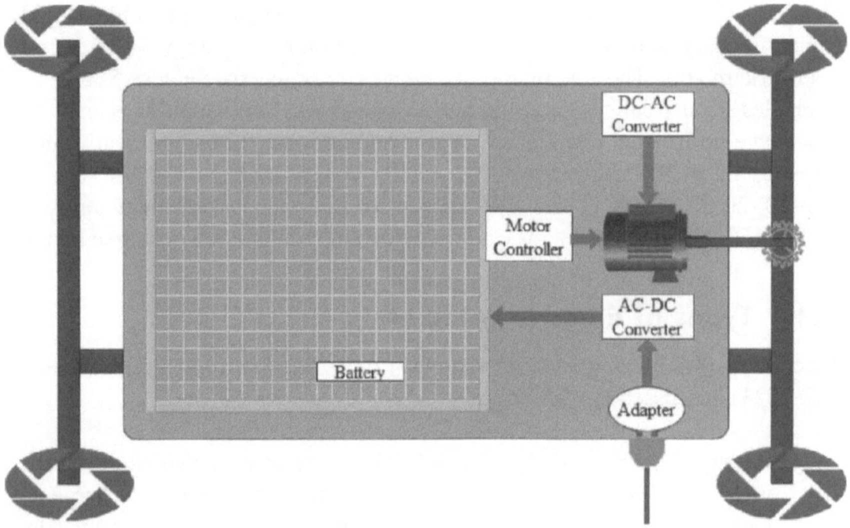

Figure 12.1 Main Parts of Electric Vehicle.

in BEVs because of their features such as efficiency, consistency, affordability, other protection features, etc., compared to other types of batteries. The other types of batteries are (1) Lead-acid batteries (Pb-PbO2), (2) Nickel-cadmium batteries (Ni-Cd), (3) Nickel-metal-hydride batteries (Ni-MH), (4) Zinc-bromine batteries (Zn-Br2), (5) Sodium chloride and nickel batteries (NA-NiCl), (6) Sodium sulfur batteries (Na-S).

(b) Electric Motors: Electric motors in EVs receive the electrical power from the battery and convert it into mechanical energy as the shaft of the motor rotates. The wheels attached to the shaft of the motor start to rotate.

The electric motor consists of the rotating part (rotor) and the stationary part (stator). Both the rotor and stator are made up of magnets and coiled by conducting wires. When a current-carrying conductor is placed in a magnetic field (Iqbal et al. 2021), it exerts a force that makes the rotor rotate. The torque of the motor is controlled by the current supplied to it, and the speed of the vehicle depends on the acceleration supplied by the driver.

Two types of motors are widely used in EVs. They are AC (Alternative Current) motors and DC (Direct Current) motors. Big EVs, such as cars, use AC motors, and small EVs, such as scooters, use DC motors.

The DC motor is simple and more cost-effective than AC motors. But it is less efficient and works with limited speed than AC motors. The AC motor is more efficient and operates at a higher speed compared to DC motors. An induction type of AC motor is used in EVs.

(c) Power Electronics System: The power electronics system plays an important role in BEVs because they manage the electric current flow between the battery unit and the motor. The power electronics system consists of DC-to-AC inverters, AC-to-DC converters, and other controllers. The power electronic system is the system responsible for the safe operation of BEVs. The plugged-in AC supply from the charging unit will be rectified to DC by the power electronics converters, and it will be stored in the battery unit. The stored DC charge in the battery will be inverted to AC, and it will be supplied to the AC motor by using a DC-AC inverter. In the case of small EVs, the DC-AC inverter is not needed as it has only a DC motor. A microprocessor is used to control the inverter that provides the required frequency and voltage to match the speed and torque. The power electronic system also controls and regulates the other electronic accessories in the system. Also, few controllers are used to protect the battery unit and the electric motor from damages due to over temperature, overloading, overheating, etc.

(d) Charging Port: The charging port is connected to the battery unit of EVs. A charging port is used in EVs to recharge the battery unit for uninterrupted operation. The charging port is connected to a power source available in charging units or homes. The time taken to charge fully depends on battery size, charger type, and the current state of the charge. The charging ports are classified into (i) AC charging port, (ii) DC charging port, and (iii) Wireless charging port.

  (i) AC Charging Port: AC charging ports are widely used as they are readily available in homes, workplaces, and charging stations. As the AC power is converted into DC, it leads to some power loss during the conversion. This is the main drawback of AC charging compared to DC. But AC charging is cheaper and simpler to charge EVs compared to DC charging. The charging rate of AC is lesser than DC.

  (ii) DC Charging Port: Most of the charging stations use DC supply to charge EVs. The battery unit in EVs is DC-compatible only. The DC input supply given directly to the battery unit will have more advantages, such as fast charging time, no power loss, more charging rate, etc. The main drawbacks of DC charging are (i) it is not readily available everywhere and (ii) the cost is higher compared to AC charging. Also, DC charging needs a special connector that can be charged to 100% within approximately 40 minutes.

  (iii) Wireless Charging Port: Wireless charging technology eliminates the use of hard electric wires for charging. It uses magnetic induction to charge the BEV's battery unit. Wireless charging technology consists of two main units: (a) a charging mat and (b) a receiver.

  The charging mat is placed on the ground, and the receiver is installed on the BEV. As soon as the BEV is parked over the charging mat, it generates a magnetic field that induces the current in the receiver. The receiver is connected to the battery unit of the BEV, charging it as soon as it receives the current. After the battery unit is fully charged, the BEVs can be moved from the charging mat. The advantages of wireless charging are safety and convenience. The disadvantages are efficiency, cost, and compatibility.

(e) Regenerative Braking Unit: In BEVs, energy management (Roy 2018) plays a vital role. Acquiring energy through all means is very important. One of the ways of acquiring energy from braking is known as regenerative braking. Regenerative braking, as shown in Figure 12.2, is designed for all types of electric vehicles. When the driver operates the brake during driving, the electric motor will operate as a generator. Regenerative braking absorbs the vehicle's motion energy (mechanical energy) into electrical energy when the vehicle is slowing down. That is, during the deceleration, the kinetic energy of the vehicle wheels will be converted into electrical energy by the generator. The braking pedal in BEVs is the triggering signal for changing the function of an electric motor into a generator. One of the important pros of regenerative braking is that it prevents battery degradation even at low battery temperatures (10°C) (Gao 2012; Islameka et al. 2023; Wong et al. 2013).

Certain disadvantages are also there when using regenerative braking. Those limitations are:

(1) The state of charge (SOC) of the battery unit should not be more than 90% and should not be less than 10%. If it is more than 90%, there is a possibility of overcharging and causing battery damage. If the SOC is less than 10%, there is a chance of over-discharging because of high battery resistance.

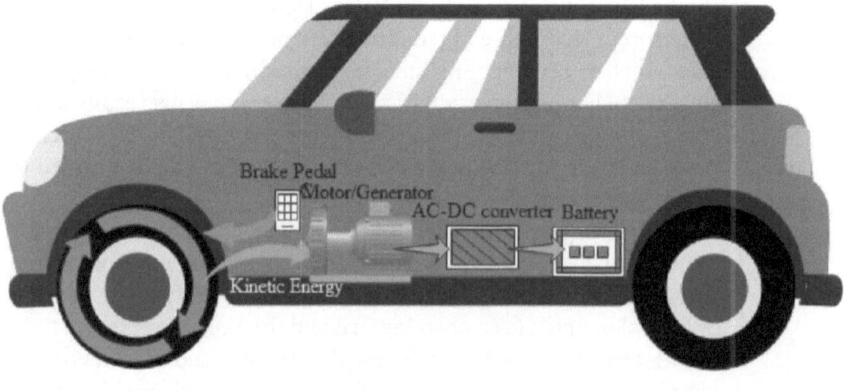

Figure 12.2 Components of Regenerative Braking.

(2) The vehicle speed should not be lesser than the threshold speed.
(3) Amount of acceleration and braking signals.
(4) Battery voltage.
(5) Specification of the electric motor.

Modern BEVs use an electronic control unit to monitor and regulate the regenerative braking.

(f) Thermal Management System (TMS): In BEVs, high temperatures cause the battery to overheat, which leads to fire. The TMS is useful to regulate the temperature of battery units, electric motors, power electronics units, and other electrical and electronic accessories. The TMS is helpful for EVs to improve their efficiency and overall performance. The components of a TMS are (i) a cooling system and (ii) a heating system.
   (i) Cooling System: It is used to reduce the electric motor temperature, battery unit, and power electronics unit either by liquid cooling techniques or air cooling techniques.
   (ii) Heating System: During cold temperature conditions, it is necessary to maintain the temperature of the battery unit. Battery units in BEVs will be maintained within the standard temperature range of 20–30°C.

(g) Infotainment System: Modern BEVs have aninfotainment system that provides entertainment for travellers and information to the driver. The infotainment system provides entertainment devices such as a video system, audio system, navigation system, touchscreen display, smartphone integration, etc. The infotainment unit needs internet for most of its functions.

The infotainment system provides information to the driver for smooth and safe driving. It also provides information about battery level, charging status, and temperature conditions of other components such as the brake system, electric motor, etc.

12.1.2 Driverless BEVs

Driverless BEVs are fully automatic vehicles capable of operating without the support of a driver. Popular driverless BEVs are manufactured by well-known automobile giants such as Tesla, Google's Waymo, General Motors, BMW, Nissan, Volkswagen, Baidu, Toyota, and Ford.

The main advantages of using driverless BEVs are increased safety, reduced traffic congestion, improved accessibility, and increased efficiency. A few challenges in driverless BEVs are technical issues, legal and regulatory issues, and cybersecurity risks.

Driverless BEVs consist of ultrasonic sensors, RADAR, LIDAR, and cameras, as shown in Figure 12.3. The higher-end ultrasonic sensors are used to detect

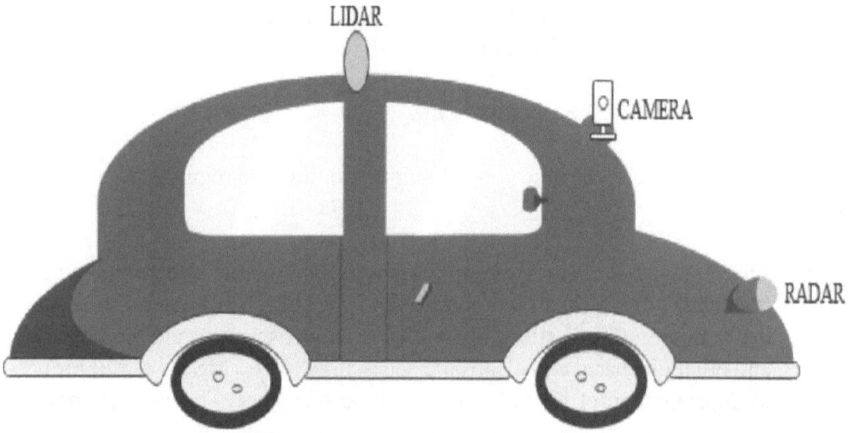

Figure 12.3 Driverless Automated Car.

nearby objects/vehicles and measure the distance from the vehicle. It protects the BEVs from damage. Ultrasonic sensors use high-frequency sound waves for detection and distance calculation.

Along with ultrasonic sensors, RADAR (Radio Detection and Ranging) is used to detect the objects as well as the location and speed of the object. It emits radio waves and receives the reflected waves from the object. The Doppler Effect determines the velocity of the object. The main advantage of RADAR is it can work in any climatic conditions.

LIDAR stands for Light Detection and Ranging. It emits laser beams to the objects and receives the reflected light from the objects. It provides a highly accurate 3D map 360°around the vehicle, which is very much necessary for driverless BEVs. It is useful to detect even small objects or obstacles in its path or surroundings.

High-resolution cameras are the essential device for the navigation of the vehicle in driverless BEVs. Cameras capture the vehicle environment and send the images/video to the Artificial Intelligence Control Unit (AICU). This AICU makes decisions about how to avoid obstacles and navigate the BEV. The AICU is the brain of the driverless BEV. The cameras also provide information about the colour, size, and texture of the objects to help the AICU to distinguish between objects. All the sensors, cameras, and other supporting devices are interconnected with the AICU and send input information. After collecting the input information, the AICU operates the BEV by sending the signals to the components such as the accelerator, brake, etc.

Some of the key advancements in electric vehicles include:

(a) Environmental Impact: The environmental benefits of electric vehicles are well documented. EVs do not emit tailpipe emissions and produce less overall greenhouse gas emissions compared to gasoline-powered vehicles.

(b) Battery Technology: The development of better and more efficient batteries has been a major driver of the growth in EVs. Today, lithium-ion batteries are the standard for EVs due to the development in battery chemistry and manufacturing, which has led to longer ranges and quicker charging times.
(c) Range: The range of electric vehicles has increased significantly over the years. The latest typical EVs are capable of travelling more than 200 miles per single charge, which makes them more sensible for daily usage.
(d) Affordability: As production volumes have increased and battery costs have come down, the cost of electric vehicles has become more affordable, making them available to more consumers.
(e) Charging Infrastructure: Public charging stations are becoming more common, and fast-charging technology allows quick top-ups.
(f) Performance: Electric vehicles are no longer seen as slow and unresponsive. Most electric cars can now outperform their gasoline-powered vehicles for their acceleration and top speed.
(g) Design: Electric cars are no longer seen as quirky and futuristic. Many EVs now have sleek and modern designs that are indistinguishable from their gasoline-powered counterparts.

12.2 HYBRID ELECTRIC VEHICLES (HEV)

HEVs are the technology that has the best chance of minimizing fuel consumption and greenhouse gas emissions. They are vehicles that have both an internal combustion engine (ICE) and an electric motor. Regenerative braking can be used to charge the battery system, which powers the electric drive. The engine may also be utilized to provide charging of the battery pack.

HEVs are designed to improve fuel efficiency while simultaneously reducing pollution. The amount of fuel burnt by the engine can be reduced to a certain extent by using the electric motor to power the vehicle. By using features like stop-start technology, the engine can be made to turn off when the vehicle is idling, and thereby, the engine is additionally enhanced to be more efficient.

12.2.1 EV vs. HEV

An EV is a vehicle running solely on electric power stored in batteries or other energy storage devices, while an HEV is powered by the combination of an electric motor and a gasoline engine. HEVs typically use the electric motor at low speeds and switch to gasoline engines at higher speeds.

In general, EVs have larger batteries and can travel longer distances on electricity alone than HEVs. However, HEVs are more fuel-efficient than traditional gasoline-powered vehicles and can switch between electric and gasoline power, making them more versatile for longer trips.

12.2.2 Why HEV?

There are numerous explanations why someone could decide to purchase an HEV, including fuel efficiency, environmental advantages, superior performance, and lower running expenses.

(1) Fuel Economy: HEVs often have better fuel economy than conventional ICE cars since the electric motor helps the ICE push the car, using less gasoline in the process.
(2) Environmental Advantages: Since HEVs emit fewer emissions than conventional ICE vehicles, they can aid in lowering greenhouse gas emissions and air pollution.
(3) Less Expensive to Operate: Throughout their lifetime, HEVs can save drivers money on fuel costs. They may also require less maintenance than conventional ICE vehicles.
(4) Performance Advantages: Due to the additional power of the electric motor, HEVs frequently offer better acceleration and torque than conventional ICE vehicles.

12.2.3 Types of Hybrid Electric Vehicles

HEVs come in different types:

(a) Series Hybrid: In this type of HEV, the engine is used to generate electricity to power the electric motor. The engine does not directly drive the wheels.
(b) Parallel Hybrid: This is the most common type of HEV, where the engine and the electric motor are connected to the same transmission and can power the vehicle separately or together.
(c) Plug-In Hybrid: This type of HEV can be charged from an external power source, such as a wall socket. It has a comparatively larger battery pack, which aids in travelling a longer distance on electric power alone.

Table 12.1 compares all types of Hybrid Electric Vehicles.

Table 12.1 Comparison of All Types of HEVs

Description	*Series Hybrid*	*Parallel Hybrid*	*Plug-In Hybrid*
Fuel Efficiency	Better fuel efficiency in stop-and-go driving conditions (due to electric motor) Less efficient at higher speeds	A balance between fuel efficiency and performance Optimal fuel consumption is attained by making the ICE and electric motor work together	Best fuel efficiency among all three models But less efficient in hybrid mode

Description	*Series Hybrid*	*Parallel Hybrid*	*Plug-In Hybrid*
Cost	More expensive to manufacture than parallel HEVs due to the larger battery and electric motor	More affordable to manufacture than the other two	More expensive due to the large battery capacity and complex power train
All-Electric Range	Has a shorter range than the other two types of HEVs Electric motor is not the primary power source	Has a shorter range than plug-in type but typically longer than series-type HEVs	Has the longest range of the three types It can drive for a certain distance on electric power alone, after which the ICE is utilized
Charging Infrastructure	Not required as the battery is charged through regenerative braking and the ICE	Not required but optional	Requires access to charging infrastructure to fully utilize the all-electric range and reduce reliance on the ICE

12.2.4 Parts of Hybrid Electric Vehicles

The key components (Alternative Fuels Data Center: How Do Hybrid Electric Cars Work?, April 30, 2023) of a hybrid electric vehicle (Geng et al. 2021) are illustrated in Figure 12.4.

(a) ***Internal Combustion Engine (ICE)***
The majority of vehicles have an ICE, which is a conventional engine that produces power by burning gasoline or diesel fuel. In a hybrid car, the ICE is frequently smaller and more effective than in a conventional automobile, and it powers the car alongside an electric motor. When the electric motor is not in use or when more power is needed, the ICE is in charge of supplying the car with power. It can drive the wheels directly since it is connected to the transmission system.

(b) ***Transmission***
Transmission in HEVs represents a power transfer from the engine or electric motor to the vehicle's wheels. Energy transmission (Saiteja et al. 2022) involves the power flow between the vehicle's internal combustion engine (ICE), electric motor, battery, power control unit (PCU), and electric generator. The ultimate objective of this energy transmission is to efficiently manage the flow of power between these components to optimize the fuel economy and thereby reduce emissions.

There are two types of transmissions used in HEVs:

(i) Automatic Transmission: Most HEVs use an automatic transmission, which is similar to the transmission found in traditional gasoline-powered

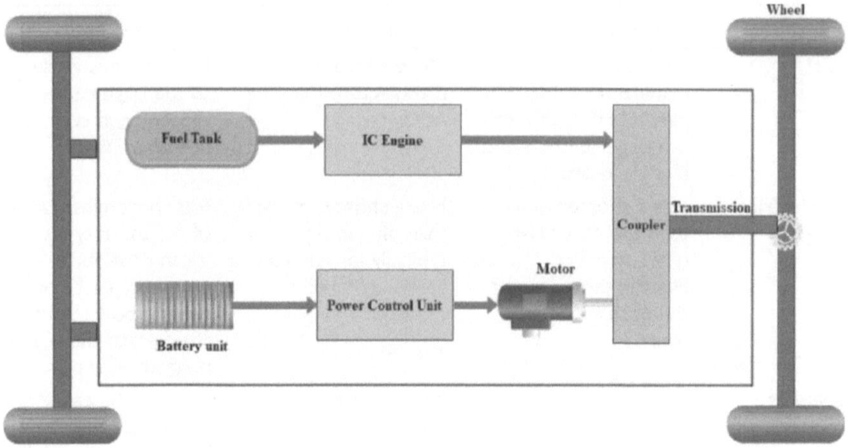

Figure 12.4 Typical Block Diagram of a Hybrid Electric Vehicle (Upadhyay et al. 2021).

vehicles. The power is transmitted from the engine/electric motor with the help of a torque converter, planetary gear sets, and clutches to transfer power from the engine or electric motor to the wheels. The automatic transmission can be a continuously variable transmission (CVT) or a traditional stepped automatic transmission.

(ii) Manual Transmission: Some HEVs use a manual transmission, which allows the driver to manually select the gear ratios. However, manual power transmissions are less common in HEVs compared to automatic transmissions.

There are two modes of energy transmission in HEVs:

(1) Electric Mode: In this mode, the electric motor (Sharma et al. 2020) is the primary source of power, and the vehicle operates solely on electrical power from the battery. This mode is typically used for low-speed driving or during enforced stops due to traffic signals or heavy traffic.
(2) Hybrid Mode: In this mode, the ICE and electric motor work together to provide power to the vehicle's wheels. The PCU manages the power flow between the IC Engine, electric motor, and battery to optimize fuel consumption and reduce emissions. The electric motor provides additional torque to assist the ICE during the acceleration task and deceleration. The regenerative braking system generates electrical power to recharge the battery (Kebriaei et al. 2015).

The power transmission between the ICE and the electric motor can be performed with the aid of various methods, including:

(a) Power Split Device: This device allows the ICE and electric motor to work together to drive the vehicle's wheels. It consists of a planetary gear set and two electric machines that work in tandem to power the wheels.
(b) Two-Mode Hybrid Transmission: This transmission use two electric motors and two planetary gear sets to power the wheels. It allows the vehicle to operate in various modes like electric, hybrid, or extended range (ICE acts as a generator for recharging the battery).

By effective power flow management between these components, HEVs can operate more efficiently with lower emissions than traditional gasoline-powered vehicles.

(c) Electric Motor/Generator

The vital component of the HEVs, the electric motor, works with the ICE for powering the wheels of the vehicle. The electric motor powered by the vehicle's battery provides extra torque and energy to the wheels during acceleration, allowing the ICE to operate more efficiently.

Several types of electric motors are used in HEVs, including:

(i) AC Induction Motor: This is the simplest and most commonly used type of electric motor in HEVs. It consists of a stator and a rotor that spins inside the stator. The AC induction motor, though it is typically less efficient than other types of electric motors, is less expensive and has fewer components, making it a popular choice for HEVs.
(ii) Permanent Magnet Synchronous Motor: This type of motor uses permanent magnets on the rotor, which provide a stronger magnetic field and improve efficiency. The permanent magnet synchronous motor is more efficient than the AC induction motor but is more expensive.
(iii) Switched Reluctance Motor: This is a newer type of electric motor that uses a simple rotor design and is less expensive than other types of engines. However, it has lower power output and less efficiency.

The electric motor in a typical HEV is smaller and more efficient than the motor in a traditional gasoline-powered vehicle. It is also designed to operate simultaneously in conjunction with the vehicle's ICE, thereby providing supplemental power during acceleration and allowing the ICE to operate more efficiently during steady-state driving.

(d) Electric generator or Motor Generator Unit (MGU)

It is a major important device that works together with the vehicle's internal combustion engine (ICE) and electric motor to generate and manage electrical power. The purpose of the electric generator is to convert the mechanical energy of the ICE into electrical energy and also to recharge the vehicle's battery or power the motor.

There are two types of electric generators commonly used in HEVs (Mustafi and Nirendra Nath, 2022):

(i) Starter-Generator: This type of generator is typically used in mild hybrid systems. This device is integrated with the ICE and is used primarily to start the engine and provide additional torque during acceleration. The starter generator also generates electrical energy during deceleration, which helps recharge the vehicle's battery.

(ii) Motor-Generator: This type of generator is used in full hybrid systems and is designed to operate independently of the ICE. This unit is integrated with the transmission and is accountable for generating electrical energy during times of deceleration and braking. This energy is then used to recharge the vehicle's battery or power the electric motor.

(e) Battery

Energy storage is an important part of maintaining supply and demand in an optimal manner (Prabhansu and Nayan Kumar, 2023). The batteries are used in a hybrid vehicle to store the energy and power the electrical motor. It is a high-capacity rechargeable lithium-ion battery that typically can be charged by ICE or regenerative braking.

Several types of batteries are commonly used in Hybrid Electric Vehicles (HEVs) include:

(i) Lithium-ion (Li-ion) Batteries: These are the most common type of battery used in HEVs. They are lightweight, have a high energy density, and can deliver a lot of power quickly. They are also durable and have a long lifespan.

(ii) Nickel-Metal Hydride (NiMH) Batteries: These were the most common type of battery used in HEVs before Li-ion batteries became more popular. They are also lightweight and have a good energy density, but they are less efficient than Li-ion batteries and have a shorter lifespan.

(iii) Lead-Acid Batteries: These are the oldest type of battery technology used in vehicles, and they are still used in some HEVs. They are heavy and have a low energy density. But, it is inexpensive and has a long lifespan.

(iv) Solid-State Batteries: This is a new type of battery technology still being developed, but it shows promise for use in HEVs. They are lightweight, have a high energy density, and are very efficient.

(f) Power Control Unit (PCU)

The PCU is a critical component in Hybrid Electric Vehicles (HEVs). It is responsible for managing the flow of power between the vehicle's internal combustion engine (ICE), electric motor, and battery. The PCU uses sensors to monitor the vehicle's performance and adjusts the power delivery as needed to optimize fuel efficiency and reduce emissions.

The PCU consists of several key components, including:

(a) Inverter: The DC power from the vehicle's battery is converted into AC power by using the inverter to drive the electric motor.
(b) DC-DC Converter: The DC-DC converter is responsible for stepping down the voltage of the high-voltage battery to power the low-voltage systems in the vehicle, such as the lights and audio system.
(c) Control Module: The control module is the brain of the PCU. It monitors the vehicle's performance and adjusts the power delivery as needed to optimize fuel efficiency and reduce emissions.

The PCU works by coordinating the flow of power between the ICE, electric motor, and battery based on the vehicle's driving conditions. For example, during acceleration, the PCU might use power from the ICE and the electric motor to provide the necessary power. During deceleration, the PCU might use the regenerative braking system to generate electricity and recharge the battery.

(g) Regenerative Braking System

The regenerative braking system used in Hybrid Electric Vehicles (HEVs) is similar to that of electric vehicles.

In some HEVs, the regenerative braking system can also be used to provide additional power to the vehicle's electric motor during acceleration. This further improves fuel efficiency and performance, as the system reduces the amount of energy that needs to be supplied by the vehicle's internal combustion engine.

12.3 HYDROGEN FUEL CELL VEHICLE (HFCV)

A Hydrogen Fuel Cell Vehicle (HFCV) is a type of electric vehicle that uses hydrogen gas to generate electricity to operate the electric motors. These HFC vehicles are manufactured with and without battery units. HFC vehicles with battery units store the excess electricity generated and the energy due to the regenerative braking system. The HFCV consists of a well-designed Hydrogen tank to store hydrogen gas. The hydrogen gas can be fuelled from the fuel station.

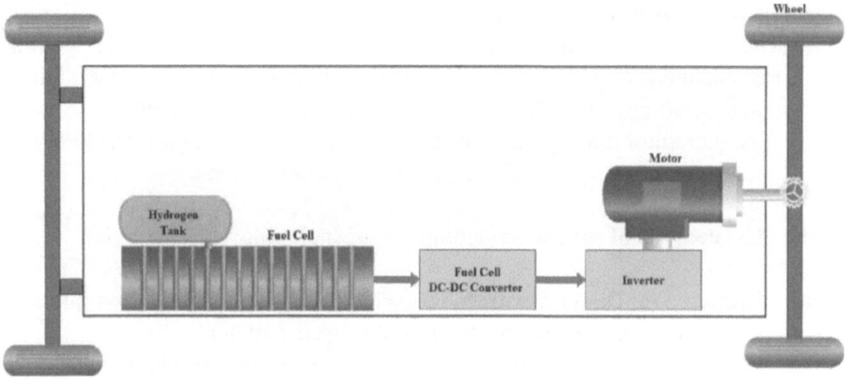

Figure 12.5 Essential Parts of a HFCV.

(a) Hydrogen Fuel Cell

The hydrogen fuel cell unit is important in a HFCV. It is similar to batteries used in electric vehicles. This fuel cell converts the chemical reaction into electricity. It consists of two platinum electrodes, an anode and cathode, and a polymer electrolyte membrane, as shown in Figure 12.6 (Todd Omatick et al. 2022).

The hydrogen gas is passed to the anode of the fuel cell, where it splits into H+ ions (protons) and electrons. The H+ ions will travel through the membrane and enter the cathode. Meanwhile, the electrons travel through a circuit (Dash et al. 2022), where they generate an electric current and excess heat. The oxygen from the atmosphere enters the cathode. The oxygen combines with the H+ ions and the electrons which are left out of the circuit and produce water (H_2O). The remaining H_2 will be recycled. The generated waste by-products in this process are water and heat.

The chemical reactions in a fuel cell are:

At Anode side	:	$2H_2 => 4H^+ + 4e^-$	by oxidation reaction
At Cathode side	:	$O_2 + 4H^+ + 4e^- => 2H_2O$	by reduction reaction
Net reaction	:	$2H_2 + O_2 => 2H_2O$	by oxidation-reduction (redox) reaction

A hydrogen fuel cell unit consists of many fuel cells connected in series because each fuel cell produces less than 1.16 volts of electricity.

(b) Power Control Unit

The power control unit controls the power flow from the Hydrogen fuel cells to other components, such as motors and battery units, and maintains the constant voltage and current of the fuel cell unit.

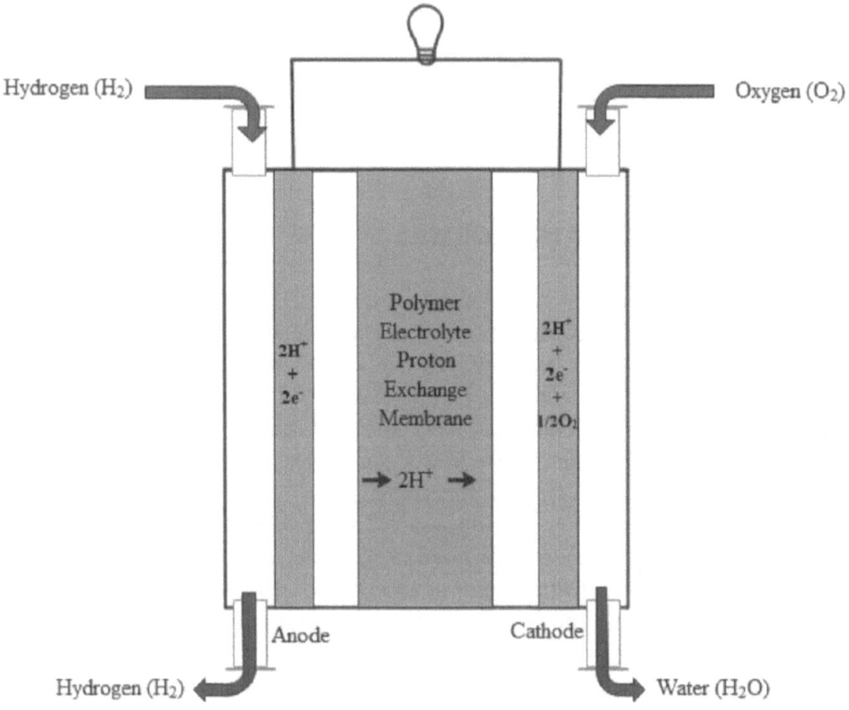

Figure 12.6 Hydrogen Fuel Cell.

(c) *Electric Motor*

The electric motor converts the electrical energy supplied by the fuel cell to mechanical energy that is driving the wheels.

(d) *Battery Unit*

The battery unit is useful to store the excess energy generated. It can also supply additional power to the vehicle in some cases, like uphill driving. It also stores energy due to regenerative braking.

(e) *Regenerative Braking*

Regenerative braking converts the kinetic energy developed during the braking of the vehicle to electrical energy stored in the battery.

(f) *Water Management System*

Water as a waste product is generated during the process of converting chemical energy to electrical energy. Here, the water will be stored in a tank or disposed of as soon as it is produced. The Water Management System ensures that the fuel cell is hydrated.

(g) Vehicle Control System

The vehicle control system monitors and controls the fuel cell stack, battery unit, electric motor, transmission, and safety features. It also monitors the vehicle's performance and communicates with the driver through a display or other interface.

12.4 CONCLUSION AND FUTURE SCOPE

Automobiles in the world will be dominated by Battery Electric Vehicles (BEV) soon due to their environmental impact, energy efficiency, and lower operating costs. Battery Electric Vehicles (BEVs) use rechargeable batteries. Hybrid Electric Vehicles (HEVs) can be preferred by users because of their advantages, such as fuel efficiency, better acceleration and torque, and less noise. Hydrogen Fuel Cell Vehicles (HFCV) work with the use of a fuel cell. The advantages of HFCV are zero emissions, long range, and fast refuelling.

In the future, solar-powered electric vehicles will be designed. Solar panels will be installed on the vehicle's roof or hood to generate electricity from the sun to help power the vehicle's electric motor or accessories. However, this is not a common feature of most EVs and is usually only found in specialized or experimental vehicles. Solar power alone is generally not sufficient to fully power an EV, but it can supplement the vehicle's power needs and help improve its overall efficiency.

REFERENCES

Alternative Fuels Data Center. How Do Hybrid Electric Cars Work? Accessed April 30, 2023. https://afdc.energy.gov/vehicles/how-do-hybrid-electric-cars-work.

Dash, Santanu Kumar, Suprava Chakraborty, Michele Roccotelli, and Umesh Kumar Sahu. Hydrogen Fuel for Future Mobility: Challenges and Future Aspects. Sustainability 14, no. 14 (July 6, 2022): 8285. https://doi.org/10.3390/su14148285.

Gao, Yimin. Vehicle Dynamics and Performance. In Electric, Hybrid, and Fuel Cell Vehicles, edited by Amgad Elgowainy, 67–90. New York, NY: Springer, 2012. https://doi.org/10.1007/978-1-0716-1492-1_799.

Geng, Cong, Dawen Ning, Linfu Guo, Qicheng Xue, and Shujian Mei. Simulation Research on Regenerative Braking Control Strategy of Hybrid Electric Vehicle. Energies 14, no. 8 (April 15, 2021): 2202. https://doi.org/10.3390/en14082202.

InsideEVs. Nissan Reveals LEAF E-Plus: 62 KWh Battery, 226-Mile Range. Accessed April 10, 2023. https://insideevs.com/news/341958/nissan-reveals-leaf-e-plus-62-kwh-battery-226-mile-range/.

Iqbal, Atif, Shaikh Moinoddin, and Bhimireddy Prathap Reddy. Electrical Machine Fundamentals with Numerical Simulation Using MATLAB/SIMULINK. Hoboken, NJ: Wiley, 2021.

Islameka, Metha, Bentang Arief Budiman, Firman Bagja Juangsa, and Muhammad Aziz. Energy Management Systems for Battery Electric Vehicles. In Emerging Trends in

Energy Storage Systems and Industrial Applications, 113–150. Elsevier, 2023. https://doi.org/10.1016/B978-0-323-90521-3.00006-5.

Kebriaei, Mohammad, Abolfazl Halvaei Niasar, and Behzad Asaei. Hybrid Electric Vehicles: An Overview. In 2015 International Conference on Connected Vehicles and Expo (ICCVE), 299–305. Shenzhen: IEEE, 2015. https://doi.org/10.1109/ICCVE.2015.84.

Limon, Mohammad Hossain, Binoy Debnath, and A. B. M. Mainul Bari. Exploration of the Drivers Influencing the Growth of Hybrid Electric Vehicle Adoption in the Emerging Economies: Implications towards Sustainability and Low-Carbon Economy. Sustainable Operations and Computers (April 2023). https://doi.org/10.1016/j.susoc.2023.04.002.

Mustafi, Nirendra Nath. An Overview of Hybrid Electric Vehicle Technology. In Engines and Fuels for Future Transport, edited by Gautam Kalghatgi, Avinash Kumar Agarwal, Felix Leach, and Kelly Senecal, 73–102. Singapore: Springer, 2022. https://doi.org/10.1007/978-981-16-8717-4_5.

Omatick, T., D. Zhang, M. S. Blais, M. Mossolly, P. Renzi, R. Kurz, S. Freund, S. Harvey. Machinery and Energy Systems for the Hydrogen Economy, 251–304. The Netherlands: Elsevier, 2022. https://doi.org/10.1016/B978-0-323-90394-3.00009-6.

Prabhansu, and Nayan Kumar, eds. Emerging Trends in Energy Storage Systems and Industrial Applications. London: Academic Press, 2023.

Roy, Souvik. Energy Management Systems for Electric Vehicles and Microgrids. International Journal of Students' Research in Technology & Management 6, no. 1 (January 3, 2018): 1–5. https://doi.org/10.18510/ijsrtm.2018.611.

Saiteja, Pemmareddy, and B. Ashok. Critical Review on Structural Architecture, Energy Control Strategies and Development Process towards Optimal Energy Management in Hybrid Vehicles. Renewable and Sustainable Energy Reviews 157 (April 2022): 112038. https://doi.org/10.1016/j.rser.2021.112038.

Sanguesa, Julio A., Vicente Torres-Sanz, Piedad Garrido, Francisco J. Martinez, and Johann M. Marquez-Barja. A Review on Electric Vehicles: Technologies and Challenges. Smart Cities 4, no. 1 (March 15, 2021): 372–404. https://doi.org/10.3390/smartcities4010022.

Sharma, Snigdha, Amrish K. Panwar, and M. M. Tripathi. Storage Technologies for Electric Vehicles. Journal of Traffic and Transportation Engineering (English Edition) 7, no. 3 (June 2020): 340–361. https://doi.org/10.1016/j.jtte.2020.04.004.

Types of Electric Vehicles. Accessed April 11, 2023. https://e-amrit.niti.gov.in/types-of-electric-vehicles.

Upadhyay, Abhishek, Mihir Dalal, Naman Sanghvi, Vaibhav Singh, Sheeja Nair, Ionut Cristian Scurtu, and Cristian Dragan. Electric Vehicles over Contemporary Combustion Engines. IOP Conference Series: Earth and Environmental Science 635, no. 1 (January 1, 2021): 012004. https://doi.org/10.1088/1755-1315/635/1/012004.

Wong, Y. S., and C. C. Chan. Vehicle Energy Storage: Batteries. In Transportation Technologies for Sustainability, edited by Mehrdad Ehsani, Fei-Yue Wang, and Gary L. Brosch, 1082–1102. New York, NY: Springer, 2013. https://doi.org/10.1007/978-1-4614-5844-9_808.

Chapter 13

Application of Solar Energy in Sustainable Mobility

K. Balamurugan, T. Jerry Alexander, V. Karthikeyan

13.1 INTRODUCTION

The sun is a natural primary source of renewable and clean energy available for life on Earth. The sun is expected to continue emitting energy for billions of years, making solar energy a virtually inexhaustible power source. The sun is a sustainable energy source with the potential to provide clean, reliable, and affordable energy to meet our growing energy demands while mitigating the impacts of climate change.

Solar energy is a renewable energy source harnessed in various ways, such as plant photosynthesis, solar thermal energy, and photovoltaic (PV) cells. There are multiple applications of solar energy pertaining to sustainable transportation, including solar-powered vehicles like bikes, cars, buses, trains, boats, ships, etc., and solar-powered charging stations. One of the applications is its utilization in the solar-powered electric vehicle. With this application, one can preserve fossil fuels by limiting their usage to a certain extent.

13.2 SOLAR-POWERED ELECTRIC VEHICLES (SPEVs)

Solar-electric and solar-powered electric vehicles are similar because they use solar energy to power their electric motors. However, there is a distinction between the two terminologies.

A solar electric vehicle is a type of electric vehicle that has solar panels mounted over itself, either on the roof or hood. These solar panels capture sunlight and convert it into electricity, which is utilized to charge the vehicle's batteries or power its electric motor directly. In short, it generates its own electricity.

On the other hand, a solar-powered electric vehicle is powered by electricity from the grid or a charging station, but it is charged using solar energy. It means that solar energy does not directly power the vehicle. However, the energy used to charge the vehicle's batteries comes from solar panels, either on the vehicle itself or elsewhere.

DOI: 10.1201/9781003464556-13

Solar-powered electric vehicles are vehicles that utilize the energy from the sun to power their electric motors, either directly or indirectly. One of the ways to power these EVs using solar energy is by installing solar panels on the vehicle's rooftop, which can charge the vehicle's battery pack while it is parked or being driven (Can Emir, 2022). The second way is to use solar-powered charging stations, where EVs can be charged using electricity generated by solar panels. By doing this, sustainable mobility can be achieved.

These solar EVs can replace traditional gasoline-powered vehicles if sustainable transportation is attained. Thereby, the following is made feasible:

- Utilization of renewable energy sources – reduces dependence on fossil fuels (Holechek et al., 2022)
- Reduced carbon emissions due to fuel
- Energy independence (up to a certain degree)
- Reduced transportation/handling costs

However, solar-powered EVs have some limitations, such as the cost of the solar panels and the limited range of the vehicle, which depend on the size and efficiency of the solar panels. Additionally, solar-powered charging stations may only be available in some locations, which could limit the practicality of long-distance travel using a solar-powered EV.

13.2.1 Components of a Solar-Powered Electric Vehicle

The critical components to creating a renewable and sustainable mode of transportation are solar panels, batteries, electric motors, charge controllers, inverters, and power management systems, as shown in Figure 13.1.

(a) Solar Panels

Solar panels can be a valuable addition to an electric vehicle and may be limited to a particular type of SPEV. They are photovoltaic panels that convert sunlight into electricity. They are mounted on the vehicle's roof or body and are connected

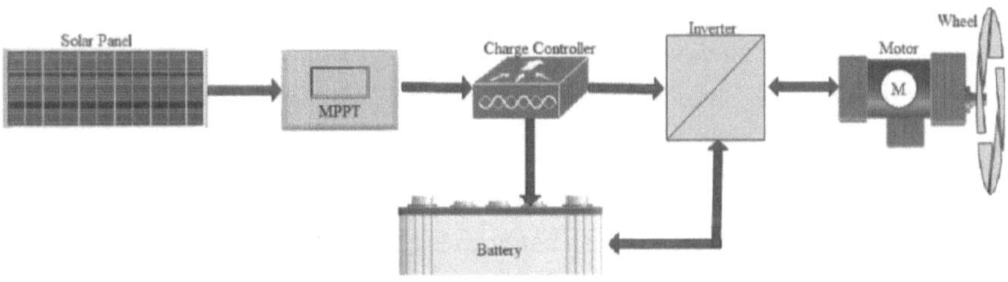

Figure 13.1 Components of a Solar-Powered Electric Vehicle.

to a battery storage system. The solar panels' energy can be utilized to power the vehicle's electric motor or charge its battery.

The amount of energy solar panels can generate on a vehicle is limited by the size of the panels and the amount of sunlight they receive. In most cases, the energy generated by the solar panels will not be enough to power the vehicle on its own, so the vehicle will still need to be charged from an external source, such as a charging station (which may be solar-powered) or a wall outlet (Manivannan and Kaleeswaran, 2016).

The weight and cost of the solar panels can be a limiting factor in designing a solar-powered electric vehicle.

Several types of solar panels are used in a solar-powered electric vehicle, but the most common types are:

(i) Monocrystalline Solar Panels: These panels are made from a single silicon crystal, which makes them highly efficient and durable. They are known for their high conversion efficiency, making them the prominent type of solar panel used in electric vehicles.
(ii) Polycrystalline Solar Panels: Polycrystalline-type solar panels are made up of multiple silicon crystals and are slightly less efficient than monocrystalline panels. Nevertheless, they are still highly efficient and famous for solar-powered electric vehicles.
(iii) Thin-Film Solar Panels: The thin film panels are made by depositing a thin layer of photovoltaic material onto a substrate. They are less efficient than crystalline silicon panels but are lighter and more flexible, making them easier to integrate into the design of a vehicle.
(iv) Amorphous Silicon Solar Panels: These panels are constructed from non-crystalline silicon and are highly flexible and lightweight. They are less efficient than crystalline silicon panels but can be integrated into curved surfaces, making them a good choice for electric vehicles with unique designs.

(b) Battery Storage System

The battery storage system consists of a set of batteries that store the energy generated by the solar panels. The batteries are usually lithium-ion, which powers the electric motor that propels the vehicle.

Solar-powered electric vehicles typically use a rechargeable battery system to store energy generated by the solar panels. The most common types of battery storage systems used in solar-powered electric vehicles are (Kartikay et al., 2021):

(i) Lithium-ion Batteries: These batteries are the most popular type of battery used in electric vehicles, including solar-powered ones. They are lightweight, have a high energy density, and are relatively safe and long-lasting.

Lithium-ion batteries can be charged quickly, which makes them ideal for use in EVs.

(ii) Nickel-Metal Hydride (NiMH) Batteries: These batteries are a more mature technology than lithium-ion batteries and are still used in some electric vehicles, although they are becoming less common. NiMH batteries have a lower energy density than lithium-ion batteries but are more affordable and can last longer.

(iii) Lead-Acid Batteries: These batteries are an older technology and are used less frequently in electric vehicles. They are heavy, have low energy density, and require regular maintenance. However, they are affordable and can be helpful for some niche applications.

(iv) Solid-State Batteries: This newer technology is still developing but can be a game-changer for electric vehicles. Solid-state batteries have a high energy density, but they are lightweight and safer than lithium-ion batteries. They are still in the experimental stage, but several companies are working on developing them for use in electric vehicles.

The type of battery storage system used in a solar-powered electric vehicle will depend on factors such as energy density, weight, cost, and safety. Lithium-ion batteries are currently the most preferred choice for electric vehicles

(c) Electric Motor

The motor is the device that converts electrical energy into mechanical energy to drive the vehicle. It is connected to the battery storage system and controlled by an electronic controller.

Electric motors used in solar-powered electric vehicles are typically high-efficiency brushless DC (BLDC) motors. These motors use permanent magnets to generate the rotating magnetic field that drives the motor. They are controlled by an electronic controller that converts the direct current (DC) from the battery to the alternating current (AC) the motor needs.

BLDC motors are preferred in electric vehicles because they are highly efficient and have a high power-to-weight ratio. They also have a longer lifespan and require less maintenance than traditional brushed motors. Additionally, they provide smooth and quiet operation, which is ideal for electric vehicles.

The size and power of the electric motor used in a solar-powered electric vehicle will depend on factors such as the vehicle's weight, the desired speed and acceleration, and the range required. In general, electric motors for electric vehicles range in power from a few kilowatts to several hundred kilowatts.

It's worth noting that some solar-powered electric vehicles use multiple electric motors, with one motor powering each wheel or axle. This configuration, known as a "hub motor" or "in-wheel motor" system, can provide better traction and control and more efficient regenerative braking. However, designing and manufacturing can also be more complex and expensive.

(d) Charge Controller

The charge controller is an electronic device that is used to regulate the current flow from the solar panels to the battery storage system. It ensures that the batteries are charged efficiently and safely.

A charge controller is essential in a solar-powered electric vehicle as it regulates the current flow between the solar panels and the battery. The charge controller prevents the battery from overcharging or undercharging, which can damage the battery and reduce its lifespan.

Charge controllers used in solar-powered electric vehicles are typically Maximum Power Point Tracking (MPPT) controllers. The MPPT controllers optimize the output power from the solar panels by continuously monitoring the voltage and current of the panels and adjusting the current and voltage parameters to match the battery's requirements (Shiba et al., 2023).

MPPT controllers are more efficient than other charge controllers as they can convert more solar energy into usable electricity for the battery. This efficiency is significant in solar-powered electric vehicles where space and weight are at a premium.

In addition to MPPT controllers, some solar-powered electric vehicles may also use pulse-width modulation (PWM) controllers. PWM controllers work by turning the power on and off at a high frequency, which controls the energy flowing between the solar panels and the battery.

The type of charge controller used in a solar-powered electric vehicle will depend on factors such as the size of the solar panel array, the battery's capacity, and the vehicle's energy requirements. The charge controller must be sized appropriately to ensure it can handle the maximum current and voltage generated by the solar panels and protect the battery from damage.

(e) Inverter

The inverter is a device that converts Direct Current (DC) electricity from the battery storage system into Alternating Current (AC) electricity, which powers the vehicle's accessories, such as lights, air conditioning, and entertainment systems.

The inverter is a crucial component in an SPEV that converts the DC power from the battery to AC power that can be utilized by the electric motor. The inverter also controls the speed and torque of the motor and manages the regenerative braking system that captures energy from the motor during braking and sends it back to the battery.

The inverter used in a solar-powered electric vehicle is typically a high-efficiency, three-phase, pulse-width modulation (PWM) inverter. PWM inverters are highly efficient and provide precise control over the output voltage and frequency, making them ideal for use in electric vehicles.

The inverter must be sized appropriately for the motor's power rating and the voltage and current requirements of the battery. In addition, the inverter must be able to handle the peak power demands of the vehicle, such as during acceleration or hill climbing.

Some solar-powered electric vehicles may also use a bi-directional inverter, allowing energy to flow from the battery to the motor during normal operation and back to the battery during regenerative braking. This bi-directional functionality can extend the vehicle's range and improve its efficiency.

Hence, the inverters ensure the battery's DC power conversion into usable AC power for the electric motor, enabling the vehicle to move efficiently and with a longer range of transportation.

(f) Power Management System

This electronic system manages the power flow between the solar panels, battery storage system, and electric motor. It optimizes available power and ensures that the vehicle operates efficiently.

A power management system (PMS) is an essential component in a solar-powered electric vehicle that coordinates the energy flow between the solar panels, battery, electric motor, and other systems in the vehicle. The PMS ensures energy is used efficiently and optimizes the vehicle's performance while extending its range.

The PMS used in a solar-powered electric vehicle typically includes several subsystems, including:

(i) Battery Management System (BMS): The BMS monitors the battery's state of charge, temperature, and voltage and manages the charging and discharging of the battery.
(ii) DC-DC Converter: The DC-DC converter regulates the voltage and current flowing between the vehicle's solar panels, battery, and other DC loads.
(iii) Inverter: As mentioned earlier, the inverter converts the DC power from the battery into AC power that the electric motor can use.
(iv) Motor Controller: The motor controller manages the speed and torque of the electric motor, ensuring that it operates efficiently and safely.
(v) Regenerative Braking System: The regenerative braking system captures energy from the motor during braking and sends it back to the battery, increasing the vehicle's range and efficiency.

The PMS must be able to handle the dynamic energy requirements of the vehicle, ensuring that energy is used efficiently while maintaining safe and reliable operation. The PMS also plays a critical role in protecting the battery from damage due to overcharging, undercharging, or excessive discharge.

13.3 SOLAR-POWERED TRAINS

Trains are an important and unavoidable transport for the public. Trains have different locomotives, such as steam locomotives, diesel-electric locomotives, electric locomotives, gas turbine-electric locomotives, hybrid locomotives, and Maglev trains. Steam locomotives use steam generated by heating water with coal, oil, or wood to power the engine. Diesel-electric locomotives use diesel engines to generate electricity to power electric motors. Electric locomotives use only electricity that is provided separately to run the train. Gas turbine-electric locomotives use a gas turbine to generate electricity to power the electric motors. Hybrid locomotives use two engines, diesel and electric, to provide better fuel efficiency and lower emissions. Maglev trains are the new technology trains that use magnetic levitation to lift the train above the track and propel it forward. They do not have wheels and are powered by electricity.

Solar-powered trains are a type of train that uses solar energy to power its locomotive. The advantages of using solar-powered trains are (a) a reduction in greenhouse gas emissions and (b) a reduction in operating costs (Chouhan, 2020).

Solar panels can be installed on the roof of the train. The solar panel receives photons from the sun and converts them into electricity, which is then collected by the electric system in the train. Even though the roof of the train has a solar panel, the energy generated by the solar panel will not be sufficient to run the locomotive. So, hybrid systems can be installed to avoid this problem. Hybrid systems combine solar power and other energy sources, such as diesel or electric power (Jia et al., 2020).

The disadvantages of solar-powered trains are (a) limited space available on the train's roof for solar panels and (b) the installation cost.

13.4 SOLAR-POWERED AIRCRAFT

Solar-powered aircraft use solar energy to power their engines and other electrical systems. Solar panels can be installed on the outer portion of aircraft. The solar panel captures solar energy from the sun and converts it into electrical energy to power the aircraft. The main advantage of using solar-powered aircraft is it avoids carbon emissions or noise pollution. Also, it is cheaper to operate in the long run as they don't require fuel.

There are two main types of solar-powered aeroplanes: (a) unmanned aerial vehicles (UAVs) and (b) manned aeroplanes. UAVs are typically smaller and have limited range and payload capabilities (Chu et al., 2021), while manned solar-powered aeroplanes are larger and can carry passengers and cargo.

The disadvantages of solar-powered aircraft are (a) the very limited space available on the aeroplane for solar panels, (b) the weight of the solar panels, which can reduce the aeroplane's payload capacity, and (c) the efficiency of solar panels due to weather conditions, such as clouds and low light (Noth and Siegwart, 2007).

13.5 SOLAR-POWERED CHARGING STATIONS

The advantages of using solar energy over conventional electric power are:

a) Readily available throughout the year
b) No cost for the source
c) Environmentally friendly
d) Easy installation

The use of electric vehicles (EVs) is increasing drastically because of people's awareness of polluting the environment. EVs have a battery unit to store the electric energy. One of the main limitations of the EVs is charging the battery. The driver must use the charging station to charge the EV's battery. If the number of electric vehicles increases, there is a need to increase the charging station. This automatically increases the energy consumption from the electricity supplier (government or non-government). It may create a demand for electricity that, in turn, affects the industry, residences, etc. Installing the solar panel in the charging station is necessary to overcome this problem. Even though solar energy is insufficient to power more vehicles, it will support the charging station to provide an uninterrupted supply. Figure 13.2 shows the example charging station, the Baka-charging station.

13.5.1 Components of Solar-Powered Charging Stations

A solar charging station for EVs consists of solar panels, a DC-to-AC converter, a battery unit, and EV charging equipment, as shown in Figure 13.3. The solar panel receives the photon energy from the sun's rays. The photons

Figure 13.2 Baka-Charging Station (Source: Wikimedia Commons).

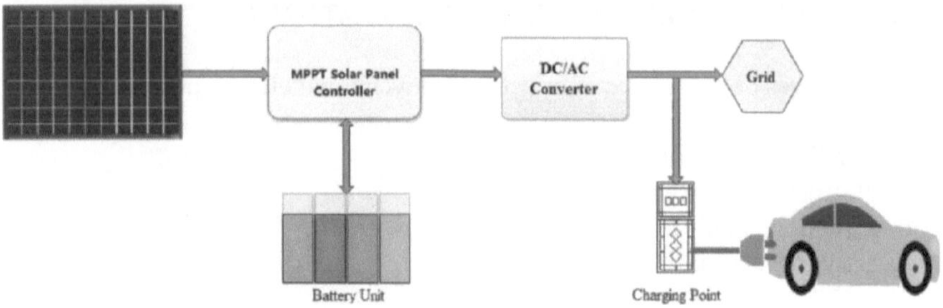

Figure 13.3 Components of a Solar-Powered Charging Station.

hit the atoms in the semiconductor and release the electrons that cause electricity. The solar panel capacity can be up to 50 kW for a medium-sized solar charging station.

Maximum Power Point Tracking (MPPT) is a type of controller used to optimize the efficiency of solar panels. MPPT continuously monitors the voltage and current output of the solar panel and adjusts the same to maintain its maximum power point (MPP). The MPP is the point at which the solar panel produces full power, which can increase the efficiency of the solar power system by up to 30%. The generated electricity is then converted into AC supply by the converter (Divyapriya et al., 2021). The specification of converter capacity should be matched with the capacity of the solar panels. For a 50 kW solar panel, the DC-to-AC converter should be capable of handling 50 kW. The battery unit in the charging station stores excess electric charges generated during the day and helps the charging station during the required time. The AC supply is then provided to the EVs through the charging point. Some charging stations are connected to the grid to transfer unused energy when no EVs exist.

In a few solar-based charging stations, an additional monitoring and control system performs the real-time monitoring of the entire performance of the solar charging station. Also, along with the MPPT controller, it controls the whole power generation process. It monitors and controls the solar panel's output, the battery unit's charge level and charging status, etc. For the improvement of solar power generation, some of the factors to be considered are (a) the location of the solar panel installation, (b) the orientation of the solar panel, (c) the tilt angle of the solar panel, and (d) the type of battery used. The solar panel's location, orientation, and tilt angle should be technically fixed to receive more solar energy. Automatic tilting of the solar panel according to the intensity of sunlight will also improve the output efficiency of the solar power. This automated tilting technique works like the sunflower plant. Sensors are used to determine the intensity of the sunlight, and according to the intensity level, the entire solar panel will tilt to receive more sunlight. The selection of the battery unit should also play an essential role in the safety of the charging station, and the battery unit should be environmentally friendly (Alrubaie et al., 2023).

Most charging stations only partially depend on solar power due to limitations such as capacity, climatic conditions, etc. They all use hybrid structures, such as combining public electricity with solar energy to provide an uninterrupted supply to EVs. Even though the solar system has many advantages, such as a reduction in carbon emissions, etc., the main disadvantage is the cost of installing a solar system in the charging station.

13.6 SOLAR-POWERED ROADS

Solar roads are a promising technology for future generations. A highly efficient, uninterrupted power supply can be given to EVs almost all the time. Implementing solar roads is still being researched to discover real-time problems.

The roads are made up of strong photovoltaic (PV) cells that convert solar energy into electricity. These cells are embedded into the road surface, as shown in Figure 13.4, and protected by a transparent, durable material that is strong enough to support the weight of vehicles. The generated electricity can charge EVs, buildings, streetlights, etc. The excess power will be given to the grid for other uses. The main challenges in implementing solar roads are cost and durability (Dezfooli et al., 2017). PV cells are expensive. It is also necessary to maintain it properly. A robust protective layer is needed to withstand heavy loads, changing weather conditions, and vehicle wear and tear. The power generated by solar roads is connected to the grid directly, or the battery banks can be used to store the energy

Figure 13.4 Solar Road (Source: Wikimedia Commons).

for future use. The battery bank capacity will be chosen depending on the power generated. Batteries for solar roads are typically made up of lithium-ion cells, the same type of batteries used in electric vehicles and many other electronic devices (Hu et al., 2021).

There are three layers in the solar road. They are (a) the glass layer, (b) the electronic layer, and (c) the base plate layer.

(a) Glass Layer (Encapsulation Layer): This is the topmost layer of the road on which vehicles can travel. It is made of strong, transparent material, such as tempered glass or a durable polymer. This layer protects the PV cells from damage due to weather conditions, debris, and heavy loads from vehicles.
(b) Electronic Layer (PV Cell Layer): This layer contains solar photovoltaic cells, and it is under the glass layer. It includes a chip board that senses the loads and controls and monitors the fixed heating element within the surface layer. The electronic layer converts light vitality to electric energy. The solar panels are usually made of glass or durable polymer material and are designed to withstand heavy loads from vehicles and changing weather conditions.
(c) Base Plate Layer (Substrate Layer): This is the bottom-most layer of the solar layer. The purpose of this layer is to collect the electrical energy and transmit it to the power station. The substrate layer is the underlying layer that supports the encapsulation layer and PV cells. The substrate layer is typically made of strong and durable material, such as concrete or asphalt, and supports the other layers while resisting external weather conditions.

The main advantages of solar roads are:

i. Generation of renewable energy
ii. Utilization of space
iii. Less maintenance
iv. Increased safety
v. Improved durability

The main disadvantages of the solar road are:

i. High cost
ii. Lower efficiency
iii. Limited performance in harsh weather
iv. Maintenance and repair challenges
v. Safety concerns
vi. Limited scalability

13.7 SOLAR PERSONAL RAPID TRANSIT SYSTEMS

A solar personal rapid transit (PRT) system is an innovative solution for urban transportation that combines the benefits of solar power, automation, and on-demand service to provide a fast, efficient, and sustainable mode of transport for the future. It uses small, automated, electric vehicles to transport passengers quickly and efficiently between destinations within a specific area, like a university campus, business park, or amusement park (Baertsch and Swenson, 2008). These vehicles are operated on solar power generated by solar panels, which are installed on the roofs of PRT stations or the vehicles themselves.

The vehicles used in a solar PRT system are typically lightweight, four-seater pods that run on an elevated guide way network. The pods are battery-powered and can be charged using solar energy stored in the battery or from the grid during low-demand periods. The pods can be called on-demand by passengers using a smart phone app or a touch-screen interface at the PRT station.

A solar PRT system offers several advantages over traditional transportation systems. It is cost-effective, environmentally friendly, and provides excellent flexibility for passengers (Furman et al., 2021). Additionally, the small size of the pods and the on-demand nature of the system make it easier for its expansion, either up or down, to meet the changing demand.

13.8 LONG-RANGE SOLAR ELECTRIC VEHICLES

A Long-Range Solar Electric Vehicle (LSEV) is an innovative solution for sustainable transportation that combines the benefits of solar power and electric vehicles to provide a clean and efficient mode of transportation for the future (Arulious et al., 2021). The LSEV is powered by electricity from solar panels and has a range that exceeds typical electric vehicles. The range of a solar electric vehicle is determined by the size of the battery pack, the efficiency of the motor, and the amount of energy generated by the solar panels.

The range of a long-range solar electric vehicle can vary widely depending on the specific model and design, but some models have ranges of up to 640 km or more. These vehicles are often designed with larger battery packs and more efficient motors than traditional electric vehicles, allowing them to travel further on a single charge.

In addition to the larger battery packs and more efficient motors, long-range solar electric vehicles are designed with more advanced solar panels to generate more electricity from sunlight. These panels are often integrated into the body of the vehicle, such as on the roof or hood, to maximize the amount of solar energy that can be harvested (Dimitrova, 2018).

The benefits of a long-range solar electric vehicle include reduced dependence on fossil fuels, lower greenhouse gas emissions, and the ability to travel long distances without needing to stop for a charge. However, these vehicles can be more expensive than traditional electric vehicles due to their larger battery packs and advanced solar panel technology.

13.9 CONCLUSION AND FUTURE SCOPE

Solar-powered electric vehicles offer a sustainable and environmentally friendly mode of transportation. SPEVs use solar energy to power the vehicle, thereby reducing reliance on fossil fuels and the carbon footprint. Similarly, solar-powered trains, aircraft, and long-range transportation are also the applications of sustainable mobility. The solar-based charging station is also one of the applications of solar energy in sustainable mobility.

The technology used in solar-powered electric vehicles is rapidly evolving, and there is significant potential for future developments in this area. Some of the areas for future research and development include:

(a) Improving the efficiency of solar panels and lower manufacturing cost: There is ongoing research to develop solar panels that are more efficient, lightweight, and flexible, which could significantly increase the range and efficiency of solar-powered electric vehicles. At the same time, low-cost manufacturing of the panel can contribute to its development.
(b) Developing better battery technology: Improved battery technology, including higher energy density and faster charging times, could help to increase the range and reduce the cost of solar-powered electric vehicles. Also, extra care should be taken in disposing of the chemical batteries for the benefit of a green, clean, and healthy environment.
(c) Increasing the efficiency of the power management system: The power management system is a critical component of solar-powered electric vehicles. Improving its efficiency could extend the vehicle's range and improve its performance.
(d) Integrating solar charging infrastructure: Developing solar charging infrastructure, such as solar-powered charging stations, could help to increase the adoption of solar-powered electric vehicles and make them more accessible to the general public.
(e) Optimizing vehicle design: Vehicle design plays a vital role in the efficiency and performance of solar-powered electric vehicles. Further research and development could improve these vehicles' aerodynamics, weight, and overall efficiency.

The future of solar-powered electric vehicles is promising, with significant potential for improvements in efficiency, range, and accessibility. Suppose the overall manufacturing cost comes with the compromise concerning the transportation distance. With ongoing research and development, these vehicles could become a more common mode of transportation, helping reduce the environmental impact of transport and the dependence on fossil fuels, which aids in sustainable transportation.

REFERENCES

A. J. Alrubaie, M. Salem, K. Yahya, M. Mohamed, and M. Kamarol. A Comprehensive Review of Electric Vehicle Charging Stations with Solar Photovoltaic System Considering Market, Technical Requirements, Network Implications, and Future Challenges. Sustainability, vol. 15, no. 10 (2023): 8122, doi: 10.3390/su15108122.

J. A. Arulious, D. Earlina, D. Harish, P. Sakthi Priya, A. Inba Rexy, and J. S. Nancy Mary. Design of Solar Powered Electric Vehicle. Journal of Physics: Conference Series, vol. 2070, no. 1 (2021): 012105. doi: 10.1088/1742-6596/2070/1/012105.

R. Baertsch, and R. Swenson. Solar-Powered Personal Rapid Transit (PRT): Electric Vehicles Without Batteries or Congestion. Proceedings of the Solar Conference, vol. 2 (2008).

Can Emir. Here Are the Top 8 Electric Vehicles of 2022 with Solar Roofs, (June 3, 2022). https://interestingengineering.com/transportation/eight-electric-vehicles-2022-solar-roofs.

P. Chouhan. Feasibility Report of Solar Powered Train. IJRASET, vol. 8, no. 11 (2020): 337–343. doi: 10.22214/ijraset.2020.32136.

Y. Chu, C. Ho, Y. Lee, and B. Li. Development of a Solar-Powered Unmanned Aerial Vehicle for Extended Flight Endurance. Drones, vol. 5, no. 2 (2021): 44. doi: 10.3390/drones5020044.

A. S. Dezfooli, F. M. Nejad, H. Zakeri, and S. Kazemifard. Solar Pavement: A New Emerging Technology. Solar Energy, vol. 149 (2017): 272–284. doi:10.1016/j.solener.2017.04.016.

Z. Dimitrova. Optimal Designs of Electric Vehicles for Long-Range Mobility. MATEC Web of Conferences, vol. 234 (2018): 02001. doi: 10.1051/matecconf/201823402001.

S. Divyapriya, A. Amudha, and R. Vijayakumar. Design of Solar Smart Street Light Powered Plug-in Electric Vehicle Charging Station by Using Internet of Things. Journal of the Institution of Engineers (India): Series B vol. 102, no. 3 (2021): 477–486. https://doi.org/10.1007/s40031-021-00548-y.

B. Furman, L. Ramasubramanian, S. McDonald, R. Swenson, J. Fogelquist, Y. Chiao, A. Pape, and M. Cruz. Solar-Powered Automated Transportation: Feasibility and Visualization, (2021, December). *Mineta Transportation Institute Publications*. https://doi.org/10.31979/mti.2021.1948.

J. L. Holechek, H. M. E. Geli, M. N. Sawalhah, and R. Valdez. A Global Assessment: Can Renewable Energy Replace Fossil Fuels by 2050? Sustainability, vol. 14, no. 8 (2022): 4792. doi: 10.3390/su14084792.

H. Hu, D. Vizzari, X. Zha, and R. Roberts. Solar Pavements: A Critical Review. Renewable and Sustainable Energy Reviews, vol. 152 (2021): 111712. doi: 10.1016/j.rser.2021.111712.

L. Jia, J. Ma, P. Cheng, Y. Liu. Perspective of Solar Energy-Powered Road and Rail Transportation in China. CSEE Journal of Power and Energy Systems, vol. 6, no. 4 (2020). doi:10.17775/CSEEJPES.2020.02040.

P. Kartikay, K. Mokurala, B. Sharma, R. Kali, N. Mukurala, D. Mishra, A. Kumar, S. Mallick, J. Song, and S. H. Jin. Recent Advances and Challenges in Solar Photovoltaic and Energy Storage Materials: Future Directions in Indian Perspective. *Journal of Physics: Energy*, vol. 3, no. 3 (2021): 034018. doi: 10.1088/2515-7655/ac1204.

S. Manivannan, and E. Kaleeswaran. Solar Powered Electric Vehicle. In 2016 First International Conference on Sustainable Green Buildings and Communities (SGBC), 1–4. Chennai, India: IEEE. (2016). https://doi.org/10.1109/SGBC.2016.7936074.

A. Noth, and Siegwart. Designing Solar Airplanes for Continuous Flight. Environmental Research (2007): 1–17.d. doi: 10.1117/2.1200906.1649. https://www.semanticscholar.org/paper/Design-of-Solar-Powered-Airplanes-for-Continuous-Noth/97687b192c202581bcb59297b72e233c89f62215

S. M. Shariff, M. S. Alam, F. Ahmad, Y. Rafat, M. S. J. Asghar, and S. Khan. System Design and Realization of a Solar-Powered Electric Vehicle Charging Station. IEEE Systems Journal, vol. 14, no. 2 (2020): 2748–2758. doi: 10.1109/JSYST.2019.2931880.

M. Shiba, A. Hussien, S. Abouel-Seoud, and A. Sultan. Innovative Design and Performance Analysis of Solar-Powered Electric Vehicles for a Promising Approach for Green Transportation. Journal of International Society for Science and Engineering 2023, April 29. https://doi.org/10.21608/jisse.2023.205733.1070.

Chapter 14

State Estimation of Lithium-ion Battery for Electric Vehicle Application

Types, Design Methods, and Future Trends

Roushan Kumar

14.1 INTRODUCTION

State estimation of lithium-ion batteries for electric vehicle (EV) applications [1] is a critical aspect of battery management systems. It involves estimating key parameters such as state of charge (SoC), state of health (SoH), and remaining useful life (RUL) to ensure optimal performance, efficiency, and safety of the battery pack [2]. Several techniques and algorithms are utilized for state estimation in EV batteries [3], such as electrochemical models, equivalent circuit models, Kalman filters, particle filter models, and different machine learning models. Electrochemical models are based on the fundamental electrochemical processes occurring within the battery [4]. They describe the complex behaviour of lithium-ion batteries and provide insights into the internal dynamics. Electrochemical models use mathematical equations and physical principles to represent the battery's electrochemical reactions, diffusion processes, and other phenomena [5]. These models can offer high accuracy but are computationally intensive and require detailed knowledge of battery chemistry. Equivalent circuit models provide a simplified representation of the battery's electrical behaviour. They consist of resistors, capacitors, and voltage sources that approximate the battery's internal electrochemical processes. Equivalent circuit models are computationally efficient and widely used for real-time applications [6]. However, they may lack accuracy compared to electrochemical models, especially under varying operating conditions.

Kalman filters are recursive estimation algorithms that combine measurements from sensors with a mathematical model of the system to estimate the true state. Extended Kalman filters (EKF) and unscented Kalman filters (UKF) are commonly employed for state estimation of lithium-ion batteries [7]. These filters incorporate measurements such as voltage, current, and temperature along with the battery model to estimate SoC, SoH, and other parameters. Kalman filters are computationally efficient and provide good estimation accuracy but require accurate battery models and knowledge of system uncertainties. Particle filters, also known as Monte Carlo filters, are another class of recursive estimation algorithms

DOI: 10.1201/9781003464556-14

used for state estimation. They use a set of particles (sample points) to represent the probability density function of the battery state. Particle filters can handle nonlinear and non-Gaussian systems and are suitable for online estimation [8]. However, they can be computationally demanding and require careful tuning of the particle set size and resampling strategies.

Machine learning approaches, such as artificial neural networks, support vector machines, and random forests, have gained attention in battery state estimation [9]. These techniques can learn complex relationships from training data and estimate battery states based on input features such as voltage, current, temperature, and time. Machine learning-based state estimation methods offer potential improvements in accuracy and can adapt to changing battery behaviours. However, they require large amounts of training data and may lack interpretability [10]. These state estimation techniques are applied in combination with real-time measurements from sensors, such as current sensors, voltage sensors, and temperature sensors, to estimate the battery's state variables. The choice of technique depends on factors such as accuracy requirements, computational constraints, available information, and the specific application context. Overall, state estimation plays a crucial role in maximizing the performance, reliability, and longevity of lithium-ion batteries in electric vehicle applications. It enables accurate monitoring and control, facilitates efficient battery utilization, and enhances overall system safety [11]. Ongoing research and advancements in state estimation techniques continue to contribute to the development of advanced battery management systems for electric vehicles.

14.2 STATE ESTIMATION TYPES

The states and parameters of SOC, RUL, SOH, etc., are crucial for battery management systems to monitor, estimate, and control the behaviour of the battery, ensuring its safe and optimal operation, as shown in Figure 14.1. It covers all the different state estimation types used in battery management systems for electric vehicle applications.

14.2.1 State of Charge

SOC stands for "State of Charge." It is a critical parameter used to indicate the current level of charge or remaining capacity of a battery relative to its full capacity. SOC is typically expressed as a percentage, with 0% representing a fully discharged battery and 100% indicating a fully charged battery [12]. Battery management systems (BMS) monitor and estimate the SOC to provide essential information about the available energy in the battery and to ensure the battery's safe and efficient operation. Accurate SOC estimation is crucial for proper battery management, as it helps prevent overcharging, over-discharging, and other potentially

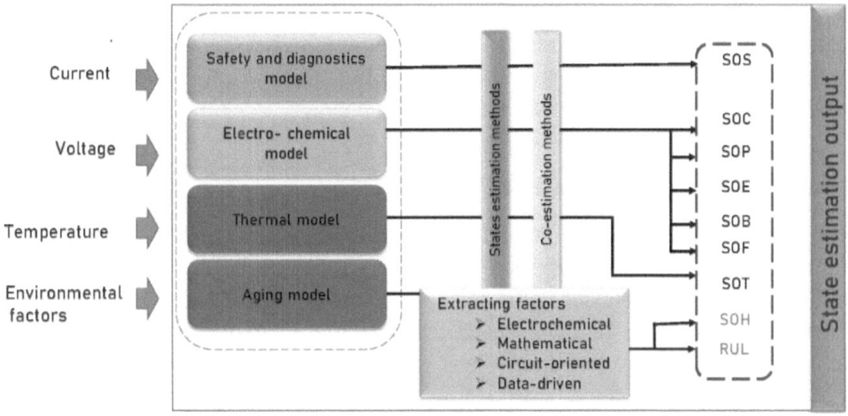

Figure 14.1 Different State Estimation Types Used in a Battery Management System.

damaging operating conditions. The SOC is used in various applications, such as electric vehicles, portable electronic devices, and renewable energy storage systems [13]. In electric vehicles, knowing the SOC helps drivers understand the remaining driving range, while in renewable energy systems, it aids in optimizing energy storage and utilization. SOC estimation is usually achieved through advanced algorithms and techniques, such as Coulomb counting, voltage-based methods, Kalman filters, and various model-based approaches.

14.2.2 State of Health

SOH stands for "State of Health." It is a crucial parameter that indicates the overall health or condition of a battery, specifically how well it is performing compared to its original or expected performance when it was new. As batteries undergo charge-discharge cycles and age over time, their performance can degrade due to factors such as chemical changes, capacity loss, increased internal resistance, and mechanical stress [14]. SOH provides an assessment of these degradation effects and helps determine the remaining useful life of the battery. A high SOH value indicates that the battery is in good condition and still capable of delivering close to its original capacity and performance. Conversely, a low SOH value suggests that the battery has experienced significant degradation, resulting in reduced capacity and performance. Monitoring the SOH is essential in battery management systems (BMS) as it allows operators and users to make informed decisions regarding battery maintenance, replacement, or reconditioning. It helps prevent unexpected failures, ensures safety, and optimizes the use of batteries in various applications such as electric vehicles, renewable energy storage, and portable electronics. The estimation of SOH often involves complex algorithms that analyze data collected during the battery's lifetime, including charge and discharge cycles, temperature, and other operating conditions [15]. By accurately assessing the SOH, battery operators can determine the remaining useful life and take appropriate actions to maximize the performance and reliability of the battery system.

14.2.3 Remaining Useful Life

RUL stands for "Remaining Useful Life." It is a critical parameter used to estimate the remaining operational lifespan of a battery or any other system/component before it reaches the end of its useful life. In the context of batteries, RUL provides an estimate of how much longer the battery can perform its intended functions effectively before its capacity and performance degrade to the point where it becomes unsuitable for its intended application [16]. Accurate RUL estimation is essential for proactive maintenance, replacement planning, and optimizing the usage of the battery or the system it powers. It helps prevent unexpected failures and downtime, reduces operational risks, and ensures safety and reliability. The estimation of RUL is a challenging task, especially for complex systems like lithium-ion batteries, as it depends on various factors such as charge-discharge cycles, operating conditions, environmental factors, and battery chemistry. Researchers and engineers use advanced modelling techniques, machine learning algorithms, and data-driven approaches to estimate the RUL with reasonable accuracy. By monitoring and predicting the RUL of batteries, battery management systems (BMS) and operators can make informed decisions about when to replace a battery, schedule maintenance, or adjust system operations to maximize the remaining useful life and overall system performance. This proactive approach also helps in optimizing the total cost of ownership and improving the efficiency of battery-dependent applications, such as electric vehicles, renewable energy storage, and industrial energy systems.

14.2.4 State of Temperature

Temperature management is a critical aspect of a BMS to ensure the safe and efficient operation of batteries. Proper temperature control is essential because temperature significantly impacts battery performance, safety, and lifespan. BMS employs various strategies to manage the temperature of the battery, some of which include sensing, protection mechanism, cooling system, heating system, thermal regulation, modelling, thermal balancing, and control. BMS uses temperature sensors placed at strategic locations within the battery pack to continuously monitor the temperature [17]. These sensors provide real-time temperature data, allowing the BMS to respond to temperature changes promptly. If the temperature exceeds safe operating limits, the BMS activates thermal protection mechanisms. This may include reducing the charging or discharging current or completely disconnecting the battery from the load or charging source to prevent overheating and potential safety hazards. Some battery packs incorporate cooling systems, such as liquid cooling or forced air cooling, to dissipate excess heat and maintain the temperature within an optimal range. Cooling systems are particularly useful in high-power applications like electric vehicles. In cold environments, BMS may include heating elements to keep the battery temperature within the optimal range. Heating ensures the battery maintains sufficient performance and capacity even in low-temperature conditions.

Thermal regulation during charging/discharging is adjusted in the BMS through charging and discharging rates based on temperature measurements to prevent overheating or performance degradation. Charging and discharging at controlled rates help manage the temperature during high-power or fast-charging scenarios [18]. BMS may utilize thermal models to predict and estimate temperature changes during different operating conditions. These models assist in proactive thermal management, enabling the BMS to take preventive measures to control temperature fluctuations. In multi-cell battery packs, the BMS may employ temperature-balancing techniques. These techniques redistribute the heat dissipation within the pack by controlling individual cell temperatures, ensuring all cells operate within a safe temperature range. In some applications, the BMS may incorporate environmental control systems that regulate the ambient temperature around the battery, further aiding in temperature management. Effective temperature management in a BMS helps optimize battery performance, extends battery life, and ensures safe and reliable operation. It is especially critical in high-power and fast-charging applications, where temperature variations can have a significant impact on battery health and safety [19].

14.2.5 State of Balance

SOB stands for "State of Balance" in the context of battery management systems and battery packs. SOB refers to the condition of individual cells or modules within a battery pack in terms of their state of charge or voltage. In a multi-cell battery pack, it is essential to ensure that all cells or modules have similar SOC or voltage levels to maintain optimal performance and prevent issues related to cell imbalances. If there are significant differences in the SOC or voltage levels among cells or modules in a battery pack, it can lead to various problems such as reduced capacity, overcharging, over-discharging, reduced efficiency, and reduced lifespan. Reduced capacity deals with cells with lower SOC or voltage, which may limit the overall capacity of the battery pack. Overcharging or over-discharging deals with imbalanced cells, which can result in some cells reaching their voltage limits (overcharging) while others may be at risk of being over-discharged, both of which are harmful to battery health [20]. Reduced efficiency targets the imbalanced cells, which can lead to energy losses during charging and discharging processes. Reduced lifespan covers cells that are consistently overcharged or over-discharged, which may experience accelerated aging, leading to a shorter overall battery lifespan. To manage the SOB in a battery pack, the BMS uses balancing techniques. Balancing involves redistributing charges among individual cells or modules to bring them to a similar state of charge. Balancing can be achieved through various methods, such as active balancing (transferring charge between cells using electronic circuits) or passive balancing (using resistors to discharge cells with higher SOC). By maintaining a balanced SOC among cells or modules, the BMS ensures that the battery operates optimally, maximizing capacity, efficiency, and overall lifespan. Proper balancing is crucial for the safe and reliable operation of battery packs in various applications, including electric vehicles and energy storage systems [21].

14.2.6 State of Power

"State of Power" (SOP) refers to the power capability or power delivery state of the battery or battery pack at a specific point in time. It represents the maximum power the battery can provide at a given moment, considering its SOC, temperature, and other operating conditions. The SOP is an essential parameter for applications that require high-power performance, such as EVs and hybrid electric vehicles (HEVs). In these applications, the battery needs to deliver sufficient power to accelerate the vehicle, climb steep hills, or perform other demanding tasks. Battery management systems continuously monitor the SOP and manage power delivery to ensure that the battery can meet the power demands of the vehicle or the system it powers [22]. If the power demand exceeds the battery's current capability, the BMS may limit the power output to prevent overheating or damage to the battery. In addition to SOP, BMS also monitors and manages other key battery states, such as SOC, SOH, SOT, and SOE, to ensure the safe and optimal operation of the battery and maximize its performance and longevity. By accurately monitoring and controlling these states, the BMS ensures that the battery operates efficiently and reliably in various applications.

14.2.7 State of Energy

The "State of Energy" (SOE) represents the total amount of energy stored in a battery or battery pack at a specific point in time. It is typically expressed in watt-hours (Wh) or kilowatt-hours (kWh) and provides an estimate of the battery's remaining energy capacity. SOE is closely related to the SOC, but there is a subtle difference between the two [23]. SOC indicates the percentage of the total battery capacity that has been utilized. A SOC of 100% means the battery is fully charged, and a SOC of 0% means the battery is fully discharged. SOE, on the other hand, represents the actual amount of energy remaining in the battery, measured in watt-hours or kilowatt-hours. It takes into account not only the battery's current SOC but also its total energy capacity. The difference between SOC and SOE becomes apparent when a battery operates at different voltage levels or under varying load conditions. For example, a battery with a 50% SOC can have different energy levels depending on its voltage and the current draw. SOE is a crucial parameter for managing battery usage, especially in applications like electric vehicles (EVs) and energy storage systems. By monitoring the SOE, the BMS can provide accurate estimates of the remaining driving range in EVs or the available energy in energy storage systems. This information is valuable for users and operators to plan their activities, optimize battery usage, and prevent unexpected depletion of energy during critical operations. Accurate estimation and management of SOE, along with other battery states like SOC and SOH (State of Health), are essential for ensuring the safety, performance, and longevity of batteries in various applications.

14.2.8 State of Safety

"State of Safety" (SOS) is a widely recognized term in the context of battery management systems. In general, battery safety is a critical aspect of battery management, especially in high-power applications such as electric vehicles and energy storage systems. Battery safety encompasses various measures and considerations to prevent hazardous events, such as overheating, thermal runaway, fires, and explosions. Important aspects of battery safety include thermal management, under and over-voltage protection, cell balancing, fault detection and diagnostics, safety standards, and guidelines [13]. Thermal management involves monitoring and controlling the temperature of the battery to prevent overheating and thermal runaway. Over-current and over-voltage protection-implementing measures protect the battery from excessive currents or voltages that could lead to damage or safety hazards. Cell balancing ensures that cells within a battery pack are balanced in terms of SOC and voltage to prevent overcharging or over-discharging of individual cells. Fault detection and diagnostics detect potential faults or abnormalities in the battery and take appropriate actions to mitigate risks. Short circuit protection implements safeguards to prevent short circuits that could lead to high currents and safety hazards. Battery safety standards and guidelines, such as those developed by organizations like IEC (International Electro Technical Commission) and ISO (International Organization for Standardization), provide essential frameworks to ensure the safe design, manufacturing, and use of batteries in various applications [24].

14.3 STATE ESTIMATION METHODS

14.3.1 Electrochemical Models

Electrochemical models are one of the key methods used for state estimation in battery management systems to accurately predict and monitor the state of lithium-ion batteries. These models describe the electrochemical processes that occur inside the battery, allowing the BMS to estimate important battery states such as SOC, SOH, SOP, etc. The commonly used electrochemical models for state estimation are the single particle model, dual particle model, pseudo two-dimensional model, electrochemical impedance spectroscopy, electrochemical thermal model, reduced order electrochemical model, and finite element model [25]. The single particle model is a simplified electrochemical model that represents each electrode particle within the battery as a single pseudo-particle. It is widely used in applications that require fast computational times and can provide reasonably accurate SOC estimations. The dual particle model is an extension of the single particle model that considers separate pseudo-particles for lithium insertion and extraction in the anode and cathode [26]. It provides more accurate SOC and voltage predictions than the single particle model but requires slightly more computational resources.

The pseudo two-dimensional model is a more complex electrochemical model that considers particle dynamics and radial diffusion in the electrodes. It provides higher accuracy in SOC and voltage predictions compared to the single particle model and double particle model but requires more computational effort. The electrochemical impedance spectroscopy (EIS) model utilizes impedance measurements to estimate the battery's internal parameters and can be used to infer information about the battery's health and performance. The electrochemical thermal model combines the electrochemical model with a thermal model to predict the battery's SOC, SOH, and temperature simultaneously. It is essential for estimating battery performance under different operating conditions [27]. The reduced-order electrochemical model simplifies the complex physics of the battery while retaining key features that affect its state. It offers a compromise between accuracy and computational complexity. The finite element model is a numerical method that can provide detailed spatial distributions of battery states and can be used for accurate local state estimation. These electrochemical models can be combined with filtering techniques, such as extended Kalman filters or unscented Kalman filters, to perform real-time state estimation and continuously update the battery's state as it operates. The choice of the specific model depends on the application's requirements, computational resources, and the desired level of accuracy for state estimation in the battery management system.

14.3.2 Equivalent Circuit Models

Equivalent Circuit Models (ECMs) are simplified electrical models used to represent the behaviour of lithium-ion batteries in various applications, including electric vehicles and energy storage systems. These models provide a balance between accuracy and computational efficiency, making them suitable for real-time battery state estimation and control in battery management systems (BMS). ECMs are widely used because of their ability to capture essential battery characteristics while being relatively simple to implement and solve. The most common types of equivalent circuit models include the single resistor model, single resistor capacitor model, dual resistor capacitor model, Doyle fuller Newman model, and Rint model [28]. The single resistor model is the simplest ECM, representing the battery as an ideal voltage source (representing the open-circuit voltage) in series with a single resistor (representing internal resistance). It is suitable for quick estimations of battery voltage under steady-state conditions but lacks the ability to capture transient behaviour and complex dynamics. The single RC (resistor capacitor) model adds a single RC network to the single resistor model, which introduces the ability to model some transient effects, such as charging and discharging dynamics. The RC network captures some capacitive behaviour of the battery [29].

The dual RC model is an extension of the single RC model and includes two RC networks, one for the electrolyte and one for the solid-state diffusion processes. It offers improved accuracy for transient effects and can better model battery behaviour during different operating conditions. The Doyle Fuller Newman (DFN) model is a more comprehensive electrochemical model that combines an equivalent circuit with partial differential equations to simulate various electrochemical processes inside the battery, including diffusion, migration, and reactions. It is more complex and computationally intensive but provides a higher level of accuracy [30]. The Rint model estimates the internal resistance (Rint) of the battery and is often used to assess the battery's SOH. It involves measuring the voltage drop across the battery during a pulse or step in current and using Ohm's law to calculate Rint. It's important to note that equivalent circuit models are empirical and do not capture all the underlying physics of battery operation. However, they are valuable tools for battery management, state estimation, and control due to their simplicity and ability to represent important battery characteristics. The choice of the appropriate equivalent circuit model depends on factors such as the desired level of accuracy, computational resources, and specific applications. For real-time applications like BMS, simpler equivalent circuit models are often preferred due to their efficiency, while more complex models like the DFN model are used for detailed research and analysis purposes.

14.3.3 Kalman Filters

Kalman filters are commonly used as state estimation techniques in Battery Management Systems (BMS) to accurately predict the state of a lithium-ion battery, such as State of Charge (SOC), State of Health (SOH), and State of Power (SOP) [31]. Kalman filters are optimal estimation algorithms that combine measurements from various sensors with system dynamics to estimate the true state of a system, even in the presence of noise and uncertainties. The Kalman filter consists of two main steps: a prediction step and an update step. In the prediction step, the filter predicts the current state of the battery based on the previous state and system dynamics model. The system dynamics model is often derived from battery physics, such as equivalent circuit models or electrochemical models. The prediction incorporates the effect of inputs (e.g., current and voltage) and battery behaviour models to estimate the state evolution over time. In the update step, the filter incorporates sensor measurements (e.g., voltage, current, and temperature) to correct the predicted state and obtain a more accurate estimate. The Kalman filter uses the difference between the predicted and measured states to adjust the estimated state and its uncertainty based on the reliability of the sensor measurements [32].

The Kalman filter continuously iterates between the prediction and update steps, dynamically adjusting the state estimation based on new measurements and system dynamics. This process results in a more robust and accurate estimation of the battery's state, even in the presence of measurement noise and modelling errors. One of the main advantages of using Kalman filters for state estimation in BMS is their ability to handle real-time data and provide accurate and reliable state estimates with relatively low computational overhead. However, the effectiveness of the Kalman filter depends on the accuracy of the battery model used in the prediction step and the quality of the sensor measurements in the update step. Various implementations and variations of the Kalman filter, such as the Extended Kalman Filter (EKF) and Unscented Kalman Filter (UKF), are employed in BMS based on the complexity of the battery model and the available sensor data. These filters have been successfully applied to optimize battery performance, extend battery life, and enhance the safety and reliability of lithium-ion batteries in electric vehicles and other energy storage applications.

14.3.4 Particle Filters

Particle filters are widely used for state estimation in various fields, including battery management systems for electric vehicles and other battery-powered devices. BMS plays a crucial role in monitoring and controlling the battery's state of charge, state of health, and other important parameters to ensure safe and efficient operation [33]. A particle filter is a Bayesian filtering technique that allows for estimating the state of a system when there is uncertainty and noise in the measurements. It operates on the principle of representing the state probability distribution using a set of discrete samples or particles. These particles are propagated through the system's dynamics and updated based on the available measurements, eventually providing an estimate of the state. Particle filters are typically applied in BMS state estimation as battery model, particle initialization, prediction step, measurement, etc. In the battery model, first, a dynamic model of the battery is defined. This model describes how the battery's state (SOC, SOH, etc.) changes over time in response to various inputs and loads. The battery model may include factors like current-voltage characteristics, capacity degradation models, temperature effects, etc. In the particle initialization methods, the particle filter starts by initializing a set of particles representing different possible states of the battery. These particles are sampled from the initial state probability distribution. The number of particles determines the resolution of the estimation [34].

The prediction step deals with the particles that are propagated through the battery model using the system dynamics (e.g., current and voltage measurements). The goal is to predict the new state of each particle based on the current state and the system's dynamics. Once new measurements from sensors become available, the particles' weights are updated based on their likelihood compared to the measurements. Particles that match the measurements well will be assigned higher weights, while particles with poor matches will be given lower weights. After the update step, the particles' weights are normalized, and resampling is performed. The resampling step randomly selects particles with replacements based on their weights. Particles with higher weights are more likely to be selected multiple times, while particles with lower weights may be discarded. This process helps concentrate particles in regions of high probability. The final state estimate (SOC, SOH, etc.) is computed using a weighted sum of the particles' states. The particles' weights are used as importance factors in this summation, where particles with higher weights contribute more to the estimate. Particle filters are particularly useful for BMS because they can handle nonlinear and non-Gaussian models, which are common characteristics of battery behaviour. Advanced variations of particle filters, such as the Sequential Monte Carlo (SMC) filter, have been developed to improve estimation accuracy and stability in real-world applications.

14.3.5 Machine Learning Techniques

Machine learning techniques have been successfully applied to state estimation problems in various fields, including robotics, control systems, signal processing, and environmental monitoring. State estimation involves inferring the unobservable state of a system based on available measurements and prior knowledge [32]. Some machine learning techniques are commonly used for state estimation, recurrent neural networks, deep learning techniques, etc. Recurrent neural networks, specifically long- and short-term memory networks, are well-suited for time-series data and sequential state estimation problems. They can process sequential measurements over time and capture temporal dependencies, making them useful for tasks such as trajectory prediction and sensor fusion. Gaussian processes are probabilistic models that can provide estimates of the state along with uncertainty information [35, 36]. They are useful for problems with sparse or irregularly sampled data, where they can interpolate and extrapolate the state estimates with uncertainty bounds. Figure 14.2 illustrates the different machine learning algorithms used as regression, classification, association, and control applications in battery management systems for electric vehicle applications.

Deep learning methods, such as deep neural networks, have been used for state estimation tasks in various ways. For instance, they can be used for feature extraction from sensor data, data-driven modelling of system dynamics, or

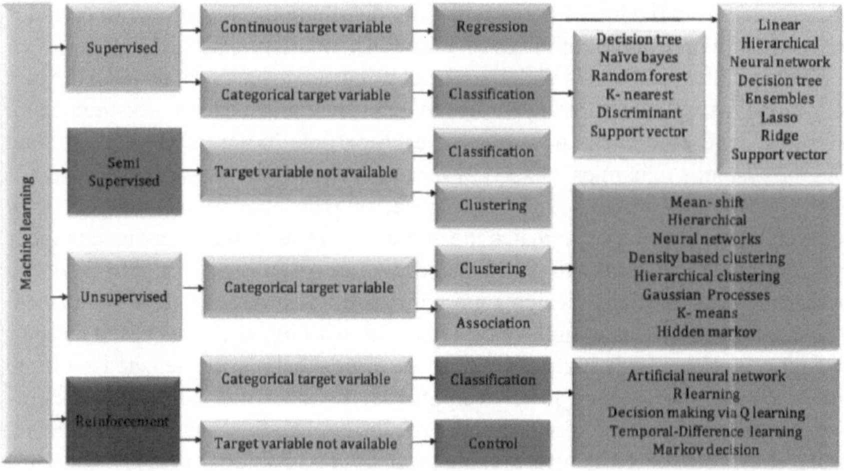

Figure 14.2 Different Machine Learning Algorithm Used as Regression, Classification, Association, and Control Applications in battery management systems.

direct mapping from measurements to state estimates [37, 38]. Deep learning techniques can handle complex relationships and learn representations from raw sensor data. Ensemble methods, like random forests and gradient boosting machines, can be used for state estimation by building an ensemble of multiple models to improve prediction accuracy and robustness [39, 40]. The choice of the machine learning technique depends on the characteristics of the state estimation problem, the system dynamics, the type and quality of sensor measurements, and the available training data. Additionally, hybrid approaches that combine different techniques, like incorporating machine learning models into traditional state estimation methods, can often lead to improved performance in real-world applications [41, 42].

14.4 CONCLUSION AND FUTURE WORKS

State estimation of lithium-ion batteries in electric vehicle (EV) applications is a critical task for ensuring optimal battery performance, safety, and longevity. Accurate state estimation allows EVs to efficiently utilize the available energy, extend battery life, and provide reliable range predictions to the driver. Some common state estimation techniques for lithium-ion batteries in EVs are SOC, SOH, RUL, and SOF estimation. Coulomb counting is one of the simplest methods; it estimates SOC based on the net amount of charge transferred in and out of the battery. The open circuit voltage method estimates SOC by relating the battery's open-circuit voltage to its state of charge. Model-based approaches are more complex methods using battery models that consider factors like internal resistance, capacity fading, and temperature effects. Capacity estimation tracks changes in the battery's capacity over time to estimate its health. Model-based approaches use the degradation models to estimate SOH based on the battery's usage and aging characteristics. Estimates the functional state of the battery, considering factors like thermal stability, power capability, and impedance changes.

The future trends of the state estimation of lithium-ion batteries in electric vehicle include data-driven approaches, hybrid model approaches, online learning, sensor fusion, safety, and fault diagnostics. In the data-driven approaches, as more EVs are deployed, there will be an abundance of data available from real-world driving conditions. Data-driven approaches using machine learning techniques like neural networks or Gaussian processes will become more prevalent for state estimation tasks. These methods can capture complex relationships and improve estimation accuracy. Future trends will likely involve the combination of physics-based models with data-driven approaches to leverage the strengths of both. Hybrid models can provide more robust and accurate estimates by compensating for the limitations of each type of model. Battery behaviour can change over time due to aging and other factors. Future state estimation systems will likely incorporate online learning and adaptation mechanisms to continuously update models and adapt to changing battery characteristics. As sensor technology advances, state estimation will benefit from sensor fusion techniques that combine information from multiple sensors (current, voltage, temperature, etc.) to provide more robust and reliable estimates. Future state estimation systems may include fault detection and diagnosis capabilities to identify abnormal battery behaviour and ensure safety.

REFERENCES

[1] R. Kumar, R. K. Pachauri, P. Badoni, D. Bharadwaj, U. Mittal, and A. Bisht, "Investigation on parallel hybrid electric bicycle along with issuer management system for mountainous region," *J. Clean. Prod.*, vol. 362, p. 132430, 2022, doi: 10.1016/j.jclepro.2022.132430.

[2] S. Jafari, Z. Shahbazi, Y. C. Byun, and S. J. Lee, "Lithium-ion battery estimation in online framework using extreme gradient boosting machine learning approach," *Mathematics*, vol. 10, no. 6, 2022, doi: 10.3390/math10060888.

[3] K. Saqli, H. Bouchareb, M. Oudghiri, and N. K. M'Sirdi, "Critical review of ageing mechanisms and state of health estimation methods for battery performance," *Smart Innov. Syst. Technol.*, vol. 163, pp. 507–518, 2020, doi: 10.1007/978-981-32-9868-2_43.

[4] L. Gaines, "Lithium-ion battery recycling processes: Research towards a sustainable course," *Sustain. Mater. Technol.*, vol. 17, p. e00068, 2018, doi: 10.1016/J.SUSMAT.2018.E00068.

[5] J. Bi, J. C. Lee, and H. Liu, "Performance comparison of long short-term memory and a temporal convolutional network for state of health estimation of a lithium-ion battery using its charging characteristics," *Energies*, vol. 15, no. 7, 2022, doi: 10.3390/en15072448.

[6] A. Kumar, R. Kumar, A. Jain, A. Pandey, S. Thapliyal, and A. Sharma, *Automatic street lighting system*, vol. 624. Springer, Singapore, 2018.

[7] R. Kumar, A. Kumar, M. K. Gupta, J. Yadav, and A. Jain, "Solar tree-based water pumping for assured irrigation in sustainable Indian agriculture environment," *Sustain. Prod. Consum.*, vol. 33, pp. 15–27, 2022, doi: 10.1016/j.spc.2022.06.013.

[8] S. Khaleghi, D. Karimi, S. H. Beheshti, M. S. Hosen, H. Behi, M. Berecibar, and J. Van Mierlo, "Online health diagnosis of lithium-ion batteries based on nonlinear autoregressive neural network," *Appl. Energy*, vol. 282, no. PA, p. 116159, 2021, doi: 10.1016/j.apenergy.2020.116159.

[9] K. Das, R. Kumar, and A. Krishna, "Supervised learning and data intensive methods for the prediction of capacity fade of lithium-ion batteries under diverse operating and environmental conditions," *Water Energy Int.*, vol. 66, no. 1, pp. 53–59, 2023.

[10] R. Kumar, N. J. Ahuja, and M. Saxena, "Improvement and approval of impediment recognition and activity for power window," in *Intelligent communication, control and devices*, Springer, Singapore, 2018, pp. 855–864.

[11] R. Kumar, P. K. Dwivedi, D. Praveen Reddy, and A. S. Das, "Design and implementation of hydraulic motor based elevator system," 2015, doi: 10.1109/IICPE.2014.7115821.

[12] R. Kumar, K. Bansal, A. Kumar, J. Yadav, M. K. Gupta, and V. K. Singh, "Renewable energy adoption: Design, development, and assessment of solar tree for the mountainous region," *Int. J. Energy Res.*, no. May, pp. 1–17, 2021, doi: 10.1002/er.7197.

[13] B. Gou, Y. Xu, and X. Feng, "State-of-health estimation and remaining-useful-life prediction for lithium-ion battery using a hybrid data-driven method," *IEEE Trans. Veh. Technol.*, vol. 69, no. 10, pp. 10854–10867, 2020, doi: 10.1109/TVT.2020.3014932.

[14] Y. Chen, Y. He, Z. Li, L. Chen, and C. Zhang, "Remaining useful life prediction and state of health diagnosis of lithium-ion battery based on second-order central difference particle filter," *IEEE Access*, vol. 8, pp. 37305–37313, 2020, doi: 10.1109/ACCESS.2020.2974401.

[15] R. Kumar, Divyanshu, and A. Kumar, "Nature based self-learning mechanism and simulation of automatic control smart hybrid antilock braking system," *Wirel. Pers. Commun.*, vol. 116, no. 4, pp. 3291–3308, 2021, doi: 10.1007/s11277-020-07853-7.

[16] J. Tian, R. Xiong, and W. Shen, "A review on state of health estimation for lithium ion batteries in photovoltaic systems," *eTransportation*, vol. 2, p. 100028, 2019, doi: 10.1016/j.etran.2019.100028.

[17] Z. Chen, H. Zhao, Y. Zhang, S. Shen, J. Shen, and Y. Liu, "State of health estimation for lithium-ion batteries based on temperature prediction and gated recurrent unit neural network," *J. Power Sources*, vol. 521, 2022, doi: 10.1016/j.jpowsour.2021.230892.

[18] R. Kumar and A. Kumar, "Design and hardware development of power window control mechanism using microcontroller," *2013 Int. Conf. Signal Process. Commun. ICSC 2013*, pp. 361–365, 2013, doi: 10.1109/ICSPCom.2013.6719813.

[19] A. Kumar, K. Bansal, D. Kumar, A. Devrari, R. Kumar, and P. Mani, "FPGA application for wireless monitoring in power plant," *Nucl. Eng. Technol.*, vol. 53, no. 4, pp. 1167–1175, 2020, doi: 10.1016/j.net.2020.09.003.

[20] A. Pal, R. Kumar, and V. R. Sanal Kumar, "Conceptual design of an automatic fluid level controller for aerospace applications," *Proc. IEEE Int. Conf. Soft-Computing Netw. Secur. ICSNS 2015*, 2015, doi: 10.1109/ICSNS.2015.7292433.

[21] A. S. Das, P. K. Dwivedi, A. K. Mondal, R. Kumar, R. M. Reddy, and A. Kumar, "Storage optimization of automated storage and retrieval systems using breadth-first search algorithm," 2017, doi: 10.1007/978-981-10-2999-8_18.

[22] A. G. Siani, M. Mousavi Badjani, H. Rismani, and M. Saeedimoghadam, "State of health estimation of lithium-ion batteries based on the CC-CV charging curve and neural network," *IETE J. Res.,* 2021, doi: 10.1080/03772063.2021.1908853.

[23] A. Barragán-Moreno, E. Schaltz, A. Gismero, and D. I. Stroe, "Capacity state-of-health estimation of electric vehicle batteries using machine learning and impedance measurements," *Electron.*, vol. 11, no. 9, 2022, doi: 10.3390/electronics11091414.

[24] Q. Lin, Z. Xu, and C. M. Lin, "State of health estimation and remaining useful life prediction for lithium-ion batteries using FBELNN and RCMNN," *J. Intell. Fuzzy Syst.*, vol. 40, no. 6, pp. 10919–10933, 2021, doi: 10.3233/JIFS-201952.

[25] R. Kumar, N. J. Ahuja, M. Saxena, and A. Kumar, "Modelling and simulation of object detection in automotive power window," *Indian J. Sci. Technol.*, vol. 9, no. 43, 2016, doi: 10.17485/ijst/2016/v9i43/104393.

[26] Y. Che, D. I. Stroe, X. Hu, and R. Teodorescu, "Semi-supervised self-learning-based lifetime prediction for batteries," *IEEE Trans. Ind. Informatics*, vol. 19, no. 5, pp. 6471–6481, 2022, doi: 10.1109/TII.2022.3206776.

[27] J. Hong, Z. Wang, W. Chen, L. Wang, P. Lin, and C. Qu, "Online accurate state of health estimation for battery systems on real-world electric vehicles with variable driving conditions considered," *J. Clean. Prod.*, vol. 294, p. 125814, 2021, doi: 10.1016/j.jclepro.2021.125814.

[28] S. Chen, H. Zhang, L. Zeng, Y. Fan, L. Chang, and Y. Zhang, "Secondary structural ensemble learning cluster for estimating the state of health of lithium-ion batteries," *ACS Omega*, vol. 7, no. 20, pp. 17406–17415, 2022, doi: 10.1021/acsomega.2c01589.

[29] R. Kumar, N. Gupta, D. Bharadwaj, D. Dutt, and A. Joshi, "Design and development of electronic clutch control unit for manual transmission," *Mater. Today Proc.*, vol. 69, no. 2, pp. 240–245, 2022, doi: 10.1016/j.matpr.2022.08.470.

[30] H. Sharma, R. Kumar, A. Kumar, M. K. Gupta, and V. Kaundal, "Implementation of HMI using VISUAL BASIC. NET for induction hardening process," 2020, doi: 10.1007/978-981-13-8618-3_96.

[31] P. Li, Z. Zhang, Q. Xiong, B. Ding, J. Hou, D. Luo, and S. Li, "State-of-health estimation and remaining useful life prediction for the lithium-ion battery based on a variant long short term memory neural network," *J. Power Sources*, vol. 459, p. 228069, 2020.

[32] A. Kumar, P. Sharma, M. K. Gupta, and R. Kumar, "Machine learning based resource utilization and pre-estimation for network on chip (NoC) communication," *Wirel. Pers. Commun.*, vol. 102, no. 3, pp. 2211–2231, 2018, doi: 10.1007/s11277-018-5376-3.

[33] K. Rawat, D. Kumar, R. Kumar, D. Bharadwaj, V. Soti, G. Rathore, . . . and P. Singh, "Structural optimisation of go kart chassis with basic electronic driver assistance systems". *In 2023 2nd international conference on paradigm shifts in communications embedded systems, machine learning and signal processing (PCEMS)*, IEEE, 2023, April, pp. 1–6.

[34] J. Yadav, S. K. Kurre, A. Kumar, and R. Kumar, "Nonlinear dynamics of controlled release mechanism under boundary friction," *Results Eng.*, vol. 11, p. 100265, 2021, doi: 10.1016/j.rineng.2021.100265.

[35] E. Ezemobi, A. Tonoli, and M. Silvagni, "Battery state of health estimation with improved generalization using parallel layer extreme learning machine," *Energies*, vol. 14, no. 8, pp. 1–15, 2021, doi: 10.3390/en14082243.

[36] M. K. Gupta, R. Kumar, V. Verma, and A. Sharma, "Robust control based stability analysis and trajectory tracking of triple link robot manipulator," *J. Eur. Syst. Autom.*, vol. 54, pp. 641–647, 2021.

[37] A. Jain, A. Kumar, A. P. Shukla, H. Alshazly, H. Elmannai, A. D. Algarni, and J. Yadav, "Smart communication using 2D and 3D mesh network-on-chip," *Intell. Autom. Soft Comput.*, vol. 34, no. 3, 2022.

[38] R. Kumar, M. Kumar Gupta, A. Kumar, P. Sharma, and R. Deorari, "Analysis of electronic clutch control unit for manual transmission vehicle oriented toward safety," *Mater. Today Proc.*, vol. 69, no. 2, pp. 240–245, 2022, doi: 10.1016/j.matpr.2022.08.473.

[39] K. Das, R. Kumar, and A. Krishna, "Analyzing electric vehicle battery health performance using supervised machine learning," *Renew. Sust. Energy Rev.*, vol. 189, p. 113967, 2024.

[40] K. Das, and R. Kumar, "Electric vehicle battery capacity degradation and health estimation using machine-learning techniques: A review," *Clean Energy*, vol. 7, no. 6, pp. 1268–1281, 2023.

[41] K. S. Rawat, and R. Kumar, "Insight into the novel low cost green air pollution monitoring and control systems: A technological solution from concept to market," *J. Clean. Prod.*, vol. 427, p. 139218, 2023. https://doi.org/10.1016/j.jclepro.2023.139218

[42] K. Das, and R. Kumar, "Assessment of electric two-wheeler ecosystem using novel pareto optimality and TOPSIS methods for an ideal design solution," *World Electr. Veh. J.*, vol. 14, no. 8, p. 215, 2023.

Chapter 15

Social, Economic, and Environmental Issues Associated with Sustainable Transportation

Francis Luther King M., E. Elayarani

15.1 INTRODUCTION

The concepts of sustainability, sustainable development, and sustainable transportation are attracting an increasing amount of attention. There have been a significant number of articles, studies, and publications devoted to the topic of sustainability, and a significant number of communities are now engaged in sustainable planning initiatives. The extent and scale of these difficulties, as well as the consequences they have for transportation planning and policy, are only just starting to be investigated. The interest in these matters might be attributed to a number of different sources. The developing knowledge that human activities have considerable environmental repercussions that may impose economic, social, and ecological costs is the origin of the concern about sustainability. Air pollution on a global scale, the long-lasting consequences of man-made toxins, the degradation of natural resources such as fresh water and fisheries, and the transboundary character of a large number of environmental issues are all factors that underline the need to take a holistic approach to analyzing the effects of human activity. The concept of sustainability places a focus on the interconnected aspects of human activities and, as a result, highlights the need to coordinate planning efforts across a variety of industries, authorities, and community organizations. Instead of waiting for problems to escalate into crises, sustainability planning acts as development's equivalent of preventative medicine by identifying potential issues ahead of time and devising solutions to such issues. A primary aim of sustainable development is to strike a healthy equilibrium between competing economic, social, and ecological considerations. It is possible that skeptics may reason that sustainable planning is nothing more than a rebranding of comprehensive planning. This may be true, but many different governments have not done a very good job planning in this area. The idea of sustainability offers a structure and a set of instruments for long-term, all-encompassing planning. This kind of planning takes into account the intricate connections that exist beyond traditional geographical and temporal boundaries.

DOI: 10.1201/9781003464556-15

15.2 CONCEPT OF SUSTAINABILITY

It is widely accepted that sustainability promotes circumstances that are beneficial to the ecosystem, the economy, and society as a whole without jeopardizing the quality of life of people who will live in the future. Despite this, history reveals that the legacies left behind by previous civilizations have a significant impact on the circumstances that exist in subsequent societies. It is important that the worth (utility) of any and all types of assets (including money, real land, infrastructures, natural resources, and knowledge) handed down to subsequent generations be at least equivalent. The introduction of environmental capital, in particular ecosystems, into this frame of reference is what might be considered novel. The incorporation of the idea of sustainability into this temporal framework necessitates the use of three primary pillars: society, the economy, and the environment.

15.2.1 Society

Achieving social justice. In this context, "conditions favouring a distribution of resources among the current generation based upon comparative levels of productivity and the promotion of equality of opportunities" (https://transportgeography.org/contents/chapter4/transportation-sustainability-decarbonization/#:~:text=Social%20equity.,promotion%20of%20equality%20of%20opportunities) refers to the circumstances that encourage such a distribution of resources. This suggests that people, institutions, or companies are free to follow the path of their choosing and get benefits commensurate with the risks and efforts they take. In most cases, the most difficult aspect of the idea of sustainability is attempting to provide a definition for social equality. It is not to be confused with equality of result, sometimes known as socialism, which is an ideology in which discriminatory practices are applied in favour of one socioeconomic group and against another with the professed purpose of rectifying perceived inequities.

Transportation systems that are environmentally friendly should be safe, should not be harmful to people's health, and should cause as little disruption as possible in their communities. Additionally, access and equality are two crucial concepts, since the goal of transportation should be to facilitate access to products and services for the greatest number of people feasible.

15.2.2 Economy

The study of economic efficiency focuses on the conditions that provide the opportunity for higher levels of economic efficiency in terms of the use of resources and labor. It lays a strong focus on a company's skills, competitiveness, production flexibility, and the delivery of goods and services that are in line with the demands of the market. Under these circumstances, there must be unrestricted distribution of the components necessary for production, and the marketplaces must be accessible to business transactions.

Transportation has a role in economic expansion and development, as well as job opportunities. It needs energy for operations and the materials for the modes and the infrastructure, but those resources may be utilized more effectively. Users of the transportation system should be responsible for all expenses, both direct and indirect, associated with their use of the system, and this should be reflected in the price structure used. It is more probable that modal choice and efficiency will be promoted within a transportation system that allows for open and fair competition. It is possible for pricing distortions and inappropriate capital allocation to occur in a system in which transportation is provided by a public or private monopoly. These issues are likely to make the system unsustainable over time.

15.2.3 Environment

A sense of duty towards the environment entails building a footprint for human activity that is smaller than the capacity of the environment to accept it. This is done to reduce the impact that humans have on the ecosystem. This encompasses the provision of resources (such as food, water, and energy) as well as the management of waste in an appropriate manner. The reduction of waste and the maximization of product and resource reuse are two of its fundamental concepts.

One potential option for achieving sustainability is to lessen the negative effects that transportation has on the surrounding environment. Pollutant emissions, unwanted noise, and reducing the harmful impacts that transportation has on the environment around it is one of the possible solutions that might be used in the quest to make sustainability a reality. Emissions of harmful pollutants, disturbances of peace caused by unwelcome noise, and shifts in weather patterns are all side effects of human mobility. It is estimated that the transportation sector is responsible for around 15% of the total emissions of greenhouse gases and 22% of the emissions of carbon dioxide. Despite this, the overall number of automobiles in the world's fleet is increasing while, at the same time, becoming less damaging to the environment. Another important strategic goal to strive towards is reducing the negative impact that the transportation sector currently has on the surrounding environment, especially as it relates to the consequences of the planning, construction, and upkeep of physical infrastructure. Reduce, repurpose, and recycle as much of the trash that is produced by transportation systems as possible. This waste includes items like vehicles, components, and packaging, among other things.

15.3 SUSTAINABLE ISSUES

One's understanding of sustainability is susceptible to being constrained if, for example, one chooses to focus attention only on the challenges posed by diminishing resource levels and increasing air pollution. This is done on the basis that these concerns pose the biggest threat to the environment over the long term and are likely to be ignored by conventional planning (Committee for a Study on Transportation and a Sustainable Environment, 1997). As a result, this action is

Figure 15.1 Sustainability Issues.

taken. However, sustainability is being considered in a wider context, which now includes the problems shown in Figure 15.1. In spite of the fact that Figure 15.1 gives the impression that every problem can be placed neatly into one of many categories when, in reality, these problems often intersect. For instance, pollution is a problem for the environment as well as for human health (a concern for the social sphere) and for the fishing and tourist sectors (concerns for the economic sphere). When it comes to sustainable planning, the realization that effects and goals often interact with one another is reflected in the fact that solutions must represent integrated analyses.

Sustainability that is defined too narrowly has a tendency to ignore many of the linkages between problems and prospects for coordinated responses. For instance, some techniques for reducing emissions of greenhouse gases may make other economic, social, and environmental issues worse, while other solutions bring many advantages [1]. A limited study does not take into account these extra repercussions, but a complete analysis can take them into account. An in-depth examination may assist in uncovering "no regrets" solutions, which contribute to the accomplishment of numerous goals and are, as a result, justifiable regardless of the value attributed to expenses such as climate change.

The tendency that occurs in decision-making to focus on goals and impacts that are straightforward to measure while dismissing those that are more difficult to quantify may be understood as a reaction to this predisposition. Concern about sustainability may be seen as a response to this propensity. Making choices that are sustainable may thus be characterized as planning that takes into account goals and results, notwithstanding how difficult it may be to measure the implications of such actions. In the beginning, people were worried about the long-term repercussions of the current pace of resource consumption, which paralleled the principles of "intergenerational justice" (also known as being fair to future generations) and ecological integrity. These ideas were mirrored in people's concerns about the environment. Nevertheless, if there is a worry about the quality of the environment in the future, it makes very little sense to dismiss the equitable and environmental consequences that are taking place during this generation in remote locations. These repercussions are having an effect on the environment. As a result, sustainability, in the end, reflects the goals of equality, ecological integrity, and human health, regardless of when or where they take place [2]. The concept of sustainability is often articulated in terms of natural capital since this is the primary focus of the academic discipline of ecological economics, which studies the monetary worth of ecological resources. The term "natural capital" refers to the value that may be derived from natural ecosystems in the form of products and services. These include climatic stability, clean air and water, and other environmental benefits. Keeping these services accessible is akin to a corporation maintaining the value of the productive assets it has. Ecological economists argue that human consumption should not deplete natural capital at a rate that is faster than the rate at which it can be replaced by human capital that is both viable and long-lasting. This is a key tenet of the sustainable development paradigm. This philosophy contends, for example, that non-renewable resources, such as petroleum, should not be depleted without appropriate development of substitutes, such as other kinds of renewable energy. Traditional economics has a propensity to ignore the non-market costs of economic activities, which may even be regarded as good economic events by indices such as gross domestic product [3]. This is because traditional economics tends to ignore the costs of economic activity that are not related to markets. Ecological economics attempts to take these costs into account in order to present a more realistic picture of the actual cost of economic activity. This is one of the primary goals of ecological economics.

This involves determining the economic value of non-market goods and services, such as the advantages that a wetland delivers in boosting water quality and supporting fishing operations. An example of this would be the benefits that a wetland gives. One illustration of this may be seen in the benefits that wetlands provide to the fishing industry. According to Daly (1996) [4], the field of sustainable economics makes a clear difference between growth (increasing quantity) and development (improved quality). It criticizes standard economic metrics such as gross domestic product, which measure the amount of market activity but not the quality of those activities, and it places emphasis on social welfare outcomes rather than merely evaluating material riches. The goal of sustainable economics, in contrast to the objective of neoclassical economics, is not to achieve ever-increasing consumption but rather sufficiency.

A conservation ethic is often reflected in sustainability, which implies that patterns of production and consumption are designed to avoid resource use and waste as much as possible. Changing the existing economic policies that favour wasteful production and consumption is necessary in order to accomplish this goal. To stimulate industry, for instance, the cost of electricity is kept as low as possible in a lot of nations so that people can afford their utilities and their cars. This demonstrates a consumerist point of view. The price of energy could go up (possibly as a result of a tax on carbon), but at the same time, programmes could be implemented to weatherize buildings, increase the fuel efficiency of vehicles, improve alternative modes of transportation, and increase industrial efficiency. This would allow manufacturers and consumers to meet their needs with less resource consumption.

15.4 TRANSPORTATION IMPACTS

The infrastructure for transportation and the activities that take place inside it have substantial consequences on the environment, including those that are outlined in Table 15.1.

Up until fairly recently, the majority of economists believed that more mobility brings net economic advantages despite the possibility that it may have detrimental impacts on society and the environment. This was the case despite the fact that increased mobility may have negative effects on the environment. However, recent research suggests that an increase in the number of miles travelled by motor vehicles above a certain ideal level may have a detrimental effect on the economy as a whole. This is due to the fact that the marginal productivity of increasing travel is dropping, and the use of vehicles imposes external expenses that might negate direct economic advantages [5, 6]. In addition, the amount of time spent travelling has been shown to have a negative impact on health. This demonstrates that planning for sustainability does not necessarily require making concessions between economic, social, and environmental goals. Rather, it is a question of developing strategies that help accomplish all of these goals over the long run

Table 15.1 Transportation Impacts on Sustainability

Economical	*Social*	*Environmental*
Traffic congestion	Inequity of impacts	Air and water pollution
Mobility barriers	Mobility disadvantaged	Habitat loss
Accident damages	Human health impacts	Hydrologic impacts
Facility costs	Community interaction	Depletion of non-renewable resources
Consumer costs	Community liveability	
Depletion of non-renewable resources	Aesthetics	

by increasing the efficiency of the transportation system. Conventional planning has a tendency to make the assumption that advancements in transportation occur in a linear fashion, with more recent, quicker means eventually displacing older, slower modes, as shown in the following diagram. This series model operates on the presumption that the more established modes are of little relevance; hence, it is not a problem, for instance, if an increase in automotive traffic causes a delay in the operation of public transportation due to congestion or if it erects a barrier to prevent pedestrian movement. From this point of view, it would be a step in the wrong direction to give emphasis to modes of transportation other than the vehicle, such as public transit or walking.

Walk → Bicycle → Train → Bus → Automobile → Improved automobiles

Sustainability represents a parallel paradigm, which presupposes that each mode may serve a purpose and works towards creating balanced transportation networks that put each mode to use in the context of what it does most effectively.

As seen in the following diagram, advancement in transport necessitates the enhancement of all existing forms of transportation, not simply the most recent method. For instance, increasing the number of people who walk and ride bicycles, bolstering support for public transportation, and limiting the use of automobiles in crowded metropolitan areas are all potential initiatives that may prove to be quite helpful. This does not imply that enhanced transportation necessarily results in quicker travel or more distance; rather, improvements may boost comfort and safety, give cost savings, or even lessen the overall need for travel.

Walk → Improved walking conditions
Bicycle → Improved cycling conditions
Train/Bus → Improved public transit service
Automobile → Improved automobile travel conditions

15.5 ECONOMIC, ENVIRONMENTAL, AND SOCIAL IMPACTS

The population has risen over time, cities have expanded, and globalization and free trade have boosted the flow of people and products across regional and international boundaries; as a result, our nation's transportation infrastructure and systems have developed significantly. At the local, regional, and global levels, the use of automobiles, trucks, buses, subways, trains, aeroplanes, ships, and ferries to transport ourselves and our products has substantial ramifications for the amount of energy and material resources used, the amount of environmental pollution caused, the amount of noise produced, and the amount of land that is occupied. The huge amount of energy that was used by transportation ultimately led to substantial detrimental effects on the surrounding ecosystem. To begin, one of the primary contributors to climate change is transportation. Climate change is driven by so-called greenhouse gases (GHGs), which are gases that trap heat reflected from the surface of the globe in the lower atmosphere, resulting in the greenhouse effect. Carbon dioxide is the major greenhouse gas, and it is responsible for around two-thirds of the climate change that is caused by humans. Injuries and deaths that may be traced back to a variety of transportation-related mishaps are among the most troubling aspects of this industry's adverse impact on society. As a result, it was essential to develop a thorough framework of socioeconomic analysis to tackle these problems as quickly as humanly feasible.

15.5.1 Environment

Although road freight transit is necessary for the growth of the economy, it is detrimental to both the natural world and human health due to the negative impact it has on both. Before recent times, the primary goal of the planning that goes into goods transportation operations was to reduce expenses as much as possible see, for example, [7, 8]. As a result of an expanding global population because of growing concern for the environment, suppliers of logistics services and goods carriers have begun to pay greater attention to the negative externalities that are caused by their activities.

Trucks, responsible for a considerable share of freight transportation at the local and regional levels, are responsible for the emission of a huge quantity of pollutants. Even though advances have been made in both transportation technology and fuels over the years, the majority of trucks are still powered by diesel engines, which are important contributors to the emission of nitrogen oxides (N_2O), particulate matter (PM), and carbon dioxide (CO_2). There is a correlation between prolonged exposure to smog that is based on N_2O and particulate matter, which has been related to a broad variety of health concerns. Greenhouse gases, sometimes known as GHGs, are major contributors to the warming trend seen on a worldwide scale. The combustion of fossil fuels results in the majority of the greenhouse gas emissions that are produced by the transportation industry. These factors create alterations in the atmosphere and disturbances in the climate, both detrimental to the natural and built surroundings and pose dangers to human health. In the traditional sense, greenhouse gases were not considered to be pollution. The carbon dioxide equivalent, often known as CO_2e and abbreviated as CO_2, is a unit of measurement used to assess the amount of global warming that a certain kind and quantity of greenhouse gas (GHG) may create by comparing it to the amount or concentration of CO_2 that is functionally equivalent. Determining which greenhouse gases need to be included in the carbon footprint is a fundamental obstacle that must be overcome. According to Wright, Kemp, and Williams (2011) [9], there is a chance that a considerable amount of emissions may be determined by the measurement of CO_2 and CH_4, which are the two most prominent forms of human greenhouse gases. This finding suggests that there is a correlation between these two kinds of human-produced greenhouse gases and the quantity of emissions that are produced. According to Demir (2011) [10], the amount of CO_2 emissions that are created by a vehicle are exactly proportional to the amount of gasoline that the vehicle uses. This is the case even if the vehicle is fuel efficient. This, in turn, is reliant on a wide variety of parameters that are tied to the vehicle, the surroundings, and aspects that are related to traffic, such as the vehicle's speed, load, and acceleration. On the other hand, CH4 emissions are reliant on a broad range of nuanced aspects of the dynamics of combustion as well as the sort of emission control systems that are put into place. This is the case regardless of whether or not fossil fuels are being burned.

15.5.2 Societal Challenges

Society is progressing towards the nomadic rather than the sedentary; however, there still seems to be a disparity between the affluent and the less fortunate. However, a reliable evaluation of the transportation sector's real environmental consequences still faces numerous unknowns. The same is true for evaluating sustainable transport policy in a variety of sizes, from the neighbourhood to the world. For almost ten years, people have known that the current state of transportation is unsustainable. The supply and use of transport will be affected by a number of emerging social and technical changes in the 21st century. Currently, the demands put on transportation are shifting due to a number of societal factors. These include growing salaries, more leisure time, advances in communication technology, an ageing population, and the decline of the conventional family. Changes in economic activity are occurring because of developments in industrial methods, transportation, and the expansion of consumer markets. As traffic deaths. injuries, and congestion continue to climb, many are beginning to wonder if our present transport systems can survive the long haul. Whether or not our present transport networks can survive beyond the next half century is called into doubt by these tendencies [11].

There is a lack of information regarding the transportation networks. Interactions between dynamic behaviour and capacity restrictions might cause unanticipated turbulence in a transport system. Policymaking may sometimes lead to unexpected behaviour. At the same time as objectives, metrics, and societal acceptability need to be fine-tuned, there is a lack of insights into how players in transport react to policy initiatives, especially cost measures.

Because of this, transportation studies must take into account other factors influencing mobility and communication, such as those in the economy, government, and technology. The natural environment is also directly mentioned; it is heavily impacted by land use, transportation, and transportation infrastructure, as well as new technological solutions, but only little impacted in the other direction. It is the same for social integration (or lack thereof) in different regions.

Some of the most influential factors in the increasing mobility of families and people are the trend towards individualization and the demand for suburban living. There seems to be a continuous shift.

15.6 SUSTAINABLE TRANSPORTATION INDICATORS

When evaluating sustainability [12, 13], measurable indicators are used so that trends may be tracked, areas and activities may be compared, decisions regarding policies and planning may be evaluated, and performance targets may be set. It is essential that aims and methods do not get confused when choosing the signs to use.

A society's aims might be thought of as its highest goals as a whole. While objectives are the steps that will get you closer to achieving a goal, goals are the destination point. Sometimes, those in charge of making decisions prioritize less complicated outcomes and objectives above those that are more difficult to evaluate. Walking and cycling improvements, telework, and more accessible land use may enhance access without increasing traffic or mobility. On the other hand, initiatives to increase traffic and mobility may reduce access by making conditions for pedestrians worse or by establishing patterns of dispersed land use.

15.6.1 Traditional Measures of Transportation

Quality focuses primarily on the performance of motor vehicles in traffic [14].

Here are a few instances:

- The least congested roads get the highest degree of service.
- Typical travel times: greater is considered to be better.
- There should be a balance between parking convenience and cost, with more convenience and lower cost being preferred.
- Accident frequency per vehicle mile: lower is preferable.

These metrics tend to be at odds with the pursuit of sustainable transportation since they prioritize automobile usage. They may defend measures like adding lanes and parking lots, which lead to more cars on the road and fewer people walking, biking, or using public transportation. This worsens the transportation situation for those who do not have access to a car by increasing demand for resources, pollution, and land use.

They disregard increasing vehicle mileage as a risk factor and reduced vehicle travel as a potential solution to transportation issues by considering consequences per vehicle mile rather than per person. According to this view, it is OK if the number of car accidents per person rises as long as the number of miles driven rises at the same rate.

Since more miles driven in safe circumstances results in lower collision rates per mile driven, increasing vehicle travel may be seen as a safety approach.

15.6.2 Basic Indicators of Environmental Stability

Some assessments of sustainable transport use a limited number of indicators based on readily accessible data to simplify the analytical process. Here are a few instances:

Transport sustainability concerns:

- It is preferable to use less fossil fuel and produce fewer carbon dioxide emissions from transportation.
- The lower a car's pollutant output, the better.
- Lower annual miles driven per person are preferable.
- If more people use public transport, that is a good thing.
- The fewer people hurt or killed in car accidents, the better.
- Reduce your land use for transportation, please.
- Aesthetic condition of roadways: when individuals value their surroundings, they are more likely to take good care of them.

However, indicators that are too simplistic may not be able to provide useful planning assistance. They often prioritize solutions that address only one or two goals (like alternative fuel vehicles) while devaluing those that provide modest but multiple benefits (like mobility management strategies and more accessible land use), which can have significant effects (like improving community liveability and equity).

15.6.3 Global Indicators of Sustainable Transportation

Sustainable transport metrics that are all-encompassing consider many different effects. Indicators representing the whole scope of sustainability aims should be included. Quality and number of travel alternatives, mode split, quality and quantity of non-motorized transport, etc., are all markers of transport variety. It might be best to utilize only one such signal, or if more are employed, to give each one a lesser weight. Some indicators can be broken down further by factors such as income, employment, gender, age, physical ability, minority status, location (city, suburb, rural), day of the week, time of day, travel mode (walking, cycling, transit), and purpose of trip (work, commute, vacation, shopping, etc.). The quality of transportation for people who have special needs, such as those who are unemployed, physically disabled, low-income single parents, socially disadvantaged groups, rural non-drivers, etc., may receive special consideration, as may basic access trips, which society considers to be of high value (travel to medical services, education, jobs, freight, and commercial travel).

It may be challenging to collect or assess the necessary data for certain metrics. Some rely on qualitative information conveyed through a subjective rating scheme as opposed to quantitative information conveyed by objective measures. Indicators should be chosen with data collecting costs and usability in mind. While data accessibility is a crucial consideration when choosing indicators, it should not be the primary one.

There are essential effects that are not captured by these metrics, but they must not be overlooked. One such example is the widespread belief that electric automobiles are more sustainable. However, the economic and environmental implications of electric vehicles depend on the methods used to generate power. Nuclear power production may improve sustainability in certain ways (less air pollution), but it also has unintended consequences (radiation danger, thermal pollution, terrorist concerns).

15.7 SUSTAINABLE TRANSPORTATION SOLUTIONS

Planning, management, and technological shifts are only some of the options for making transportation systems greener. While options may seem to be either/or, the vast majority of studies show that many approaches are required to accomplish sustainability targets. The value put on environmental and social repercussions like global warming and inequalities is ambiguous, but 'no regrets' solutions help accomplish a mix of economic, environmental, and social goals and may be adopted regardless. However, there are sometimes heated discussions over which methods are best, which should get the most funding, and when each should be put into action.

Even when sustainability policies net good effects, they can cause friction between various groups and areas. Companies and areas that generate energy, for instance, stand to lose money as a result of increased energy efficiency measures. Transition costs and disputes exist even if they are economic transfers rather than genuine expenses.

Therefore, in order to effectively prepare for sustainability, it may be important to take steps to facilitate the transition to a more resource-efficient economy and make attempts to compensate those who stand to lose. This has significant repercussions for several sectors of the transportation business, including the oil and automotive sectors, both of which stand to lose if certain sustainability laws are put into place [15–16].

15.8 CONCLUSION

When attempting to plan for sustainable transportation, there are a number of concerns that come up, some of which include how to define and quantify sustainability, what criteria should be used to evaluate goals and objectives, and what sort of decision-making process should be applied.

To achieve sustainability, we need planning that is both more in-depth and more holistic. This planning should take into account a broad variety of economic, social, and environmental factors, including the intangibles. Having a wide range of people's opinions and preferences heard and considered is essential in sustainability planning.

In general, sustainability advocates for land use patterns that are less car-dependent and more compact, as well as for market reforms that lead to more varied and cost-effective modes of transportation. As a result of these changes, the economy operates more efficiently, fewer resources are used, the environment is impacted less negatively, and the mobility of those who do not drive is enhanced.

REFERENCES

[1]. Litman, T. (1999a) 'Reinventing transportation; exploring the paradigm shift needed to reconcile sustainability and transportation objectives', Transportation Research Record 1670, Transportation Research Board, www.trb.org, pp. 8–12.

[2]. Jansson, A., Hammer, M., Folke, C., and Costanza, R. (Eds.) (1994) 'Investing in natural capital: the ecological economics approach to sustainability', International Society for Ecological Economics, Island Press, Washington, DC.

[3]. Daly, H., & Cobb, J. (1989) For the Common Good, Beacon, Boston, MA.

[4]. Daly, H. (1996) Beyond Growth; Economics of Sustainable Development, Beacon Press, Boston, MA.

[5]. Boarnet, M. (1997) 'New highways & economic productivity: interpreting recent evidence', Journal of Planning Literature, 11(4), 476–486.

[6]. Helling, A. (1997) 'Transportation and economic development; a review', Public Works Management & Policy, 2(1), 79–93.

[7]. Crainic, T. G. (2000) 'Service network design in freight transportation', European Journal of Operational Research, 122(2), 272–288.

[8]. Forkenbrock, D. J. (2001) 'Comparison of external costs of rail and truck freight transportation', Transportation Research Part A: Policy and Practice, 35(4), 321–337.

[9]. Williams, I., Kemp, S., Coello, J., Turner, D. A., & Wright, L. A. (2012) 'A beginner's guide to carbon footprinting', Carbon, 3(1), 55–67.

[10]. Demir, E., Bektas, T., & Laporte, G. (2011) 'A comparative analysis of several vehicle emission models for road freight transportation', Transportation Research Part D: Transport and Environment, 6(5), 347–357.

[11]. Nijkamp, P., & Black, W. R. (2002) 'Introduction: pathways to sustainable transport and basic themes', in W. Black & P. Nijkamp (Eds.), Social Change and Sustainable Transport, Indiana University Press, Bloomington, IN.

[12]. Litman, T. (2003a) Sustainable Transportation Indicators, VTPI, www.vtpi.org.

[13]. Centre for Sustainable Transportation (CST). (2001) Sustainable Transportation Performance Indicators Project, www.cstctd.org.

[14]. Litman, T. (2003b) 'Measuring transportation: traffic, mobility and accessibility', ITE Journal, 73(10, 28–32, www.ite.org.

[15]. Litman, T., & Burwell, D. (2006) 'Issues in sustainable transportation', International Journal of Global Environmental Issues, 6(4), 331–347.

[16]. Forkenbrock, D. J. (1999) 'External costs of intercity truck freight transportation', Transportation Research Part A: Policy and Practice, 33(7–8), 505–526.

Chapter 16

Separation of Ternary Azeotropic Mixture of Ethanol, Isopropanol, and Methyl Ethyl Ketone – Challenges for Future Fuels and Solvents

Saurab Suman, Raj Kumar, Chandan Swaroop Meena, Nawal Kishor Banjara, and Shashikant Yadav

16.1 INTRODUCTION

Distillation is the most widely used separation technique in industries. In industries, about 95% of energy is consumed by distillation, but still, this method is widely used because of its various advantages [1]. Isopropanol (IPA), methyl ethyl ketone (MEK), and ethanol (EtOH) are widely used as intermediate solvents in many chemical processes [2, 3]. A ternary azeotropic mixture of these three solvents is generated in the Fischer-Tropsch synthesis process [4, 5]. This process indirectly converts natural gas into transportation fuels [6].

An azeotropic mixture can have either a higher or lower boiling point than the individual components. Conventional distillation methods cannot change their azeotropic composition [7, 8]. They require a very large number of stages or a very high reflux ratio to achieve separation because their relative volatility is very close to unity [9]. Therefore, it is uneconomical to use conventional distillation for their separation because of the very close boiling points [5].

Advanced distillation methods, such as homogeneous azeotropic distillation (including extractive distillation), heterogeneous azeotropic distillation, distillation with salt effects, and PSD, are used for their separation [10–14]. These methods promote separation by shifting and eventually removing this azeotropic point using different approaches. Extractive distillation is one such method in which an entrainer is added to enhance the relative volatility of the mixture and shift the azeotropic composition [15–17]. It uses a high boiling solvent, which is fed near the top of the column, and it dissolves one of the components preferentially from the feed [18, 19].

PSD is a viable alternative in which columns operate at different pressures to separate azeotropic mixtures [20, 21]. High-purity streams are obtained from one end of the column, and streams from the other end are recycled, which have compositions near the azeotropic composition. PSD uses a change in pressure to shift the azeotropic point to facilitate separation [22–24]. The change in pressure can

DOI: 10.1201/9781003464556-16

significantly alter the azeotropic composition of the mixture and alter the relative volatility of the mixture [25, 26]. When the pressure decreases, the azeotropic point shifts towards a higher value of the lighter components and increases the relative volatility [5, 27]. The capital and operating costs in PSD depend on operating pressure. For the optimum design of PSD for ternary azeotropic mixtures, the feasibility and existence of distillation boundaries are studied by means of residual curves [28].

If the chemical components of the mixture attract each other and the activity coefficients are less than one, then maximum boiling azeotropes are formed. However, if the repulsive forces between the components are large and the activity coefficients are greater than one, then minimum boiling azeotropes are formed [29, 30]. The IPA-MEK-EtOH mixture also forms a minimum boiling azeotrope.

The PSD is useful for separating both the minimum and maximum boiling azeotropes. In the case of minimum boiling azeotropes, pure products are obtained from the bottom of the column, and distillates are recycled. However, in the case of maximum boiling azeotropes, pure products are obtained from the top of the column, and bottoms are recycled to the columns [19, 31]. It is obvious that recycling bottom products requires more energy than recycling distillate because bottoms usually have higher boiling points. Therefore, maximum boiling azeotropes should require higher energy and cost due to recycling of the bottom product. However, previous studies have found that there is little effect on energy consumption in both systems [30].

Figure 16.1 shows a detailed classification of PSD. On the basis of operating modes, the PSD can be divided into three types: continuous PSD, batch PSD, and semi-continuous PSD [32]. Continuous PSD is the most widely used. It consists of two columns operating at different pressures. The feed column depends on the feed concentrations. If the concentration is above the azeotropic point, the feed is fed into a high-pressure column, and if it is under the azeotropic point, it is fed into a low-pressure column. The batch process is preferred in small-scale

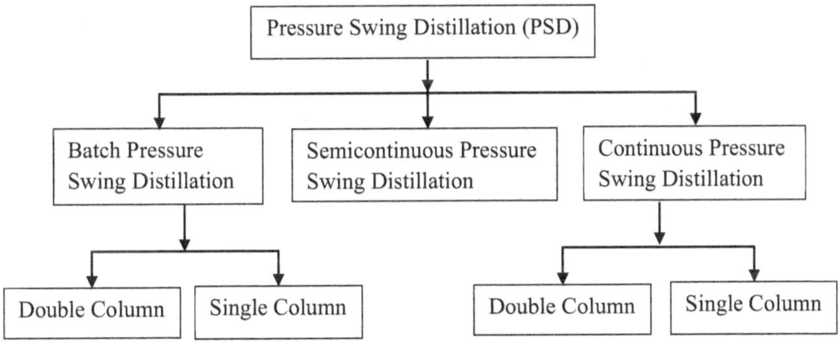

Figure 16.1 Classification of PSD Process.

production, pharmaceutical, and fine chemicals industries [33, 34]. Despite being a batch process, it has a higher energy demand than a continuous process [35, 36]. But the overall structure is simple, and the total cost is low. Semi-continuous PSD requires only a single column and, therefore, has a low capital cost. In addition, it has shorter downtimes when the mixture to be separated is changed [26].

Another classification can be based on the number of components in the mixture to be separated: binary and multi-component PSD [32]. A binary azeotropic mixture can be separated using continuous PSD to save energy consumption [37, 38]. However, multi-component PSD has not been well-studied by researchers. After the increase in the number of columns, the reflux requirements also increased significantly [9, 39].

PSD has certain advantages over other methods available for the separation of azeotropic mixtures. The major advantage is that it did not use an additional component called an entrainer for the separation of azeotropic mixtures [26, 40]. The use of entrainer increases the operating cost of the process due to the cost of the entrainer. In addition, the entrainer acts as an impurity for the mixture and, therefore, needs to be separated at the end of distillation [41]. Therefore, the PSD avoids further complicating the separation process. PSD operation requires low equipment cost because it requires a smaller number of distillation columns than is required for conventional azeotropic distillation. Furthermore, the energy saving in PSD is also high.

PSD has a major drawback in that it can only be used with pressure-sensitive azeotropes [32]. This operation requires complex process control and automation. One of the major costs is maintaining column pressures below atmospheric pressure. In the case of the minimum boiling azeotropes, the products are obtained from the bottom of the column, and the top distillates are recycled. Therefore, other contaminants are more likely to enter the bottom of the column, usually those with high boiling points.

16.2 PROCESS SIMULATION

The MEK-IPA-EtOH mixture forms two binary azeotropes of MEK-EtOH and MEK-IPA, which are highly pressure-sensitive in nature. Table 16.1 shows the chemical formula and boiling points of these three components. They have very close boiling points and cannot be separated by conventional distillation. Therefore, there is a need to achieve the separation by an alternative economical method.

To study the pressure dependence of these azeotropic mixtures, different plots are generated in DWSIM version 6.6.2. The interactions of all solvents are studied using the non-random two-liquid (NRTL) model. It relates the activity coefficients of a compound with mole fractions in the concerned liquid phase. Figure 16.2 is the P-x-y plot of ethanol/MEK binary azeotropic mixture at room temperature. From this plot, it is clear that at about 0.13 bar pressure and 0.2 mole fraction of ethanol, this binary mixture is forming an azeotropic mixture.

Table 16.1 Components to Be Separated and Their Boiling Points

Component	*Chemical formula*	*Boiling point, °C*
Iso-propanol	$CH_3CH(OH)CH_3$	82.5
Methyl ethyl ketone	$CH_3COCH_2CH_3$	79.64
Ethanol	CH_3CH_2OH	78.37

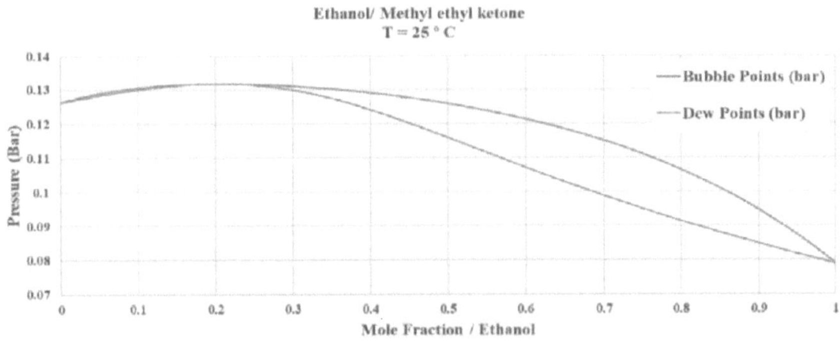

Figure 16.2 P-X-Y Diagram of the Ethanol/MEK Mixture at 25 °C.

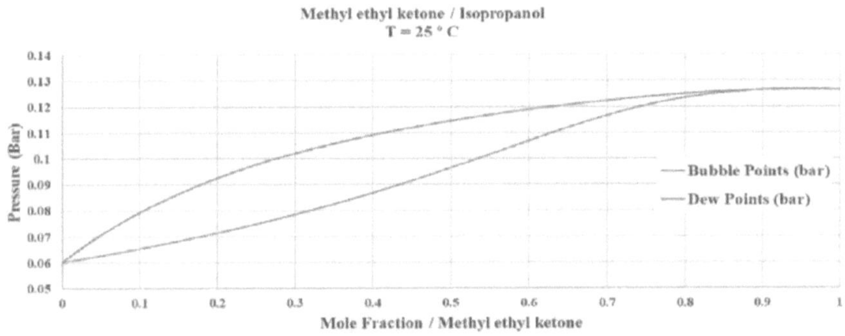

Figure 16.3 P-X-Y Diagram of the MEK/Isopropanol Mixture at 25 °C.

Similarly, the MEK/isopropanol binary mixture also forms an azeotropic mixture at normal conditions. Figure 16.3 is the P-x-y plot of the MEK/isopropanol binary azeotropic mixture at room temperature. From this plot, it is clear that at a pressure of about 0.125 bar, this binary mixture is forming an azeotropic mixture.

Figure 16.4 and Figure 16.5 show the effect of pressure on the MEK/IPA azeotropic mixture. Comparing both plots, it is clear that the azeotropic composition, which is about 0.6 mole fraction of MEK at 1.01325 bar, is almost eliminated at

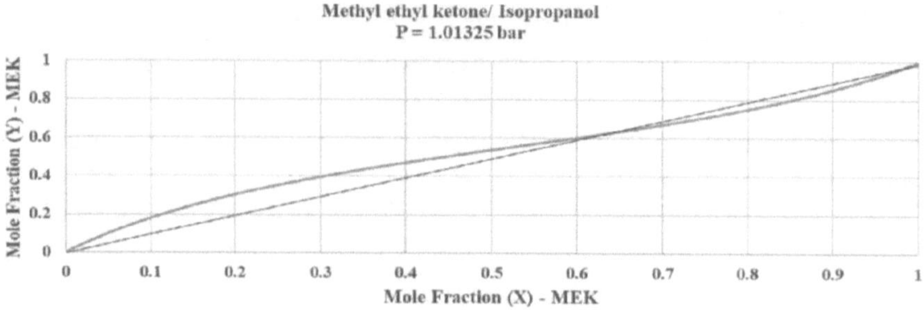

Figure 16.4 X-Y Plot of the MEK/Isopropanol Mixture at 1.01325 Bar.

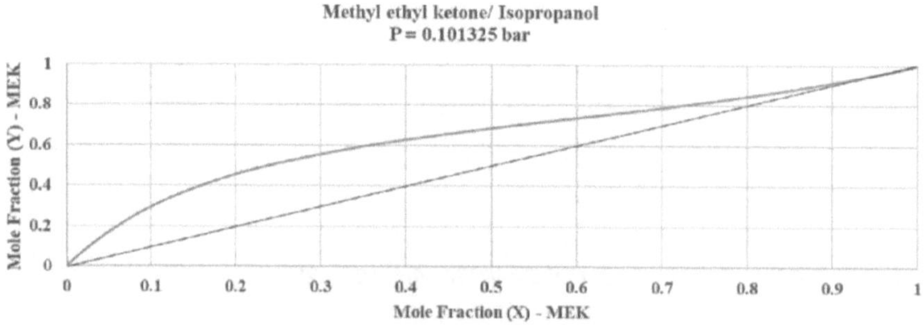

Figure 16.5 X-Y Plot of the MEK/Isopropanol Mixture at 0.101325 Bar.

0.101325 bar. This shows that if the column is operated at 0.101325 bar, a separation of more than 95% of the MEK can be achieved, which is about 60% if the column is operating at 1.01325 bar. Therefore, at a pressure below 0.101325 bar, no distillation boundary/restriction exists.

This separation process of the ternary azeotropic mixture of MEK-IPA-EtOH is simulated using DWSIM version 6.6.2. The triple column pressure-swing distillation (TCPSD) technique is used to separate this ternary azeotropic mixture. Two different separation sequences are feasible for this ternary mixture but with different pressure combinations. From the literature, the residue curve maps (RCMs) of this ternary mixture are studied at 1.01325 bar and 0.101325 bar [5]. The properties of azeotropes and residue curves are usually described using triangular diagrams [9]. The different distillation boundaries present in their triangular diagrams suggest two separation sequences.

These figures indicate that a ternary azeotropic mixture of isopropanol, methyl ethyl ketone, and ethanol can be separated with the separation sequence as MEK-IPA-EtOH if the first distillation column in the process flowsheet is operated at 4.053 bar [5]. Separation sequence means that the first column will give MEK, the second column will give IPA, and the last column will give EtOH. Since it is a minimum boiling azeotropic mixture, pure products are obtained from the bottom of the column. The distillate from the last column is recycled to the first column for efficient and economical separation. The reverse is true for maximum boiling azeotropes.

Another separation sequence is the IPA-MEK-EtOH, in which the first distillation column in the process flowsheet should be operated at 0.101325 bar [5]. In this arrangement, the IPA/MEK azeotrope almost disappeared. So, the first column yields more IPA. The distillate from the last column is recycled to the second column instead of the first to maintain IPA purity at the bottom of the first column.

16.2.1 Simulation for MEK-IPA-EtOH Separation Sequence

The flowsheet for the separation of the ternary azeotropic mixture of MEK-IPA-EtOH is prepared and simulated using DWSIM version 6.6.2. In this case, the separation sequence achieved is MEK-IPA-EtOH. The C5 system of unit is taken with pressure in bar and temperature in °C. The NRTL thermodynamic model is used to describe the interactions of mixture components. Figure 16.6 shows the DWSIM flowsheet for this separation sequence.

In this, feed (S-1) consists of 15% isopropanol, 70% MEK, and 15% ethanol. It is fed with a 1000 kg/h mass flow rate to a high-pressure distillation column (HPDC-1) at 4.053 bar pressure and 40°C temperature. The column is made with the cape-open unit operation available in DWSIM. HPDC-1 is a 34-stage column with feed entering at the 12th stage. It uses a total (liquid product) condenser and total (liquid product) reboiler. The column pressure is maintained at 4.053 bar. From this, bottoms-1 (S-3) with 90% MEK is obtained. The pressure of the

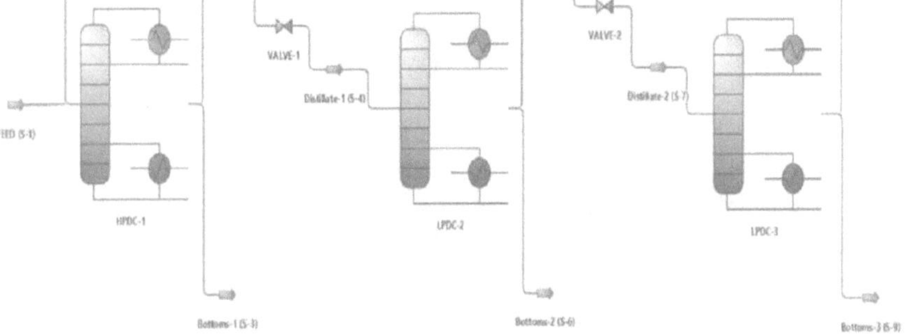

Figure 16.6 The MEK-IPA-EtOH Separation Sequence Flowsheet in DWSIM.

distillate-1 (S-2) stream is decreased with the help of valve-1 to 1.01325 bar, and then the distillate (S-4) stream is fed into the 22nd stage of the low-pressure distillation column (LPDC-2). It is a simple distillation column with 50 stages. The column pressure is maintained at 1.01325 bar. From this, bottoms-2 (S-6) stream with 90% isopropanol purity is achieved.

The distillate (S-5) stream from the LPDC-2 is fed to valve-2 to obtain the distillate-2 (S-7) stream with a pressure of 0.4053 bar. This is fed into the 16th stage of a 30-stage low-pressure distillation column (LPDC-3). The column pressure is maintained at 0.4053 bar. From this, bottoms-3 (S-9) with 99.1% ethanol purity is achieved. The distillate-3 (S-8) obtained from LPDC-3 is recycled into HPDC-1. Recycle logical operator recycle-1 is introduced with a mass flow tolerance of 36 kg/h. The stream pressure is increased to 0.4053 bar by valve-3, and then the recycled stream distillate-3 (S-11) is fed into the 26th stage of the HPDC-1 column. All percentage compositions mentioned are on the basis of mass fractions. Table 16.2 shows the master property table obtained from DWSIM software for the necessary data chosen for the table. Figure 16.7 shows the specifications of the three distillation columns in the IPA-MEK-EtOH separation sequence. It includes the number of stages and the feed point locations.

16.2.2 Simulation for IPA-MEK-EtOH Separation Sequence

The flowsheet to separate the ternary azeotropic mixture of IPA-MEK-EtOH is prepared and simulated using DWSIM version 6.6.2. In this, the separation sequence achieved is IPA-MEK-EtOH. C5 system of unit is taken with pressure in bar and temperature in °C. The NRTL thermodynamic model is used to describe the interactions of mixture components. Figure 16.8 shows the DWSIM flowsheet for this separation sequence.

Table 16.2 Master Property Table for the MEK-IPA-EtOH Separation Sequence Flowsheet

Parameters	*Feed (S-1)*	*Bottoms-1 (S-3)*	*Bottoms-2 (S-6)*	*Bottoms-3 (S-9)*
Temperature, °C	40	125.975	82.0032	56.6989
Pressure, bar	4.053	4.053	1.01325	0.4053
Mass flow, kg/h	1000	826.864	72.9357	98.5191
Volumetric flow, m^3/h	1.27841	1.2088	0.100727	0.130194
Density (mixture), kg/m^3	782.222	684.037	724.09	756.711
Mixture molar fraction				
Isopropanol	0.15	0.0976313	0.9	0.000996383
Methyl ethyl ketone	0.7	0.9	0.00104741	0.00800362
Ethanol	0.15	0.00236866	0.0989526	0.991

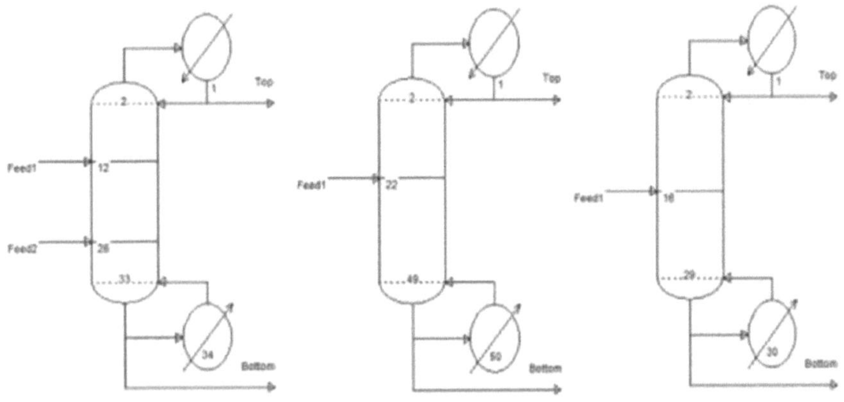

Figure 16.7 Specifications of All Columns in the MEK-IPA-EtOH Separation Sequence.

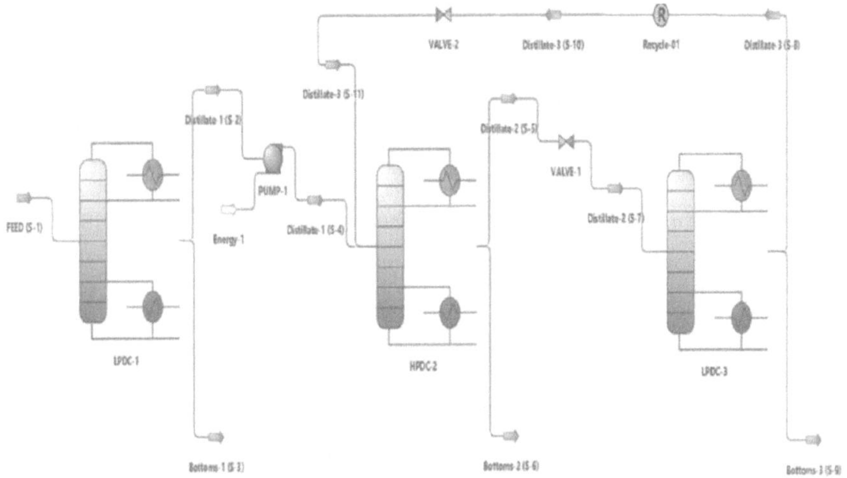

Figure 16.8 The IPA-MEK-EtOH Separation Sequence Flowsheet in DWSIM.

In this, feed (S-1) consists of 15% isopropanol, 70% MEK, and 15% ethanol. It is fed with 1000 kg/h mass flow rate into the low-pressure distillation column (LPDC-1) at 0.101325 bar and 40 °C. The column is made with the cape-open unit operation available in DWSIM. LPDC-1 is a 77-stage column with feed entering at the 19th stage. It uses a total (liquid product) condenser and total (liquid product) reboiler. The column pressure is maintained at 0.101325 bar. From this, bottoms-1 (S-3) with 99.3% isopropanol purity is achieved. The pressure of the distillate-1 (S-2) stream is increased with the help of pump-1 to 5.87685 bar, and then the distillate (S-4) stream is fed into the 22nd stage of the high-pressure distillation column (HPDC-2). It is a simple distillation column with 35 stages. The column pressure is maintained at 5.87685 bar. From this, a bottoms-2 (S-6) stream with 99.9% MEK purity is achieved.

The distillate (S-5) stream from the HPDC-2 is fed to valve-1 to obtain the distillate-2 (S-7) stream with a pressure of 0.4053 bar. This is fed into the 13th stage of the 32-stage low-pressure distillation column (LPDC-3). The column pressure is maintained at 0.4053 bar. From this, bottoms-3 (S-9) with 99.1% ethanol purity is achieved. The distillate-3 (S-8) obtained from LPDC-3 is recycled into HPDC-2. The recycle logical operator recycle-01 is introduced with a mass flow tolerance of 36 kg/h. The stream pressure is reduced to 0.101325 bar by valve-2, and then the recycled stream distillate-3 (S-11) is fed into the 17th stage of the HPDC-2 column. Table 16.3 shows the master property table obtained from DWSIM software for the necessary data chosen for the table.

Figure 16.9 shows the specifications of three distillation columns in the IPA-MEK-EtOH separation sequence. The separation sequence IPA-MEK-EtOH is

Table 16.3 Master Property Table for the IPA-MEK-EtOH Separation Sequence Flowsheet

Parameters	*Feed (S-1)*	*Bottoms-1 (S-3)*	*Bottoms-2 (S-6)*	*Bottoms-3 (S-9)*
Temperature, °C	40	33.8318	147.012	56.668
Pressure, bar	0.101325	0.101325	5.87685	0.4053
Mass flow, kg/h	1000	136.492	759.06	104.275
Volumetric flow, m^3/h	3869.79	0.176408	1.15799	0.137791
Density (mixture), kg/m^3	0.258412	773.731	655.497	756.76
Mixture molar fraction				
Isopropanol	0.15	0.993	1.74052E-06	1.79044E-07
Methyl ethyl ketone	0.7	7.56012E-13	0.999	0.00899982
Ethanol	0.15	0.007	0.000998259	0.991

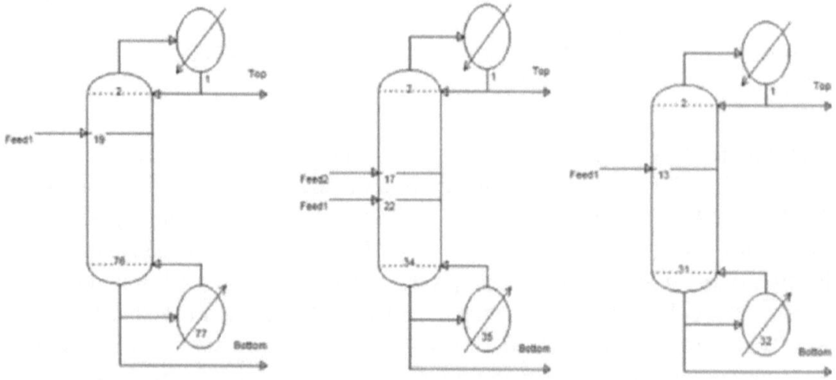

Figure 16.9 Specifications of All Columns in the IPA-MEK-EtOH Separation Sequence.

found to be more economical than the other. In further study, only the IPA-MEK-EtOH separation sequence is considered, and analysis for the components used is performed. Different components of the flowsheet (pump, valves, etc.) have been given certain inputs. Pump-1 is used to increase the pressure of distillate-1 (S-2) from 0.101325 bar to 5.87685 bar. The pump requirement is found to be 0.229519 kW for a head increase of 5.77553 bar for the distillate-1 (S-2) stream.

Valve-1 and valve-2 are used in the process flowsheet to decrease the pressure of the distillate-2 (S-5) and disillate-3 (S-10) streams to the desired pressures, according to the columns in which they are entering. The valve-1 produces a pressure drop of 5.47155 bar and a temperature drop of -78.4752 °C. The valve-2 produces a pressure drop of 0.303975 bar and a temperature drop of -19.2576 °C.

Recycle-01 is the recycle operator introduced in the distillate-3 (S-8) stream coming from LPDC-3, which is recycled into HPDC-2 to maintain the economy and efficiency of the process. It is given a mass flow tolerance of 36 kg/h and uses 100 iterations for the simulation of the recycling stream.

16.3 FEASIBILITY STUDY

Any plant design is considered acceptable if it has sufficient capability to operate under different conditions to earn profits. To assess the feasibility of any project, profitability analysis is widely used [42]. Profitability analysis helps make the process economically competitive and guides in evaluating the financial standing of any industry [43].

Capital investment is broadly divided into two types: fixed capital investment and working capital investment. Fixed capital investments such as building, piping, and utilities are invested only once during commissioning and before the plant starts production [44, 45]. It is independent of the plant operation rate. Working capital is the investment that is required for the daily operation and production of the industry. It is assumed to remain the same throughout the plant life and is recovered at the end of plant life [43, 46]. The cost estimates are made for the purpose of computing the payback period of the plant. The delivered equipment cost is calculated using the Marshall and Swift (M&S) cost index and the column specifications obtained from the simulation. A detailed estimation has been made in the next section.

16.4 RESULTS AND DISCUSSIONS

Two simulations for the PSD have been made by using DWSIM version 6.6.2. It is designed to efficiently separate the azeotropic mixture of IPA-MEK-EtOH. Flowsheets consisted of three distillation columns operating at different pressures. From the bottom of each column, one pure component is obtained. Table 16.2 and Table 16.3 give detailed information about the parameters and results of various streams for MEK-IPA-EtOH and IPA-MEK-EtOH separation sequence simulation, respectively.

Comparing the results of these two separation sequences from Table 16.2 and Table 16.3, it can be observed that the product purity and the mass flow rates of the costlier solvents (ethanol and isopropanol) are higher in the second separation sequence, i.e., the IPA-MEK-EtOH separation sequence. Each pure component obtained from the IPA-MEK-EtOH sequence has a higher purity than the other sequence for the same feed composition and mass flow rate. Figure 16.7 and Figure 16.9 can be used to compare the number of stages used for the three columns in both sequences. They also favour the IPA-MEK-EtOH separation sequence.

Table 16.4 gives the results of various columns for both separation sequences. The LPDC-2 column in the MEK-IPA-EtOH separation sequence flowsheet uses a very high reflux ratio of 8 to obtain just 90% pure isopropanol. If the reflux ratio is decreased, the number of stages required exceeds 100, which is unrealistic in real-life situations. The minimum stages suggested by the DWSIM in accordance with Fenske's equation are more than 200 for reflux ratios below 7. Each column in the MEK-IPA-EtOH sequence uses a higher reflux ratio than the IPA-MEK-EtOH sequence and gives a lower purity than the IPA-MEK-EtOH sequence.

Another criterion for deciding the best separation sequence is the comparison of the total annual cost (TAC) for operating the distillation columns. Considering carbon steel as the column material, DWSIM version 6.6.2 has given annual costs for all the columns based on various operating parameters (shell, internals, reboiler, condenser, etc.), shown in Table 16.5. Process uptime is assumed to be 8016 h/year by DWSIM for these estimates.

Table 16.4 Comparison of Different Parameters of Both Separation Sequences

Column	MEK-IPA-EtOH sequence			IPA-MEK-EtOH sequence		
	Height, m	*Reflux Ratio*	*Product Purity*	*Height, m*	*Reflux Ratio*	*Product Purity*
Column-1	34.18	2	0.9 (MEK)	51.02	1.8	0.993 (IPA)
Column-2	10.05	8	0.9 (IPA)	23.26	2.2	0.999 (MEK)
Column-3	3.79	2.1	0.991 (EtOH)	18.57	2.1	0.991 (EtOH)

Table 16.5 Total Annual Cost for Both Separation Sequences

Column	*MEK-IPA-EtOH Separation Sequence*	*IPA-MEK-EtOH Separation Sequence*
Column-1 cost, $/year	124,000	124,000
Column-2 cost, $/year	252,000	89,000
Column-3 cost, $/year	45,000	26,000
Total annual cost, $/year	**421,000**	**239,000**

The total annual cost for the MEK-IPA-EtOH separation sequence is found to be 421,000 ($/year), and for the IPA-MEK-EtOH separation sequence, it is estimated to be 239,000 ($/year). Therefore, the MEK-IPA-EtOH separation sequence is not further studied.

16.4.1 Estimation of Equipment Cost

Table 16.6 shows the specifications of the three distillation columns. These are obtained from the simulation results in DWSIM and used for cost estimation of the columns. Various methods are available for estimating equipment costs. The Marshall and Swift (M&S) index is used for capital cost estimation of the column [47]. This requires column height (H) and diameter (D) to obtain the value of index (M&S). These three parameters, along with the material factor (F_c), are used to estimate the material and installation cost for the column. The empirical correlations to calculate these two costs (in $) are [47]:

$$\text{Material cost}(\$) = \left(\frac{\text{M\&S}}{280}\right)\left(101.9D^{1.066}H^{0.82}F_c\right)$$

$$\text{Installation cost}(\$) = \left(\frac{\text{M\&S}}{280}\right)\left(4.7D^{1.55}HF_c\right)$$

The M&S index and cost calculation by these empirical correlations need column height and diameter in feet. By using these empirical correlations and taking the material factor, $F_c = 1$ (For carbon steel), the total material cost of three columns is found to be $138,634.056. Pump and valve cost can be categorized as miscellaneous costs and are usually taken as 2% of the major equipment cost [44]. With this, the miscellaneous cost is obtained as $2,772.681.

16.4.2 Depreciation Cost

Depreciation is the decrease in the value of any property or asset with time. It can also be defined as the distribution of the cost of tangible assets over their useful life period in a systematic manner [48, 49]. Several methods are available in the literature for estimating depreciation cost. Here, depreciation is calculated using

Table 16.6 Specifications of Distillation Columns

Distillation column	*Height, m*	*Diameter, m*	*M&S Index*
LPDC-1	51.02	0.64	2300
HPDC-2	23.26	0.39	800
LPDC-3	18.57	0.22	600

the straight-line method as it is the simplest of all the methods, and gives almost accurate estimates [50]. The scrap value of the equipment is considered as 30% of the original purchased cost and ten years as the service period.

By straight-line method [43],

$$\text{Depreciation} = \frac{\text{original cost}(V) - \text{scrap value}(V_s)}{\text{service period}(n)}$$

Table 16.7 gives the costs of delivered equipment, their depreciation in one year, and cost after one year by subtracting depreciation from the initial cost.

16.4.3 Fixed Capital Investment (FCI) Estimation

Various methods for estimating FCI are available in the literature [44]. The method chosen for estimating FCI is the percentage of delivered equipment cost because it requires less information and gives better estimates [51]. This estimate is made in Table 16.8 by taking the delivered equipment cost as $141,406.74. After considering the appropriate percentages of all the factors [44], the FCI is estimated as $416,442.84.

Table 16.7 Equipment Costs Before and After One Year's Depreciation

Equipment	*Cost, $*	*Depreciation Per Year (V_s = 30%)*	*Final Cost, $*
LPDC-1	122,780.99	8,594.669	114,186.321
HPDC-2	13,235.985	926.519	12,309.466
LPDC-3	2,617.081	183.196	2,433.885
Miscellaneous (Pumps and Valves)	2,772.681	194.088	2,578.593
Total	**$141,406.737**	**$9,898.472**	**$131,508.265**

Table 16.8 Estimation of Fixed Capital Investment by Percentage of Delivered Equipment Cost Method

Delivered Equipment Cost	*D (say)*	*$141,406.74*
Equipment installation	39.0 % of D	$55,148.63
Instrumentation	28.0 % of D	$39,593.89
Piping	31.0 % of D	$43,836.09
Electrical	22.0 % of D	$31,109.48
Buildings	4.6 % of D	$6,504.71
Yard improvements	10.0 % of D	$14,140.67
Service facilities	1.8 % of D	$2,545.32

Delivered Equipment Cost	*D (say)*	*$141,406.74*
Land	13.8 % of D	$19,514.13
Engineering supervision	15.0 % of D	$21,211.01
Construction expenses	9.2 % of D	$13,009.42
Contractor's fee	11.0 % of D	$15,554.74
Contingency	9.1 % of D	$12,868.01
Total fixed cost		**$416,442.84**

Table 16.9 Profit Obtained from Separated Products

Products Produced	*Amount Produced*	*Unit Cost*	*Selling Price*
Isopropanol	0.174608 m^3/h	1,250 \$/m^3	1,931,667.6 $/year
MEK	759.06 kg/h	1.22 $/kg	8,112,226.032 $/year
Ethanol	0.137791 m^3/h	2,620 \$/m^3	3,162,468.799 $/year

16.4.4 Estimation of Working Capital

In this case, the raw material is a mixture of IPA-MEK-EtOH in 15%-70%-15% mass fraction, respectively. Since it is taken as a waste from other plants or industries for their separation, its cost will be less compared to the sum of costs of these pure components. For further estimation, the cost of raw material (IPA-MEK-EtOH mixture) is assumed to be 1.48 $/kg. Therefore, for 1000 kg/h feed flow rate and taking 365 days of plant operation with 24 h/day working time, working capital is found to be 12,964,800 $/year.

16.4.5 Estimation of Net Profit

Profit can be stated as the increase in proprietorship that occurred during a given time period. It is the difference between the amount of money earned and the amount spent [52, 53]. Therefore, the gross profit can be found from the money obtained from selling the products formed/separated at the plant, which in this case is isopropanol, MEK, and ethanol.

However, net profit is what remains after subtracting the sum of all the liabilities, including stocks, from the present value of all assets. To find the net profit, the working capital is deducted from the selling price. In addition, certain taxes are also deducted from this profit to get the net profit, as shown in Table 16.9 [53]. The total profit without depreciation obtained by adding the selling price of three products is found to be 13,206,362.43 $/year. Assuming the tax on profit is 40%, then the net profit obtained is 144,937.450 $/year.

16.4.6 Estimation of Payback Period

Estimation of payback period is the minimum time required for cash flows to recover all the initial capital invested [54–56]. In this method, the interest effects are negligible. Fixed capital investment can be depreciable and non-depreciable. Land is a non-depreciable fixed capital investment. For the calculation of the payback period, only depreciable fixed capital investment is considered. The mathematical relation to finding the payback period is [44]:

$$\text{Payback period} = \frac{\text{Depreciable fixed capital investment}}{\text{avg.}\dfrac{\text{profit}}{\text{year}} + \text{avg.}\dfrac{\text{depreciation}}{\text{year}}}$$

Table 16.10 summarizes the results of the estimated costs. The payback period for the plant is found to be 2.689 years. This indicates the economic feasibility of the process.

16.5 CONCLUSION

PSD technique for the separation of a ternary azeotropic mixture of IPA-MEK-EtOH was studied. It was compared with other available distillation techniques for the separation of azeotropic mixtures. It was found economical for those azeotropic mixtures, which are highly pressure sensitive. Binary azeotropic mixtures of MEK-IPA and MEK-EtOH were studied using P-x-y and x-y plots. It was found that azeotropic compositions get eliminated at pressures below the atmosphere.

A simulation for PSD was made by using DWSIM version 6.6.2. Plant was simulated for 1000 kg/h of waste solvent mixture, with feed composition taken as 15% isopropanol, 70% MEK, and 15% ethanol. A study of triangular diagrams and residual curve maps (RCMs) from literature suggested that two separation sequences are possible for the separation of IPA-MEK-EtOH ternary mixture depending on pressure conditions: MEK-IPA-EtOH and IPA-MEK-EtOH.

Table 16.10 Result of Various Costs and Profits Estimated in the Feasibility Study

Measured Cost/Profit	*Amount, $*
Delivered equipment cost	141,406.737
Depreciation cost (annual)	9,898.472
Final cost (after depreciation)	131,508.265
Fixed capital investment	416,442.84
Working capital	12,964,800 $/year
Net profit	144,937.450 $/year

Two simulations were made for both sequences using DSWIM version 6.6.2. It is concluded that the separation sequence IPA-MEK-EtOH is more economical because it produces products with higher purity and requires smaller-sized distillation columns. Other sequences give less flowrates of desired components and comparatively less purity, as well. Each column in the MEK-IPA-EtOH sequence uses greater reflux ratios than the IPA-MEK-EtOH sequence and gives less purity compared to the IPA-MEK-EtOH sequence. The total annual cost for the MEK-IPA-EtOH separation sequence was found to be 421 (k$/year), and for the IPA-MEK-EtOH separation sequence, it was estimated to be 239 (k$/year). Also, the column heights needed for the MEK-IPA-EtOH separation sequence are unrealistic. All the remaining analyses were made on the IPA-MEK-EtOH sequence.

Cost estimation is an important component in plant design. The payback period was calculated for the profitability analysis of the plant. In this, fixed capital investment (FCI) was calculated by using the percentage of delivered equipment cost method, which was found to be $416,442.84. Then, annual depreciation was calculated using the straight-line method, and then net profit was calculated. Using all these results, the payback period for the plant was finally calculated and found to be 2.689 years. Considering the errors in the estimates, the payback period will not exceed more than three years, which makes the process economically viable.

REFERENCES

[1] T. Mahdi, A. Ahmad, M.M. Nasef, A. Ripin, State-of-the-art technologies for separation of azeotropic mixtures, Sep. Purif. Rev. 44 (2015) 308–330. doi:10.1080/15422119.2014.963607.

[2] E.R. Baümler, M.E. Carrín, A.A. Carelli, Extraction of sunflower oil using ethanol as solvent, J. Food Eng. 178 (2016) 190–197. doi:10.1016/j.jfoodeng.2016.01.020.

[3] P. Catz, W. Shinn, I.M. Kapetanovic, H. Kim, M. Kim, E.L. Jacobson, M.K. Jacobson, C.E. Green, Simultaneous determination of myristyl nicotinate, nicotinic acid, and nicotinamide in rabbit plasma by liquid chromatography – tandem mass spectrometry using methyl ethyl ketone as a deproteinization solvent, 829 (2005) 123–135. doi:10.1016/j.jchromb.2005.10.003.

[4] F. Wolke, Y. Hu, M. Schmidt, O. Korup, R. Horn, E. Reichelt, M. Jahn, A. Michaelis, Spatially-resolved reaction profiles in Fischer-Tropsch synthesis – influence of operating conditions and promotion for iron-based catalysts, Catal. Commun. 158 (2021) 106335. doi:10.1016/j.catcom.2021.106335.

[5] M. Li, Y. Ma, X. Zhang, T. Zhao, Z. Zhu, Y. Wang, Triple column pressure-swing distillation for ternary mixture of methyl ethyl ketone/isopropanol/ethanol, Chem. Eng. Trans. 61 (2017) 649–654. doi:10.3303/CET1761106.

[6] W. Ma, E.L. Kugler, D.B. Dadyburjor, Effect of Mo loading and support type on hydrocarbons and oxygenates produced over Fe-Mo-Cu-K catalysts supported on activated carbons, Stud. Surf. Sci. Catal. 163 (2007) 125–140. doi:10.1016/S0167-2991(07)80476-9.

[7] P.D.A. Bastos, F.S. Oliveira, L.P.N. Rebelo, A.B. Pereiro, I.M. Marrucho, Separation of azeotropic mixtures using high ionicity ionic liquids based on

1-ethyl-3-methylimidazolium thiocyanate, Fluid Phase Equilib. 389 (2015) 48–54. doi:10.1016/j.fluid.2015.01.011.
[8] Y. Wang, G. Bu, Y. Wang, T. Zhao, Z. Zhang, Z. Zhu, Application of a simulated annealing algorithm to design and optimize a pressure-swing distillation process, Comput. Chem. Eng. 95 (2016) 97–107. doi:10.1016/j.compchemeng.2016.09.014.
[9] Z. Zhu, D. Xu, X. Liu, Z. Zhang, Y. Wang, Separation of acetonitrile/methanol/benzene ternary azeotrope via triple column pressure-swing distillation, Sep. Purif. Technol. 169 (2016) 66–77. doi:10.1016/j.seppur.2016.06.009.
[10] J.P. Knapp, M.F. Doherty, A new pressure-swing-distillation process for separating homogeneous azeotropic mixtures, Ind. Eng. Chem. Res. 31 (1992) 346–357. doi:10.1021/ie00001a047.
[11] R. Muñoz, J.B. Montón, M.C. Burguet, J. de la Torre, Separation of isobutyl alcohol and isobutyl acetate by extractive distillation and pressure-swing distillation: Simulation and optimization, Sep. Purif. Technol. 50 (2006) 175–183. doi:10.1016/j.seppur.2005.11.022.
[12] H. Yu, Q. Ye, H. Xu, H. Zhang, X. Dai, Design and control of dividing-wall column for tert -butanol dehydration system via heterogeneous azeotropic distillation, (2015). doi:10.1021/ie504325g.
[13] F. Qasim, J.S. Shin, S.J. Cho, S.J. Park, Optimizations and heat integrations on the separation of toluene and 1-butanol azeotropic mixture by pressure swing distillation, Sep. Sci. Technol. 51 (2015) 316–326. doi:10.1080/01496395.2015.1086378.
[14] X. You, I.R. Donis, V. Gerbaud, Improved design and efficiency of the extractive distillation process for acetone-methanol with water, Ind. Eng. Chem. Res. 54 (2015) 491–501. doi:10.1021/ie503973a.
[15] C. Wang, Y. Zhuang, Y. Qin, Y. Dong, L. Liu, L. Zhang, J. Du, Optimization and eco-efficiency analysis of extractive distillation processes with different solvents for separating the ternary mixture embedding two azeotropes, Sep. Purif. Technol. 269 (2021) 118763. doi:10.1016/j.seppur.2021.118763.
[16] X. Zhu, X. Zhao, Z. Zhang, Z. Ma, J. Gao, Optimal design and control of an energy-efficient triple-side-stream quaternary extractive distillation process, Chem. Eng. Process.—Process Intensif. 167 (2021) 108510. doi:10.1016/j.cep.2021.108510.
[17] X. Shi, X. Zhu, X. Zhao, Z. Zhang, Performance evaluation of different extractive distillation processes for separating ethanol/tert-butanol/water mixture, Process Saf. Environ. Prot. 137 (2020) 246–260. doi:10.1016/j.psep.2020.02.015.
[18] Y. An, W. Li, Y. Li, S. Huang, J. Ma, C. Shen, C. Xu, Design/optimization of energy-saving extractive distillation process by combining preconcentration column and extractive distillation column, 135 (2015) 166–178. doi:10.1016/j.ces.2015.05.003.
[19] W.L. Luyben, Comparison of extractive distillation and pressure-swing distillation for acetone/chloroform separation, Comput. Chem. Eng. 50 (2013) 1–7. doi:10.1016/j.compchemeng.2012.10.014.
[20] J. Zhu, L. Hao, H. Wei, Sustainable concept design including economic, environment and inherent safety criteria: Process intensification-reactive pressure swing distillation, J. Clean. Prod. 314 (2021) 127852. doi:10.1016/j.jclepro.2021.127852.
[21] S. Yang, Q. Zhang, Y. Ma, X. Yuan, A. Zeng, Novel eco-efficient vapor recompression-assisted arrangement for minimum-boiling side-stream pressure-swing distillation system: Preheating feed stream to dew or bubble point, Sep. Purif. Technol. 257 (2021) 117920. doi:10.1016/j.seppur.2020.117920.

[22] H. Lee, C. Seo, M. Lee, J.W. Lee, CFD-aided design of internally heat-integrated pressure-swing distillation for ternary azeotropic separation constrained by pinch pressure, Appl. Therm. Eng. 195 (2021) 117198. doi:10.1016/j.applthermaleng.2021.117198.

[23] B. Shan, D. Sun, Q. Zheng, F. Zhang, Y. Wang, Z. Zhu, Dynamic control of the pressure-swing distillation process for THF/ethanol/water separation with and without thermal integration, Sep. Purif. Technol. 268 (2021) 118686. doi:10.1016/j.seppur.2021.118686.

[24] M. Ferchichi, L. Hegely, P. Lang, Economic and environmental evaluation of heat pump-assisted pressure-swing distillation of maximum-boiling azeotropic mixture water-ethylenediamine, Energy. 239 (2022) 122608. doi:10.1016/j.energy.2021.122608.

[25] J. Yang, M. Zhou, Y. Wang, X. Zhang, G. Wu, Simulation of Pressure-swing Distillation for Separation of Ethyl Acetate-Ethanol-Water, in: IOP Conf. Ser. Mater. Sci. Eng., Institute of Physics Publishing, 2017. doi:10.1088/1757-899X/274/1/012026.

[26] J.R. Phimister, W.D. Seider, Semicontinuous, Pressure-Swing Distillation (2000) 122–130.

[27] S.K. Wasylkiewicz, L.C. Kobylka, F.J.L. Castillo, Pressure Sensitivity Analysis of Azeotropes (2003) 207–213.

[28] A. Yang, W. Shen, S. Wei, L. Dong, J. Li, V. Gerbaud, Design and control of pressure-swing distillation for separating ternary systems with three binary minimum azeotropes, AIChE J. 65 (2019) 1281–1293. doi:10.1002/aic.16526.

[29] Y.H. Wang, I.L. Chien, Unique design considerations for maximum-boiling azeotropic systems via extractive distillation: Acetone/chloroform separation, Ind. Eng. Chem. Res. 57 (2018) 12884–12894. doi:10.1021/acs.iecr.8b03125.

[30] W.L. Luyben, Pressure-swing distillation for minimum- and maximum-boiling homogeneous azeotropes, Ind. Eng. Chem. Res. 51 (2012) 10881–10886. doi:10.1021/ie3002414.

[31] A. Farsi, Ö. Kayhan, G. Chehade, I. Dincer, G.F. Natarer, Multiphase flow model and experimental study of pressure swing distillation for low pressure process of hydrochloric/water separation in hydrogen production, Comput. Chem. Eng. 141 (2020) 107020. doi:10.1016/j.compchemeng.2020.107020.

[32] S. Liang, Y. Cao, X. Liu, X. Li, Y. Zhao, Y. Wang, Y. Wang, Insight into pressure-swing distillation from azeotropic phenomenon to dynamic control, Chem. Eng. Res. Des. 117 (2017) 318–335. doi:10.1016/j.cherd.2016.10.040.

[33] F. Zhao, Z. Xu, J. Zhao, J. Wang, M. Hu, X. Li, Z. Zhu, P. Cui, Y. Wang, Y. Ma, Process design and multi-objective optimization for separation of ternary mixtures with double azeotropes via integrated quasi-continuous pressure-swing batch distillation, Sep. Purif. Technol. 276 (2021) 119288. doi:10.1016/j.seppur.2021.119288.

[34] G. Modla, Pressure swing batch distillation by double column systems in closed mode, Comput. Chem. Eng. 34 (2010) 1640–1654. doi:10.1016/j.compchemeng.2010.02.037.

[35] Y. Wang, X. Yang, J. Zhao, X. Liu, D. Yao, P. Cui, L. Wang, Z. Zhu, X. Li, D. Xu, Design and comprehensive analysis of a novel pressure-swing batch distillation process for the separation of a binary azeotrope with various boiling behaviors, Sep. Purif. Technol. 251 (2020) 117329. doi:10.1016/j.seppur.2020.117329.

[36] G. Modla, A. Kopasz, P. Lang, Feasibility of Separation of Ternary Mixtures by Pressure Swing Batch Distillation, in: R.M. de Brito Alves, C.A.O. do Nascimento, E.C.B.T.-C.A.C.E. Biscaia (Eds.), 10th Int. Symp. Process Syst. Eng. Part A, Comput. Aided Chem. Eng., Elsevier, 2009: pp. 1017–1022. doi:10.1016/S1570-7946(09)70390-6.

[37] H. Luo, K. Liang, W. Li, Y. Li, M. Xia, C. Xu, Comparison of pressure-swing distillation and extractive distillation methods for isopropyl alcohol/diisopropyl ether separation, Ind. Eng. Chem. Res. 53 (2014) 15167–15182. doi:10.1021/ie502735g.

[38] A.M. Fulgueras, D.S. Kim, J. Cho, Modeling and optimization study of pressure-swing distillation for the separation process of acetone-methanol mixture with vapor-liquid equilibrium analysis, J. Chem. Eng. Japan. 49 (2016) 84–96. doi:10.1252/jcej.14we424.

[39] G. Modla, P. Lang, Separation of an acetone-methanol mixture by pressure-swing batch distillation in a double-column system with and without thermal integration, Ind. Eng. Chem. Res. 49 (2010) 3785–3793. doi:10.1021/ie9019352.

[40] L. Lü, L. Zhu, H. Liu, H. Li, S. Sun, Comparison of continuous homogenous azeotropic and pressure-swing distillation for a minimum azeotropic system ethyl acetate/n-hexane separation, Chinese J. Chem. Eng. 26 (2018) 2023–2033. doi:10.1016/j.cjche.2018.02.002.

[41] W.L. Luyben, Importance of pressure-selection in pressure-swing distillation, Comput. Chem. Eng. 149 (2021) 107279. doi:10.1016/j.compchemeng.2021.107279.

[42] D.K.S. Ng, K.S. Ng, R.T.L. Ng, Integrated Biorefineries, Elsevier, 2017. doi:10.1016/B978-0-12-409548-9.10138-1.

[43] A. Shah, N.R. Baral, A. Manandhar, Technoeconomic Analysis and Life Cycle Assessment of Bioenergy Systems, Elsevier, 2016. doi:10.1016/bs.aibe.2016.09.004.

[44] M.S. Peters, K.D. Timmerhaus, Plant Design and Economics for Chemical Engineers, 4th ed., McGraw-Hill, New York, 1991.

[45] G. Nguyen, V. Vo, Asset liquidity and venture capital investment, J. Corp. Financ. 69 (2021) 101963. doi:10.1016/j.jcorpfin.2021.101963.

[46] S. Dhole, S. Mishra, A.M. Pal, Efficient working capital management, financial constraints and firm value: A text-based analysis, Pacific Basin Financ. J. 58 (2019) 101212. doi:10.1016/j.pacfin.2019.101212.

[47] T.R. Biyanto, J.A. Rahman, S. Sarwono, R. Roekmono, H.N. Laila, A. Abdurrakhman, P.A. Darwito, Techno Economic Optimization of Petlyuk Distillation Column Design Using Duelist Algorithm, in: Procedia Eng., Elsevier Ltd, 2017: pp. 520–527. doi:10.1016/j.proeng.2017.03.083.

[48] R.M. Samaniego, J.Y. Sun, Uncertainty, depreciation and industry growth, Eur. Econ. Rev. 120 (2019) 103314. doi:10.1016/j.euroecorev.2019.103314.

[49] K.J. Liapis, D.D. Kantianis, Depreciation methods and life-cycle costing (LCC) methodology, Procedia Econ. Financ. 19 (2015) 314–324. doi:10.1016/s2212-5671(15)00032-5.

[50] C.D. Green, J.R. Grinyer, R. Michaelson, A possible economic rationale for straight-line depreciation, Abacus. 38 (2002) 91–120. doi:10.1111/1467-6281.00099.

[51] N. Zgheib, H. Takache, Recycling of used lubricating oil by solvent extraction: Experimental results, Aspen Plus simulation and feasibility study, Clean Technol. Environ. Policy. 23 (2021) 65–76. doi:10.1007/s10098-020-01893-0.

[52] J.P. Clark, Economic Evaluation, 2009. doi:10.1007/978-1-4419-0420-1_14.

[53] M.A. Al Hayek, The relationship between sales revenue and net profit with net cash flows from operating activities in jordanian industrial joint stock companies, Int. J. Acad. Res. Accounting, Financ. Manag. Sci. 8 (2018). doi:10.6007/ijarafms/v8-i3/4757.

[54] C.M. Boardman, W.J. Reinhart, S.E. Celec, The role of the payback period in the theory and application of duration to capital budgeting, J. Bus. Financ. Account. 9 (1982) 511–522. doi:10.1111/j.1468-5957.1982.tb01012.x.

[55] T.M.I. Mahlia, H.A. Razak, M.A. Nursahida, Life cycle cost analysis and payback period of lighting retrofit at the University of Malaya, Renew. Sustain. Energy Rev. 15 (2011) 1125–1132. doi:10.1016/j.rser.2010.10.014.

[56] I.L. Wong, P.C. Eames, R.S. Perera, A review of transparent insulation systems and the evaluation of payback period for building applications, Sol. Energy. 81 (2007) 1058–1071. doi:10.1016/j.solener.2007.04.004.

Index